VOLUME SIXTY SIX

ADVANCES IN
ECOLOGICAL RESEARCH

Pluralism in Ecosystem Governance

ADVANCES IN ECOLOGICAL RESEARCH

Series Editors

DAVID A. BOHAN
Agroécologie, AgroSup Dijon, INRAE,
Université de Bourgogne Franche-Comté,
Dijon, France

ALEX J. DUMBRELL
School of Life Sciences,
University of Essex,
Wivenhoe Park, Colchester,
Essex, United Kingdom

VOLUME SIXTY SIX

ADVANCES IN ECOLOGICAL RESEARCH

Pluralism in Ecosystem Governance

Edited by

JENNIFER M. HOLZER
Environmental Sustainability Research Centre,
Brock University, St. Catharines, ON, Canada

JULIA BAIRD
Environmental Sustainability Research Centre/
Geography and Tourism Studies,
Brock University, St. Catharines, ON, Canada

GORDON M. HICKEY
Department of Natural Resource Sciences,
McGill University, Montreal, QC, Canada

ELSEVIER

ACADEMIC PRESS
An imprint of Elsevier

Academic Press is an imprint of Elsevier
125 London Wall, London, EC2Y 5AS, United Kingdom
The Boulevard, Langford Lane, Kidlington, Oxford OX5 1GB, United Kingdom
525 B Street, Suite 1650, San Diego, CA 92101, United States
50 Hampshire Street, 5th Floor, Cambridge, MA 02139, United States

First edition 2022

Notices

Knowledge and best practice in this field are constantly changing. As new research and experience broaden our understanding, changes in research methods, professional practices, or medical treatment may become necessary.

Practitioners and researchers must always rely on their own experience and knowledge in evaluating and using any information, methods, compounds, or experiments described herein. In using such information or methods they should be mindful of their own safety and the safety of others, including parties for whom they have a professional responsibility.

To the fullest extent of the law, neither the Publisher nor the authors, contributors, or editors, assume any liability for any injury and/or damage to persons or property as a matter of products liability, negligence or otherwise, or from any use or operation of any methods, products, instructions, or ideas contained in the material herein.

ISBN: 978-0-323-98901-5
ISSN: 0065-2504

For information on all Academic Press publications
visit our website at https://www.elsevier.com/books-and-journals

Publisher: Zoe Kruze
Acquisitions Editor: Jason Mitchell
Developmental Editor: Paulo Federico Mendoza
Production Project Manager: Abdulla Sait
Cover Designer: Miles Hitchen

Typeset by STRAIVE, India

Contents

Amplifying muted voices

Contributors

Eerika Albrecht
LYY Institute for Natural Resources, Environment and Society, University of Eastern Finland, Kuopio, Joensuu, Finland

Daniel Amoak
Western University, London, ON, Canada

Nathan A. Badry
Department of Natural Resource Sciences, McGill University, Montreal, QC, Canada

R. Bezner Kerr
Cornell University, Ithaca, NY, United States

Timothy Chunga
SFHC, Ekwendeni, Malawi

Sarah Coose
Geography and Tourism Studies Department, Brock University, St. Catharines, ON, Canada

Kellen Copeland
Environmental Studies Graduate Program, Oregon State University, Corvallis, OR, United States

Shelby Copeland
Human Dimensions of Natural Resources Department, Colorado State University, Fort Collins, CO, United States

Thomas Cousins
School of Anthropology and Museum Ethnography, University of Oxford, Oxford, United Kingdom; Department of Sociology and Social Anthropology, Stellenbosch University, Stellenbosch, South Africa

Laifolo Dakishoni
SFHC, Ekwendeni, Malawi

Anant Deshwal
Biology Department, Bradley University, Peoria, IL, United States

Candy Donaldson
Healthy Headwaters Lab, Great Lakes Institute for Environmental Research, Department of Integrative Biology, University of Windsor, Traditional Territory of the Three Fires Confederacy, Windsor, ON, Canada

Max Duggan
Human Dimensions of Natural Resources Department, Colorado State University, Fort Collins, CO, United States

Susanne Emond
Sululiit Area Co-Management Committee (Chair, 2011–2021); Canadian Wildlife Service, Environment and Climate Change Canada, Iqaluit, NU, Canada

Stephanie Enloe
Cornell University, Ithaca, NY, United States

David Fajardo-Ortiz
German Centre for Higher Education Research and Science Studies, Hannover, Germany; KU Leuven, Leuven, Belgium

Catherine Febria
Healthy Headwaters Lab, Great Lakes Institute for Environmental Research, Department of Integrative Biology, University of Windsor, Traditional Territory of the Three Fires Confederacy, Windsor, ON, Canada

María R. Felipe-Lucia
Department of Ecosystem Services, Helmholtz Centre for Environmental Research—UFZ, Leipzig; Department of Ecosystem Services, German Centre for Integrative Biodiversity Research (iDiv), Halle-Jena-Leipzig, Germany

Samuel R. Fennell
Human Dimensions of Natural Resources Department, Colorado State University, Fort Collins, CO, United States

Tinkani Gondwe
SFHC, Ekwendeni, Malawi

Andrew Gosler
School of Anthropology and Museum Ethnography; Edward Grey Institute of Field Ornithology, Department of Zoology, University of Oxford, Oxford, United Kingdom

Jose Guzman
Human Dimensions of Natural Resources Department, Colorado State University, Fort Collins, CO, United States

Zuzana Veronika Harmáčková
Global Change Research Institute of the Czech Academy of Sciences, Brno, Czech Republic

Asier Hernandez Saez
Human Dimensions of Natural Resources Department, Colorado State University, Fort Collins, CO, United States

Gordon M. Hickey
Department of Natural Resource Sciences, McGill University, Montreal, QC, Canada

Jennifer M. Holzer
Environmental Sustainability Research Centre, Brock University, St. Catharines, ON, Canada

Roman Isaac
Social-Ecological Systems Institute, Leuphana University of Lüneburg, Lüneburg, Germany

Aaron Iverson
St. Lawrence University, Canton, NY, United States

Jessica Ives
Healthy Headwaters Lab, Great Lakes Institute for Environmental Research, Department of Integrative Biology, University of Windsor, Traditional Territory of the Three Fires Confederacy, Windsor, ON, Canada

Lingaraj G. Jayaprakash
Department of Natural Resource Sciences, McGill University, Montreal, QC, Canada

Jana Kachler
Department of Ecosystem Services, Helmholtz Centre for Environmental Research—UFZ, Leipzig; Faculty of Biological Sciences, Friedrich-Schiller-University Jena, Jena; Department of Ecosystem Services, German Centre for Integrative Biodiversity Research (iDiv), Halle-Jena-Leipzig, Germany

Penjani Kanyimbo
SFHC, Ekwendeni, Malawi

Katrina Keeshig
Healthy Headwaters Lab, Great Lakes Institute for Environmental Research, Department of Integrative Biology, University of Windsor, Traditional Territory of the Three Fires Confederacy, Windsor, ON, Canada

Daniel Kpienbaareh
Cornell University, Ithaca, NY, United States; Western University, London, ON, Canada

Georg Küstner
University of Würzburg, Würzburg, Germany

Gabby Lamarche
Biology, Acadia University, Wolfville, NS, Canada

Julia Leventon
Global Change Research Institute of the Czech Academy of Sciences, Brno, Czech Republic

Isaac Luginaah
Western University, London, ON, Canada

Esther Lupafya
SFHC, Ekwendeni, Malawi

Mark L. Mallory
Biology, Acadia University, Wolfville, NS, Canada

Berta Martín-López
Social-Ecological Systems Institute, Leuphana University of Lüneburg, Lüneburg, Germany

Vera Mayer
University of Würzburg, Würzburg, Germany

Tesfai Mehreteab
Norwegian Institute of Bioeconomy Research (NIBIO), Oslo, Norway

Pablo F. Méndez
Institute of Innovation and Knowledge Management, INGENIO (CSIC-UPV), Universitat Politècnica de València, Valencia, Spain

Abigail Mensah
Geography and Tourism Studies Department, Brock University, St. Catharines, ON, Canada

Innocent Mhoni
SFHC, Ekwendeni, Malawi

Mwapi Mkandawire
SFHC, Ekwendeni, Malawi

Tapiwa Mkandawire
SFHC, Ekwendeni, Malawi

Renata Moreno Quintero
Department of Social Sciences and Economics, Universidad Autónoma de Occidente, Cali, Colombia

Pressings Moyo
SFHC, Ekwendeni, Malawi

Petros Munthali
SFHC, Ekwendeni, Malawi

Udaya Sekhar Nagothu
Norwegian Institute of Bioeconomy Research (NIBIO), Oslo, Norway

Dane Nickerson
Human Dimensions of Natural Resources Department, Colorado State University, Fort Collins, CO, United States

Barbora Nohlová
Global Change Research Institute of the Czech Academy of Sciences, Brno; Faculty of Humanities, Charles University in Prague, Praha, Czech Republic

Hanson Nyantakyi-Frimpong
University of Denver, Denver, CO, United States

Lisa Pirie-Dominix
Canadian Wildlife Service, Environment and Climate Change Canada, Iqaluit, NU, Canada

Katja Poveda
Cornell University, Ithaca, NY, United States

Jennifer F. Provencher
Science and Technology Branch, Environment and Climate Change Canada, National Wildlife Research Centre, Raven Road, Carleton University, Ottawa, ON, Canada

Lauren Roberts
Biology, Acadia University, Wolfville, NS, Canada

Amy Schneider
Human Dimensions of Natural Resources Department, Colorado State University, Fort Collins, CO, United States

Theresa Selfa
Department of Environmental Studies, State University of New York-College of Environmental Science and Forestry, Syracuse, NY, United States

Lizzie Shumba
SFHC, Ekwendeni, Malawi

Ingolf Steffan-Dewenter
University of Würzburg, Würzburg, Germany

Lenka Suchá
Global Change Research Institute of the Czech Academy of Sciences, Brno, Czech Republic

Marley Taylor
Human Dimensions of Natural Resources Department, Colorado State University, Fort Collins, CO, United States

Yolice Tembo
Lilongwe University of Agriculture and Natural Resources, Lilongwe, Malawi

Bastian Thomsen
School of Anthropology and Museum Ethnography, University of Oxford, Oxford, United Kingdom; Human Dimensions of Natural Resources Department, Colorado State University, Fort Collins, CO; Environmental Studies Graduate Program, Oregon State University, Corvallis, OR, United States

Jennifer Thomsen
College Veterinarian Medicine and Biomedical Sciences, Colorado State University, Fort Collins, CO, United States

Jeannie Toomasie
Sululiit Area Co-Management Committee (Chair, 2011–2021), Iqaluit, NU, Canada

Simeon Vaňo
Global Change Research Institute of the Czech Academy of Sciences, Brno, Czech Republic; Department of Ecology and Environmental Sciences, Constantine the Philosopher University in Nitra, Nitra, Slovakia

Cassandra Vogel
University of Würzburg, Würzburg, Germany

Jinfei Wang
Western University, London, ON, Canada

Klara J. Winkler
Department of Natural Resource Sciences, McGill University, Ste-Anne-de-Bellevue, Canada

Introduction: Pluralism in ecosystem governance

Jennifer M. Holzer[a], Julia Baird[b], and Gordon M. Hickey[c]
[a]Environmental Sustainability Research Centre, Brock University, St. Catharines, ON, Canada
[b]Environmental Sustainability Research Centre/Geography and Tourism Studies, Brock University, St. Catharines, ON, Canada
[c]Department of Natural Resource Sciences, McGill University, Montreal, QC, Canada

> *There is a possible solution to all of this. We can try to get out of a conceptual framework that has allowed the present situation by looking at another. The Earth has produced humans in such a fashion that they have a certain flexibility in their thought patterns. There is not one language, but many. There is not one way to describe the workings of the planet, but many. And there is more than one set of explanations for the relationship between humans and the Earth.*
>
> ***Cordova (2007, p. 211)***

We live in the Anthropocene, whose name acknowledges the power of collective human society to shape ecosystems. Pluralism is a term befitting the Anthropocene. It means "diversity"—as applied to human groups and their multifarious ways of thinking and being. "Ecosystem governance" also captures the present-day power of human societies to steer the fate of ecosystems. While we are familiar with the call to save species from extinction, the disappearance of languages—a proxy for human cultures—is also occurring; fully half of the world's languages are not being passed on to children (Davis, 2007). Species pluralism is beautifully portrayed in the recent film *My Octopus Teacher* (Ehrlich and Reed, 2020), which documents the daily encounters of the filmmaker and his mollusc companion; human interactions with octopuses have led some humans to confront the possibility that the human brain is not the most highly evolved on the planet (Klein, 2021, July 13; Montgomery, 2015). In some religious communities, pluralism can mean adopting religiously stringent practices for gatherings so that even the most devout will be able to join mixed spaces. Pluralism can also be more subtle, as in recent literature on transdisciplinarity, which enjoins scientists to collaborate in mixed groups—with decision-makers, conservation groups, and locals, among others—to forge solutions to regional environmental issues.

There has been growing recognition that ecosystem governance is predicated upon human values (Jacobs et al., 2020), that the inclusion of stakeholders and rightsholders in decision-making is essential, and that expert knowledge and evidence is socially embedded and therefore not value-neutral, all of which invites dialogues on pluralism. Recent calls to conceptualize biodiversity and apply it in conservation settings using engagement with diverse knowledge and value systems (Pascual et al., 2021) have proliferated, often linked to international forums such as the Post-2020 Global Biodiversity Framework and the Inter-governmental Science-Policy Platform on Biodiversity and Ecosystem Services (IPBES) Values Assessment and Nature's Future Framework (Raymond et al., 2022). The term "inclusive conservation" has emerged "to theorize and operationalize diversity and equity in conservation science and protected areas management" (Raymond et al., 2022). Inclusive conservation builds on multiple governance approaches, such as co-management (power sharing arrangements "occurring between state-based and community-based systems", Plummer and Armitage, 2007), mosaic governance ("governance that is sensitive to the diversity and dynamics of active citizenship and which aligns with local informal networks and across scales" Buijs et al., 2016, p. 3), and convivial conservation ("'a vision, a politics and a set of governance principles that realistically respond to the core pressures of our time' by promoting 'radical equity, structural transformation and environmental justice'", Massarella et al., 2021, p. 83). However, Raymond et al. (2022) have articulated some tensions that can arise with inclusive conservation, such as power asymmetries and the conflict between recognizing and reducing plurality, and they suggest using multi- and mixed methods research designs to advance understanding to address these tensions. It appears that we are at a critical juncture in ecology and conservation in which many voices are calling to challenge views that exclude diversity or conceptualize separations between humans and societies from nature (West et al., 2020; Wyborn et al., 2021). This special issue brings together studies that employ a range of methods and approaches from diverse fields to advance understandings of pluralism in action, using ecosystem governance case studies and examples. We now turn to definitions and then introduce the themes and chapters of this special issue.

Defining and understanding pluralism

Pluralism is advocacy for the coexistence of differing views, values, and cultures to promote ecosystem integrity (OED, 2022). We define pluralism in the context of translational ecology, which is "an intentional approach in which ecologists, stakeholders, and decision-makers work collaboratively to develop and deliver ecological research that, ideally, results in improved environment-related decision making" (Enquist et al., 2017, p. 541). While definitions of pluralism across various fields (e.g. philosophy, political theory, law) restrict the term to refer to human diversity, we stretch the definition to include non-humans, extending agency to animals and landscapes. This view is aligned with contemporary thinking in post-humanist studies that seeks to break down "the myth of human exceptionalism" (Appleby and Pennycook 2017, p. 239) and pathbreaking legal precedents, such as recent decisions by the Supreme Court of Colombia to recognize rivers and related ecosystems as legal entities as a mechanism to protect them (Macpherson et al., 2020). By using the term "pluralism", we acknowledge that diversity is an important and positive force in the context of dynamic social-ecological systems (Kassam, 2010) and that greater diversity may make processes of translational ecology more effective.

Within the human realm, environmental pluralism may be described as a view in which the world is "infinitely complex, fluid, and negotiated, frustrating any attempt to impose a master narrative as an overarching, organizing frame" (Hull, 2009). Environmental pluralism seeks to build a broad political base, encourage experimentation, and strive for "higher ground" where many of the preferred outcomes of multiple stakeholders exist (Hull, 2009). It strives to overcome the polarization and paralysis symptomatic of contemporary environmental decision-making (Hull, 2009). Beyond personal values and worldviews, pluralism can also refer to ideological diversity and a "commitment to the disagreement that necessarily accompanies ideological diversity" (Brush, 2020, p. 161). Brush (2020) suggests that the complexity and multi-dimensionality of environmental challenges require intellectually and ideologically diverse responses and that this is clearly a benefit rather than a liability. Such a view is counter to calls by major environmental philosophers for a universal environmental philosophy or ethic (Brush, 2020), which may not be possible because of cultural differences (Cocks, 2017). Further, without accounting for culture, so-called pluralism might ignore the worldviews of most of the world's people. This idea has been fortified in recent years by contributions from post-colonial theory.

Post-colonial theory concerns confronting, challenging, and undoing the dominative and assimilative force of colonialism as a historical and contemporary process as well as the cultural and epistemological Eurocentrism that underwrites it (De Lissovoy, 2010; Kohn & Reddy, 2022). By incorporating decolonial views into our vision of pluralism, we acknowledge that the context and perspectives introduced thus far tend to reside in the context of Western academic thought and the colonial, exclusivist, and elitist structures embedded within it (De Lissovoy, 2010; McDowell and Hernández, 2010). This special issue aims to advance a view of pluralism that will not constitute or perpetuate barriers for anyone who wishes to engage in a dialogue about ecosystem governance.

Defining ecosystem governance

By "ecosystem", we refer generally to Tansley's 1935 definition that refers to "... the whole system ... including not only the organism-complex, but also the whole complex of physical factors forming what we call the environment of the biome—the habitat factors in the widest sense. Though the organisms may claim our primary interest, when we are trying to think fundamentally we cannot separate them from their special environment, with which they form one physical system" (as quoted in Gignoux et al., 2011). The term "ecosystem governance" implies that an ecosystem is shaped by the decisions humans make about their individual and collective activities within that system, which determine how the biosphere, hydrosphere, and geosphere are sustained, harvested, extracted, conserved, and managed, and how these decisions are made. In this context, governance is not strictly about formal institutions; rather, it includes all actors who make decisions, benefit from, or are otherwise involved in an ecosystem.

For the purpose of this special issue, we define ecosystem governance as the integrated and holistic governance of ecosystems that includes formal and informal institutions, collaboration and cooperation across different levels of government, and non-governmental and individual action. It acknowledges and addresses the integration of the social system with the ecosystem, and issues of scale, complexity, and uncertainty. It is adaptive (using experimentation and learning to responsively change) and collaborative (recognizing that governments alone are not sufficient for governance and thus convening additional actors) (Berkes, 2012; Cosens, 2010; Garmestani and Benson, 2013; Gunderson et al., 2016; Karkkainen, 2002). To be successful, ecosystem governance should operate in a transparent and equitable manner, with the goal of improving the long-term well-being of people (Adhikari and Baral, 2018; Olsen et al., 2009). We distinguish between governance and management, focusing on governance as the institutional (formal and informal) processes and structures that create a vision and direction for management (the actions and day-to-day decisions taken) (Cosens, 2013; Olsen et al., 2009).

The translational ecology approach

Translational ecology emphasizes linking ecological science to evidence-based decision-making, using cross-disciplinary research to address "the sociological, ecological, and political contexts of an environmental problem", "characterized by extended commitment to real-world outcomes", to "inform and improve decision making for environmental management and conservation" (Enquist et al., 2017, p. 541). Such an approach requires "deliberate engagement with end-users and an understanding of the social and cultural contexts in which a research project functions" (Wall et al., 2017, p. 551). Translational ecology acknowledges that while usable science exists, much of it does not get used because of implementation challenges (Wright et al., 2020). Wright et al. (2020) suggest that while decision science can help account for the uncertainty inherent in ecological systems, it is nonetheless challenging to account for the depth of complexity and uncertainty of socio-political systems.

While pluralism is not a new idea, recent and on-going trends—calls for broadening participation in ecosystem governance (e.g. Biggs et al., 2012), the rise of transdisciplinary environmental research (e.g. Brandt et al., 2013), sustainability crises such as climate change-related extreme weather events and the COVID-19 pandemic, and calls for greater diversity, equity and inclusivity in academia, government, and beyond—have prompted renewed urgency in understanding how institutions and persons diverse in their needs, views, interests, and behaviours can effectively govern ecosystems to promote sustainability and human well-being.

Summary of the special issue

This special issue features articles that highlight the benefits and challenges of pluralism in ecosystem research and governance, synthesizing work that includes scholars' own reflections. Upon request by the editors, the authors of each paper have explicitly defined pluralism for the purpose of their study as well as included relevant reflexive comments. For our own part, we, the editors, are Canadian-based social scientists and our work focuses on ecosystem governance and management. We intentionally defined "pluralism" broadly in an effort to engage diverse authors and contributions from a range of institutions, backgrounds, and geographies. One intractable challenge of editing a volume of inter-disciplinary papers is in defining common standards for studies that may be subject to distinct standards in each of their "home" disciplines. We needed to make choices, for example, about whether data sets justified conclusions, how to introduce theories from the humanities and social sciences in a manner appropriate for a journal of ecological science, and whether to include studies that included some aspects of pluralism without situating their work explicitly as such.

The resulting chapters in this volume are grouped thematically, and while these categories are not mutually exclusive, they are used to orient readers and draw out three key themes: *Spotlight on Indigenous and local knowledge*, *Amplifying muted voices*, and *Methods and approaches to foster pluralism in translational ecology*. *Spotlight on Indigenous and local knowledge* features three studies that address ecosystem and food system challenges through transdisciplinary, pluralistic ecosystem governance and knowledge production approaches. In Chapter 1, Kpienbaareh and colleagues wrangle with the question: How can agroecological research be designed to effectively engage smallholder farmers, whose farming practices rely on ecosystem services and diverse biodiversity, and who have had diverse experiences with colonialism, imperialism, and the scientific development of agriculture? The authors identify both agroecology and agroecological research as inherently pluralistic, and as processes in need of further decolonization, by being "culturally appropriate, respectful, honoring, and caring", and the research and knowledge production processes a "collective, collaborative endeavor". Their agroecological research project is considered pluralistic in that it uses participatory research with smallholder farmers, incorporating ecological data and participatory GIS and Indigenous knowledge, putting the experience of smallholder farmers and decolonial research approaches at the centre of the project. Similarly, in Chapter 2, Mallory and colleagues focus on the importance of epistemological pluralism to the successful co-management of wildlife in Nunavut, Canada. Specifically, the authors describe how government scientists have worked with the Nattivak Hunters' and Trappers' Organization for over two decades in the community of Qikiqtarjuaq to census wildlife and document local ecological knowledge to successfully develop national wildlife areas (NWAs). They focus on how international and national conservation and research goals can align with local, place-based interests and concerns through carefully designed collaboration facilitated by inclusive co-management committees. Focusing on the challenges posed by bringing together pluralities of knowledge to collaborative governance and co-management, Badry and Hickey in Chapter 3 present a literature review to identify opportunities for boundary spanning drawing on "post-human" approaches. Speaking primarily to an intended audience of non-Indigenous scholars, this paper critiques current approaches to knowledge integration between Western and traditional ecological knowledges. It proposes actor-network theory as a tool for integration, while also acknowledging the value of Indigenous knowledge integration approaches. Actor-network theory is defined as a methodological approach, or "family of tools, sensibilities, and analytical methods used to account for the webs of relations which constitute the social ...". Their chapter highlights several contexts in which Indigenous and Western worldviews seem incommensurable and suggests that actor-network theory may offer a method to "help researchers and practitioners to incorporate ontological pluralism when working across knowledge systems" in transboundary natural resource settings (Chapter 3). Together, these chapters offer working examples of projects, initiatives, and frameworks seeking to facilitate pluralistic knowledge co-production among the scientists, bureaucrats, and Indigenous and local communities governing ecosystems.

Amplifying muted voices uses ethnography and in-depth interviews in chapters that may be unconventional for an ecology journal in their methods and in the depth to which they foreground individual experience. In Chapter 4, Jayaprakash and Hickey highlight the experiences of marginal human and non-human actors in a peri-urban forest landscape in India using environmental ethnographic methods. Their research aims to "hear the voices" of marginalized forest agents to better understand the complexity of everyday forest ecosystem management and inform policy efforts to incorporate pluralism. The findings illustrate the challenge facing formal government-led forest management and conservation efforts where a plurality of forest uses and forest institutions require diverse actors to operate through multi-layered processes of institutional bricolage—the creative processes through which local actors refashion rules, practices, and norms to make institutions work for their lived realities. Invoking ethics, Thomsen and colleagues in Chapter 5 advocate for non-human species agency, welfare, and rights using a post-humanist perspective of grey wolf-human relations in the Western United States. The authors use a method known as multi-species ethnography, employing interviews with ranchers, environmentalists, and government agents "to deconstruct human exceptionalism while maintaining the core tenets of human agency ... challeng[ing] humans to ... consider nonhumans' agency in shared landscapes". Findings highlight speciesism, the tendency of interviewees to speak about wolves in the abstract, and "the use of harvest to justify the hunting, trapping, and killing of wolves ethically result[ing] in what could be considered systematic state-sanctioned murder" (Chapter 5). Their work points to the impact of power asymmetries among different human groups and posits that "if this framework can occur in practice … nonhumans could gain previously unimaginable rights …" (Chapter 5). In Chapter 6, Moreno Quintero and Selfa provide a narrative case study of how power is shaped and contested within wetland governance in Valle del Cauca, Colombia. Pluralism here refers to including varying levels of power in state and citizen institutions, with a call to expand the engagement of underrepresented groups in environmental policy processes, plural values/priorities related to the wetland, and polycentric wetland governance. The case describes how activists defined the problem, mobilized networks, and honed their messaging to stave off further wetland destruction, using these events to analyse how the state and citizens interact to manage environmental problems. The authors identify that the actions taken did not alter existing power structures or mitigate existing environmental damage They identify the need to first address social inequities, enhance legitimacy and accountability through multi-level processes, and generate new incentives to invest in ecosystem governance. Each of these chapters acknowledges the distinctness but interdependence of agents with significant power differentials, and the pluralism of institutions, while also highlighting the patchwork of formal, informal, and grey areas of governance modes and structures. They point to the need to rethink the roles and identities of marginalized actors in dominant ecosystem conservation and management paradigms.

Methods and approaches to foster pluralism in translational ecology open with Leventon and colleagues' review of participation methods (Chapter 7), a critical perspective that focuses on how broadening participation in ecosystem governance has been addressed in different disciplines (beyond ecology). The chapter argues for social plurality in participation and offers reflexive questions to guide governance design and implementation. Social plurality is defined as a "myriad of values, worldviews, beliefs and positions" being recognized and accepted to influence priorities and extends as far as "normative visions for future ecosystems and how these can be achieved". They argue that participation offers a way to bring social plurality to ecosystem governance, drawing on literature from the interdisciplinary fields of development studies and anthropology; environmental governance and planning; and sustainability transitions and transformations. They identify several challenges and issues to consider in achieving plurality through participation, and they highlight multiple spaces for participating in ecosystem governance. In Chapter 8, Febria and colleagues focus on pluralism as a way to overcome barriers to freshwater restoration efforts, requiring coordination among multiple actors across ecosystems. The authors recognize the limits of Western scientific approaches and the importance of local knowledge holders. They present a "community-responsive" framework for science-based research designed to be locally relevant as well as translatable to different contexts. The framework has three dimensions: "strategic alignment of science with decision-making contexts and needs", "holistic and decolonial approaches to pursuing science that [are] community-driven", and "partnerships across disciplines, communities, and knowledge systems to overcome barriers or mobilization of knowledge into policy and practice". The chapter describes the process of developing the framework and shares insights to how the pluralistic approach benefitted the applied research activities (both short- and long-term, benefits to whom, and trade-offs). As such, pluralism is described in terms of a research approach to determine what is studied, how, and who will benefit. In Chapter 9, Méndez and colleagues view pluralism in democratic discourse as essential for equitable and viable change processes towards sustainability in social-ecological systems. A case study of a hydraulic megaproject in the Doñana region of Spain is presented, where limited pluralism led to "discourse inertia" and ultimate project failure. They draw on the concepts of epistemic pluralism and methodological pluralism to identify the need for transdisciplinary knowledge co-production to inform the governance of sustainability change processes. Specifically, the authors apply a framework that blends approaches for understanding institutional-discursive logics to illustrate how limited pluralism emerges and can be perpetuated in ecosystem governance, and how this situation might be addressed to foster sustainability pathways. Finally, in Chapter 10 Isaac and colleagues present a systematic review of ecosystem service governance. The authors distinguish between different modes of governance—hierarchical, scientific-technical, collaborative, and the governing of strategic behaviour—and then tie these modes to anthropogenic capitals (human social, physical, and financial) involved in the co-production of nature's contributions to people. They identify that most research has looked at the interplay of natural capital with multiple anthropogenic capitals and highlight a research gap in the nature's contributions to people co-production literature, in that it tends to neglect the role of governance when assessing the interplay of anthropogenic and natural capitals. Together, the chapters in this theme provide insight to the institutional,

Fig. 1 Reaching for and integrating pluralism in ecosystem governance.

social, and research dynamics that can influence pluralism in knowledge production and on-the-ground ecosystem governance.

Fig. 1 illustrates and synthesizes the dimensions of pluralism covered by the collection of papers included in this special issue. The octopus and diver (*àla My Octopus Teacher)* signify an eye-to-eye relationship (i.e. I-Thou relationship, Buber, 1971) between human and non-human, communicating as equals, representing pluralism in its broadest sense—multi-species (i.e. post-humanist) pluralism. The octopus's tentacles grasp the three themes of this volume: (1) the importance of Indigenous and local knowledge, (2) the amplification of less-powerful voices, and (3) advancing new methods and approaches to improve pluralism in practice. The octopus and three panels are entangled in kelp, representing the many inter-connections and links between different aspects of scholarship and practice as well as the web of connections within and between social-ecological systems. Panel 1 (left) portrays a grandmother telling stories to her granddaughter, who asks questions, eager to understand her heritage, which is indivisibly linked to the place, ecology, and species of her ancestral lands. Panel 2 represents examples of less powerful individuals: an elephant, a river, agricultural labourers, and an urban worker-activist. They represent the kinds of persons whose perspectives are often not heard or are not prioritized; they often lack power and influence in governance contexts. Panel 3 shows methods and approaches to cultivate pluralism. The top half shows field research—a scientist interviewing a farmer, perhaps using ethnographic or participant-observer methods. The bottom half shows desk research, which could be anything from data analysis, systematic literature review, or image analysis to an online workshop or geographic information systems. This illustration may consist of images that are not most readily recalled when bringing to mind the term "ecosystem governance"; when we speak of governance, we may think more easily of government, scientists collecting data, voting in elections, courtroom decisions, monitoring and enforcement activities, and management, planning, and reporting processes involving suits sitting around tables with bottled water. By contrast, Fig. 1 relays images of more pluralistic governance in research and practice, suggesting that we as scientists could broaden the image of ecosystem governance in our minds as we design and conduct research.

Given the range and complexity of ecological challenges in the Anthropocene, pluralism in ecosystem governance is a fact. Top-down governance has given way, in most cases, to participatory and multi-level approaches. An array of actors is involved in decision-making and many more are affected by those decisions. While that has always been true, today we have greater awareness and understanding of social-ecological dynamics. With this knowledge comes greater responsibility. Bringing a greater diversity of perspectives to bear on ecological decision-making brings with it greater need for effective communication and collaboration. Effective communication often begins with reflexivity, parsing for oneself and explaining to others one's identity, background, affiliations, and aims (Chapters 7 and 8). Collaboration may sometimes necessitate civil disagreement and compromise. In mixed groups, some topics or perspectives may be a non-starter or be ill-received; these situations call for empathy, flexibility, and creativity. When it comes to research and knowledge co-production, doing pluralistic, translational social-ecological research is methodologically challenging. While drawing from different types of knowledge, methods, and approaches, researchers must make "careful selections of criteria and methods for the research while appreciating the limitations of the various methods ..." (Chapter 1). A more pluralistic practice of translational ecology will need to maintain rigorous standards while integrating widely differing ontologies, perspectives, and knowledge. Developing and applying new frameworks, approaches and methods will be necessary for collaborating across differences. Decolonizing science, speaking up for the less powerful and making science more collaborative can require significant additional time and effort. Advocating for this type of "slower science" can be a liability where "publish or perish" still rules, so for this and many other reasons, incorporating greater pluralism in ecosystem research and governance represents a major and ongoing challenge.

Nearly a century ago, Aldo Leopold's *land ethic* expanded the definition of "community" to include parts of the Earth beyond the human (Leopold, 1989). Today, Indigenous scholar Robin Wall Kimmerer helps us understand the meaning of reciprocity between ourselves and other species (Kimmerer, 2013). Participatory and collaborative processes may require a new ethic of inclusivity, integrating today's public conversations about equity, diversity, and decolonization. This volume offers a lens for ecosystem scientists to consider their work in relation to translational ecology and pluralism, with a focus on examples of how some scientists seek to incorporate a pluralist approach to their work and the issues arising therein. Combined, they remind us that as far as we have come, we are still in need of refining approaches, competencies—and the will—to practice pluralism.

References

Adhikari, S., Baral, H., 2018. Governing forest ecosystem services for sustainable environmental governance: a review. Environments 5 (5), 53.

Appleby, R., Pennycook, A., 2017. Swimming with sharks, ecological feminism and posthuman language politics. Crit. Inq. Lang. Stud. 14 (2-3), 239–261.

Berkes, F., 2012. Implementing ecosystem-based management: evolution or revolution? Fish. 13 (4), 465–476.

Biggs, R., Schlüter, M., Biggs, D., Bohensky, E.L., BurnSilver, S., Cundill, G., et al., 2012. Toward principles for enhancing the resilience of ecosystem services. Annu. Rev. Env. Resour. 37, 421–448.

Brandt, P., Ernst, A., Gralla, F., Luederitz, C., Lang, D.J., Newig, J., Reinert, F., Abson, D.J., Von Wehrden, H., 2013. A review of transdisciplinary research in sustainability science. Ecol. Econ. 92, 1–15. https://doi.org/10.1016/j.ecolecon.2013.04.008.

Brush, E., 2020. Inconvenient truths: pluralism, pragmatism, and the need for civil disagreement. J. Environ. Stud. Sci., 1–9.

Buber, M., 1971. I and Thou. Simon and Schuster, New York.

Buijs, A.E., Mattijssen, T.J., Van der Jagt, A.P., Ambrose-Oji, B., Andersson, E., Elands, B.H., Møller, M.S., 2016. Active citizenship for urban green infrastructure: fostering the diversity and dynamics of citizen contributions through mosaic governance. Curr. Opin. Environ. Sustain. 22, 1–6.

Cocks, S., 2017. Environmental pluralism, polar harmonies and resolution. Dialogue Universalism 4, 9–21.

Cordova, V.F., 2007. How It Is: The Native American Philosophy of VF Cordova. University of Arizona Press.

Cosens, B., 2010. Transboundary river governance in the face of uncertainty: resilience theory and the Columbia River Treaty. J. Land Resour. Environ. Law 30, 229–265.
Cosens, B.A., 2013. Legitimacy, adaptation, and resilience in ecosystem management. Ecol. Soc. 18 (1). http://dx.doi.org/10.5751/ES-05093-180103.
Davis, W., 2007. Light at the Edge of the World: A Journey Through the Realm of Vanishing Cultures. Vancouver/Toronto/Berkeley, Douglas and McIntyre.
De Lissovoy, N., 2010. Decolonial pedagogy and the ethics of the global. Discourse 31 (3), 279–293.
Ehrlich, P., Reed, J. (Directors), 2020. *My Octopus Teacher* [Film]. Off the Fence and ZDF Enterprises, in partnership with Sea Change Project.
Enquist, C.A., Jackson, S.T., Garfin, G.M., Davis, F.W., Gerber, L.R., Littell, J.A., et al., 2017. Foundations of translational ecology. Front. Ecol. Environ. 15 (10), 541–550.
Garmestani, A. S., and M. H. Benson. 2013. A framework for resilience-based governance of social-ecological systems. Ecol. Soc. 18 (1), 9. https://doi.org/10.5751/ES-05180-180109.
Gignoux, J., Davies, I.D., Flint, S.R., Zucker, J.D., 2011. The ecosystem in practice: interest and problems of an old definition for constructing ecological models. Ecosystems 14 (7), 1039–1054.
Gunderson, L.H., Cosens, B., Garmestani, A.S., 2016. Adaptive governance of riverine and wetland ecosystem goods and services. J. Environ. Manage. 183, 353–360.
Hull, R.B., 2009. Environmental pluralism. In: Callicott, J.B., Frodeman, R. (Eds.), Encyclopedia of Environmental Ethics and Philosophy. Macmillan Reference USA, Detroit, pp. 384–387.
Jacobs, S., et al., 2020. Use your power for good: plural valuation of nature—the Oaxaca statement. Global Sustain. 3,e8, 1–7. https://doi.org/10.1017/sus.2020.2.
Karkkainen, B.C., 2002. Collaborative ecosystem governance: scale, complexity, and dynamism. Virginia Environ. Law J., 189–243.
Kassam, K.S., 2010. Pluralism, resilience, and the ecology of survival: case studies from the Pamir Mountains of Afghanistan. Ecol. Soc. 15 (2), 8. http://www.ecologyandsociety.org/vol15/iss2/art8/.
Kimmerer, R., 2013. Braiding Sweetgrass: Indigenous Wisdom, Scientific Knowledge and the Teachings of Plants. Milkweed Editions.
Klein, E., 2021, Jul 13. How octopuses upend what we know about ourselves. [Audio podcast episode]. In: The Ezra Klein Show. The New York Times. https://www.nytimes.com/column/ezra-klein-podcast.
Kohn, M., Reddy, K., 2022. "Colonialism", The stanford encyclopedia of philosophy (Summer 2022 Edition). In: Zalta, E.N. (Ed.), forthcoming. https://plato.stanford.edu/archives/sum2022/entries/colonialism/.
Leopold, A., 1989. A Sand County Almanac, and Sketches Here and There. Oxford University Press, USA.
Macpherson, E., Ventura, J.T., Ospina, F.C., 2020. Constitutional law, ecosystems, and indigenous peoples in Colombia: biocultural rights and legal subjects. Transnatl. Environ. Law 9 (3), 521–540.
Massarella, K., Nygren, A., Fletcher, R., Büscher, B., Kiwango, W.A., Komi, S., et al., 2021. Transformation beyond conservation: how critical social science can contribute to a radical new agenda in biodiversity conservation. Curr. Opin. Environ. Sustain. 49, 79–87.
McDowell, T., Hernández, P., 2010. Decolonizing academia: intersectionality, participation, and accountability in family therapy and counseling. J. Fem. Fam. Ther. 22 (2), 93–111.
Montgomery, S., 2015. The Soul of an Octopus: A Surprising Exploration Into the Wonder of Consciousness. Atria Books.
OED Online, February 2022. Oxford University Press. Web.

Olsen, S.B., Page, G.G., Ochoa, E., 2009. The Analysis of Governance Responses to Ecosystem Change: A Handbook for Assembling a Baseline. GKSS Research Centre, LOICZ International Project Office, Institute for Coastal Research.
Pascual, U., Adams, W.M., Díaz, S., Lele, S., Mace, G.M., Turnhout, E., 2021. Biodiversity and the challenge of pluralism. Nat. Sustain. 4 (7), 567–572.
Plummer, R., Armitage, D.R., 2007. Charting the new territory of adaptive co-management: a Delphi study. Ecol. Soc. 12 (2), 10. [online] URL: http://www.ecologyandsociety.org/vol12/iss2/art10/.
Raymond, C.M., Cebrián-Piqueras, M.A., Andersson, E., Andrade, R., Schnell, A.A., Romanelli, B.B., et al., 2022. Inclusive conservation and the Post-2020 Global Biodiversity Framework: tensions and prospects. One Earth 5 (3), 252–264.
Wall, T.U., McNie, E., Garfin, G.M., 2017. Use-inspired science: making science usable by and useful to decision makers. Front. Ecol. Environ. 15 (10), 551–559.
West, S., Haider, L.J., Stålhammar, S., Woroniecki, S., 2020. A relational turn for sustainability science? Relational thinking, leverage points and transformations. Ecosyst. People 16 (1), 304–325.
Wright, A.D., Bernard, R.F., Mosher, B.A., O'Donnell, K.M., Braunagel, T., DiRenzo, G. V., Fleming, J., Shafer, C., Brand, A.B., Zipkin, E.F., Campbell Grant, E.H., 2020. Moving from decision to action in conservation science. Biol. Conserv., 249(June), 108698. https://doi.org/10.1016/j.biocon.2020.108698.
Wyborn, C., Montana, J., Kalas, N., Clement, S., Davila, F., Knowles, N., et al., 2021. An agenda for research and action toward diverse and just futures for life on Earth. Conserv. Biol. 35 (4), 1086–1097.

Spotlight on Indigenous and local knowledge

CHAPTER ONE

Transdisciplinary agroecological research on biodiversity and ecosystem services for sustainable and climate resilient farming systems in Malawi

Daniel Kpienbaareh[a,b], R. Bezner Kerr[a,*], Hanson Nyantakyi-Frimpong[c], Daniel Amoak[b], Katja Poveda[a], Udaya Sekhar Nagothu[d], Cassandra Vogel[e], Aaron Iverson[f], Tesfai Mehreteab[d], Isaac Luginaah[b], Ingolf Steffan-Dewenter[e], Jinfei Wang[b], Georg Küstner[e], Stephanie Enloe[a], Vera Mayer[e], Laifolo Dakishoni[g], Esther Lupafya[g], Lizzie Shumba[g], Timothy Chunga[g], Penjani Kanyimbo[g], Petros Munthali[g], Tinkani Gondwe[g], Innocent Mhoni[g], Mwapi Mkandawire[g], Tapiwa Mkandawire[g], Pressings Moyo[g], and Yolice Tembo[h]

[a]Cornell University, Ithaca, NY, United States
[b]Western University, London, ON, Canada
[c]University of Denver, Denver, CO, United States
[d]Norwegian Institute of Bioeconomy Research (NIBIO), Oslo, Norway
[e]University of Würzburg, Würzburg, Germany
[f]St. Lawrence University, Canton, NY, United States
[g]SFHC, Ekwendeni, Malawi
[h]Lilongwe University of Agriculture and Natural Resources, Lilongwe, Malawi
*Corresponding author: e-mail address: rbeznerkerr@cornell.edu

Contents

Advances in Ecological Research, Volume 66
ISSN 0065-2504
https://doi.org/10.1016/bs.aecr.2022.04.001

Abstract

How can agroecological research methods effectively engage smallholder farmers, who provide over half of the world's food supply, and whose farm management activities have significant impacts on biodiversity and ecosystem services? This question is highly relevant in Malawi where the research took place, but in other low-income countries in Africa with mostly agrarian populations, in which multi-scalar processes drive high food insecurity, alongside declining biodiversity, worsening land degradation and climate change. We analyse an innovative transdisciplinary agroecological approach that attempts to bridge the science-practice-policy gap by examining the potential of agroecological measures to enhance functional biodiversity and ecosystem services. This study involves a longitudinal, case-control and participatory research design in a region where thousands of farmers have experimented with agroecological practices, e.g., legume intercropping, composting, and botanical sprays. Innovative transdisciplinary agroecological research activities involved farmer participatory research, ecological monitoring and field experiments, social science methods (both qualitative and quantitative), participatory methodologies (public participatory Geographic Information Systems - PPGIS and scenario planning and testing) and stakeholder engagement to foster science-policy linkages. We discuss the theoretical and methodological implications of this novel transdisciplinary and participatory approach about pluralism, decolonial and translational ecological research to foster sustainability and climate resilience of tropical farming systems.

1. Introduction

Globally, the rate of food insecurity, biodiversity loss and degradation of ecosystem services has increased (IPBES, 2019; IPCC, 2021). There are nearly 1 billion people worldwide who experience hunger, many of them living in sub-Saharan Africa (FAO et al., 2019). Further, an estimated 1 million species of plants and animals are at risk of extinction within the next few decades (IPBES, 2019) which will have additional negative effects on crop productivity. This precipitous increase in food insecurity and biodiversity loss is the result of interacting threats including low crop productivity, habitat loss and alteration, direct mortality of wildlife from (over)exploitation, invasive species, emerging infectious disease, pollution and climate change (IPBES, 2019; IPCC, 2021). What is more, these threats act interactively and reinforce each other, thus making them complicated to address. These environmental challenges disproportionately impact smallholder farmers in the Global South, who paradoxically, produce the bulk of global food supply and most diverse food groups such as fruits and vegetables (Herrero et al., 2017).

Addressing food insecurity while ensuring environmental sustainability requires production systems that are resilient to climate change and other environmental challenges, while simultaneously enhancing ecosystem services and protecting biodiversity. Achieving climate resilient farming, especially in smallholder farming systems, requires a dramatic change from current production strategies that have so far made food security an illusion for smallholder farmers but have further degraded the environment and resulted in other socio-ecological challenges (Cánovas-Molina and García-Frapolli, 2021). The productivist, market-based model for agricultural production, reliant on inputs including synthetic fertilizers and pesticides, may have initially increased yields but is increasingly questioned for the costs to biodiversity and ecosystem services (IPBES, 2019). This model is also critiqued for an increasing concentration of the input and retail sectors, with costs for farmer livelihoods and a lack of consumer control over the food system (IPES-Food, 2016). In the Global South, many smallholder farmers cannot afford purchased inputs needed to maximize productivity or would prefer to use local seed varieties and planting technologies they know and trust (Rutsaert et al., 2021). Meanwhile, there is growing advocacy and preference for sustainable farming alternatives such as agroecology, a holistic approach to farming that uses ecological principles such as increasing synergies and enhancing biodiversity, and that integrates traditional ecological knowledge, alongside principles of fairness and connectivity between producers and consumers (HLPE, 2019; Wezel et al., 2020). Agroecological practices, which are adapted to a given social and environmental context, include composting, intercropping and agroforestry (Wezel et al., 2020). There is increasing evidence that agroecological approaches are more resilient to climate change (Snapp et al., 2021) while supporting food security and nutrition (Bezner Kerr et al., 2021). Yet the benefits of agroecological approaches on ecosystems services, biodiversity and livelihoods among smallholder farmers in Africa is not fully understood.

In Malawi, where the majority of the population relies mainly on agriculture, there are severe threats to ecosystems and biodiversity posed by a high rate of deforestation (Government of Malawi, 2016), driven by climate change, agricultural land expansion and other livelihood activities (Ngwira and Watanabe, 2019). About 39.7% of agricultural land in Malawi is degraded (Mbow et al., 2019), and more than 80% of the population reside in rural areas and also rely on agriculture and forest resources for their food, energy and other livelihoods needs, but where poverty is disproportionately higher (World Bank, 2019). Further, the majority of the population in Malawi is food insecure, with 40% of children under 5 years facing chronic undernutrition (National Statistical Office, 2017). Climatic events are further aggravating low productivity and inequalities, with extreme floods and droughts being indicative of the severity of climate change on agriculture and other livelihoods in the country. Recent climate change reports (IPCC, 2021), indicate that SSA will be hardest hit by negative impacts, with Malawi and other SSA countries likely experiencing higher temperatures, higher frequency and intensity of droughts and reduced yields, which will further widen poverty and inequality. Malawi's natural ecosystem forms part of the Miombo woodland ecoregion that is high in biodiversity (with over half the plants endemic to the region) but overall poorly protected, with heavy impacts by people (Ribeiro et al., 2020). The miombo ecosystems also contain valuable biodiversity that provides ecosystem services for the rural poor (Gumbo et al., 2018).

Amid the socio-ecological challenges, Malawi has largely pursued an input-intensive agriculture policy for food production to address low productivity and food insecurity through subsidies for fertilizer and hybrid seed. This input-intensive approach does not address underlying drivers of deforestation, biodiversity loss and land degradation. Significant improvements in food security or nutrition, have not been attained with these input-intensive farming strategies, with some reported productivity gains in the country likely being overestimated (Messina et al., 2017). At the same time, there is evidence that better-off households gain substantially more than poorer households when they participate in the input subsidy program (Lunduka et al., 2013), further widening existing inequalities.

As climate change intensifies and ecosystem services and biodiversity are depleted more rapidly, amid rising concerns about input-intensive agriculture in Malawi, it has become imperative to understand and integrate different knowledge systems across disciplines to ensure that the food system is climate-resilient to increase yield while enhancing ecosystem services and biodiversity. In the context of food insecurity and environmental degradation, agroecology has emerged as an alternative to transform the food system. Given the evolutionary background of agroecology from different scientific disciplines that integrates different knowledge systems (Méndez et al., 2013), sets of practices and social movements (Guzmán and Woodgate, 2013), the question arises as to how agroecological research can be designed to effectively engage smallholder farmers in other contexts, whose farming practices rely on ecosystem services and diverse biodiversity, and who have had different experiences with colonialism, imperialism and scientific development of agriculture.

Explicating the different experiences of local communities require adopting different methodologies across multiple disciplines and linking the perspectives of the community with their lived experiences and histories. Such experiences, when examined from multiple lenses, provide stakeholders in the agricultural and environmental resource management sector with the requisite information for effective policymaking. The histories of local people entail their ecological knowledge about local food systems and the environment. In this regard, translational ecologists posit that linking these ecological knowledge systems to decision-making at multiple scales by integrating ecological science with the full complement of social dimensions that underlie today's complex environmental issues will translate into tangible outcomes. Enquist et al. (2017) further outline the uniqueness of translational ecology for addressing complex issues through interdisciplinary team approaches and integrated scientist-practitioner partnerships. Thus, the creativity and context-specific knowledge of resource managers, practitioners, and decision-makers inform and enrich the scientific process and help shape use-driven, actionable science.

In this paper, therefore, we discuss transdisciplinary agroecological research implemented in Malawi that attempts to bridge the science-practice-policy gap by examining the potential of agroecological measures to ensure a climate-resilient farming approach, improve functional biodiversity and ecosystem services, enhance ecological, agronomic and social sustainability of farming among smallholder farmers. The research was implemented by an international team of researchers with diverse disciplinary backgrounds including geographers, ecologists, agronomists, sociologists, economists, and development workers, highlighting the pluralistic and transdisciplinary nature of agroecological research. An integrated, collaborative approach was adopted which includes working with local policymakers, chiefs, and traditional leaders, with the active participation of farmers.

In the next section, we present the overview of the farmer-led research project examined in this paper, followed by a discussion of agroecology as having transdisciplinary and pluralistic roots. We highlight the theories that ground our approach and discuss the methodologies used in the research and conclude with a synthesis of the entire research project.

2. Agroecological research—A transdisciplinary pluralistic agenda

Agroecology, defined as "the integrative study of the ecology of the entire food system, encompassing ecological, economic and social dimensions" (Francis et al., 2003, p. 100), entails various practices and processes that differentiate it from conventional agriculture. Agroecology is a practical approach to farming that encourages researchers, educators, and students to embrace the wholeness and connectivity of systems and stimulate a focus on the uniqueness of each place, and solutions appropriate to its resources and limitations. In this regard, agroecology is seen as an approach to transforming the food system in ways that provide environmental and social benefits.

Agroecology is subcategorized under three strands (Wezel et al., 2009)—(1) a scientific discipline—which evolved mainly from two disciplines - agronomy and ecology, but also other disciplines such as zoology and botany/plant physiology, and their applications in addressing agricultural and environmental issues; (2) agricultural practices—which applies principles that can be applied to food production in a particular context, realized through various sustainable land management practices including mulching, agroforestry, legume integration, crop rotation, mixed farming, living fences, biological corridors, microbial and botanical inputs, plants for soil cover, organic fertilizers and composting (Wezel et al., 2020). Key agroecology principles include: (i) co-production of knowledge between producers, scientists and other stakeholders, (ii) reducing the use of toxic and synthetic inputs and using natural means for controlling insect/pests and diseases,

(iii) using biological means such as organic matter to enrich the soil, (iv) conservation to minimize energy, water, nutrient and genetic resource loss, (v) farm-level and landscape-scale genetic diversification (Wezel et al., 2020); and (3) social movements—in some regions agroecology developed from agrarian social thought and movements that arose in response to rural industrialization which emerged as an unexpected outcome of the Green Revolution (Guzmán and Woodgate, 2013). Agroecology social movements envision food production systems that rely on traditional ecological knowledge, innovation, and the use of local resources (Altieri and Toledo, 2011). Agroecology aligns with food sovereignty—"the right of peoples to healthy and culturally appropriate food produced through ecologically sound and sustainable methods, and their right to define their food and agriculture systems" (Sélingué, 2007, p. 1). Thus, agroecology as a social movement empowers farmers, defends their fundamental rights such as access to land, free use of plant genetic resources, access to markets and a more inclusive political economy (Campesina, 2013).

Two additional key principles of agroecology at the food system level that relate to both transdisciplinary and pluralism is that of increasing participation of both producers and consumers in the governance of food systems, and co-production of knowledge (Wezel et al., 2020). Years of research in decolonial theories have addressed how past research approaches within local communities have failed to account for the culture of local people, thus creating a rift between researchers and community members (Datta, 2018). According to De Lissovoy (2010, p. 280) "decolonial theory is concerned with confronting, challenging, and undoing the dominative and assimilative force of colonialism as a historical and contemporary process, and the cultural and epistemological Eurocentrism that underwrites it." Integrating knowledge systems in a collaborative participatory manner allows both the research and the researchers to be decolonized (Datta, 2018) and creates a platform for consensus in identifying norms for governing ecosystems and improving agricultural systems (Tengö et al., 2014).

Levidow et al. (2014) assert that agroecology can play different roles by either conforming to the dominant food regime or by helping to transform this dominant food regime through strategic empowerment. One way that farmers are empowered is by allowing room for active participation in the design and implementation of farming strategies using locally available resources such as manure, compost, and a deep understanding of the nature of agroecosystems in local areas and the principles by which such agroecosystems function (Altieri, 2000). Agroecological transitions move beyond the farm scale, to consider unequal power dynamics in the food system, such as how agricultural research is done, and how to address social and economic inequities of the food system (Wezel et al., 2020).

The foregoing discussions reveal that addressing food system challenges from the agroecological standpoint requires approaching research from a transdisciplinary and pluralistic perspective. Transdisciplinary research is defined as studies conducted by researchers from different disciplines working collaboratively to co-create new conceptual, theoretical, methodological, and translational innovations that integrate and transcend discipline-specific approaches to address a common problem (Arnold, 2013). Transdisciplinarity in agroecology stems from its evolution from diverse disciplines and the integration of knowledge systems—both western scientific and traditional—to co-produce knowledge that helps to understand the food system and transform it (Francis et al., 2008; Méndez et al., 2013). The complexity and diversity of disciplinary dimensions that explain agroecology highlight the need for a pluralistic approach to research it. Pluralism describes the integration of different theories and methodologies to explain a complex research topic (Given, 2008). Investigating how agroecology can be made more climate-resilient to boost food security while simultaneously protecting ecosystem services and biodiversity is a complex research question that cannot be addressed with a single theory or methodology (Spash, 2009). In the next section of this paper, we will describe the farmer-led agroecological research that designed to encompass various discipline, methods, and theories.

3. Theories underpinning our approach: Indigenous knowledge, political ecology, and decolonizing research

Our research is informed by a political ecology approach to studying Indigenous knowledge (Goldman et al., 2011; Rocheleau et al., 1996). Many communities in Malawi maintain a body of Indigenous knowledge because of the longevity of their residence and an approach that emphasizes maintaining the sustainable uses of local resources (Bezner Kerr et al., 2018; Mweta and Juma, 2020). Following Berkes (2017, pp. 7–8), we define Indigenous knowledge as “a cumulative body of knowledge, practice, and belief, evolving by adaptive processes and handed down through generations by cultural transmission, about the relationship of living beings (including humans) with one another and with the environment." Indigenous knowledge is not static but is dynamic and adapted to ongoing socio-ecological changes. Any approach to studying Indigenous knowledge must be careful not to value formal scientific approaches over local knowledge (Berkes, 2017). A political ecology approach to studying Indigenous knowledge is therefore useful in attempting to break down the imbalances of power that shape knowledge production (Zanotti et al., 2020).

The work reported here is also informed by the theory of decolonizing knowledge production and the research process. In response to what many see as Western academic oppression, Indigenous researchers and community partners are increasingly calling for research to be decolonized (Bronen and Cochran, 2021; Datta, 2018). Decolonizing research is a process of conducting research with local communities that places Indigenous voices and epistemologies at the centre of the research process (Bronen and Cochran, 2021). This approach calls for research to be culturally appropriate, respectful, honouring, and caring for local communities (Zanotti et al., 2020). It connects participants and researchers with all parts of the study process as a collective collaborative endeavour (Zanotti et al., 2020). It critically examines the underlying assumptions that inform the research and challenges the widely accepted belief that Western methods and ways of knowing are the only objectives, true science. Decolonizing research is based on the assumption that holding Western beliefs and methods as the only true science marginalizes Indigenous methods and ways of knowing (Bronen and Cochran, 2021; Tengö et al., 2014). A growing body of scholarship has demonstrated how decolonizing research can help regain control over local ways of knowing and being, and ways in which research can be used for social change (Tengö et al., 2014). By focusing here on the decolonization of knowledge production and the research process, we do not seek to reject scientific methods and theories. Rather, we argue that these methods and theories may be adapted if deemed appropriate and beneficial by local communities.

In our work, research methods have been selected to create opportunities to learn through community experiences and co-learn with farmers, as has been done previously in Malawi (e.g., Bezner Kerr and Chirwa, 2004; Nyantakyi-Frimpong et al., 2017). Working closely with local farmers, key attributes of the ways we have attempted to decolonize the research process include collective data collection, data analysis, and ownership of research findings. The following section shows how we have operationalized these concepts and theories on the ground in rural Malawi.

4. Farmer-led agroecological research in Malawi using scenarios for biodiversity and ecosystem services (FARMS4Biodiversity)

Farms4Biodiversity was a transdisciplinary, multi-scalar research project designed to understand the relationships between biodiversity conservation, ecosystem services and food security using participatory methodologies under scenarios of land-use change in the Global South. The project arose following a large scale agroecology project (Kansanga et al., 2020), during which time farmer interest in ecological pest management strategies was identified. This project was implemented in 24 communities in Northern Malawi (2019–22). We selected 24 farms across a 500 km^2 area, that spanned two independent gradients: the first a continuum of a semi-natural area in the surrounding landscape, and the second a continuum of village-level exposure to training in agroecological farming practices, described as agroecology villages and non-agroecology villages (approximately half of each type). This independent-gradient design allowed for assessment of the influence of socio-ecological dynamics within communities on farming experiences and agroecological transitions, as well as understanding how these practices and semi-natural habitats drive pest pressure, the presence of natural enemies and pollinators, and access to compost materials. A growing body of literature suggests that agroecological practices may provide a buffer for farmlands against biodiversity loss and build resilience against climate change (Snapp et al., 2021), the need to compare agroecological and non-agroecological farming contexts. While agroecology is seen as a movement that envisions local knowledge, little is known about whether local-level practices have landscape-level benefits when large numbers of farmers in an area adopt such agroecological practices. For instance, there may be thresholds beyond which the landscape cannot support high levels of biodiversity or ecosystem services.

FARMS4Biodiversity adopted a farmer-led participatory, transdisciplinary, pluralistic methodological approach to bridge the science-policy-practice gap, using farmer participatory research in ecological surveys, socio-economic surveys, scenario models and participatory scenario planning and spatial participation. The project consisted of four interrelated work packages (WPs) outlined in Fig. 1.

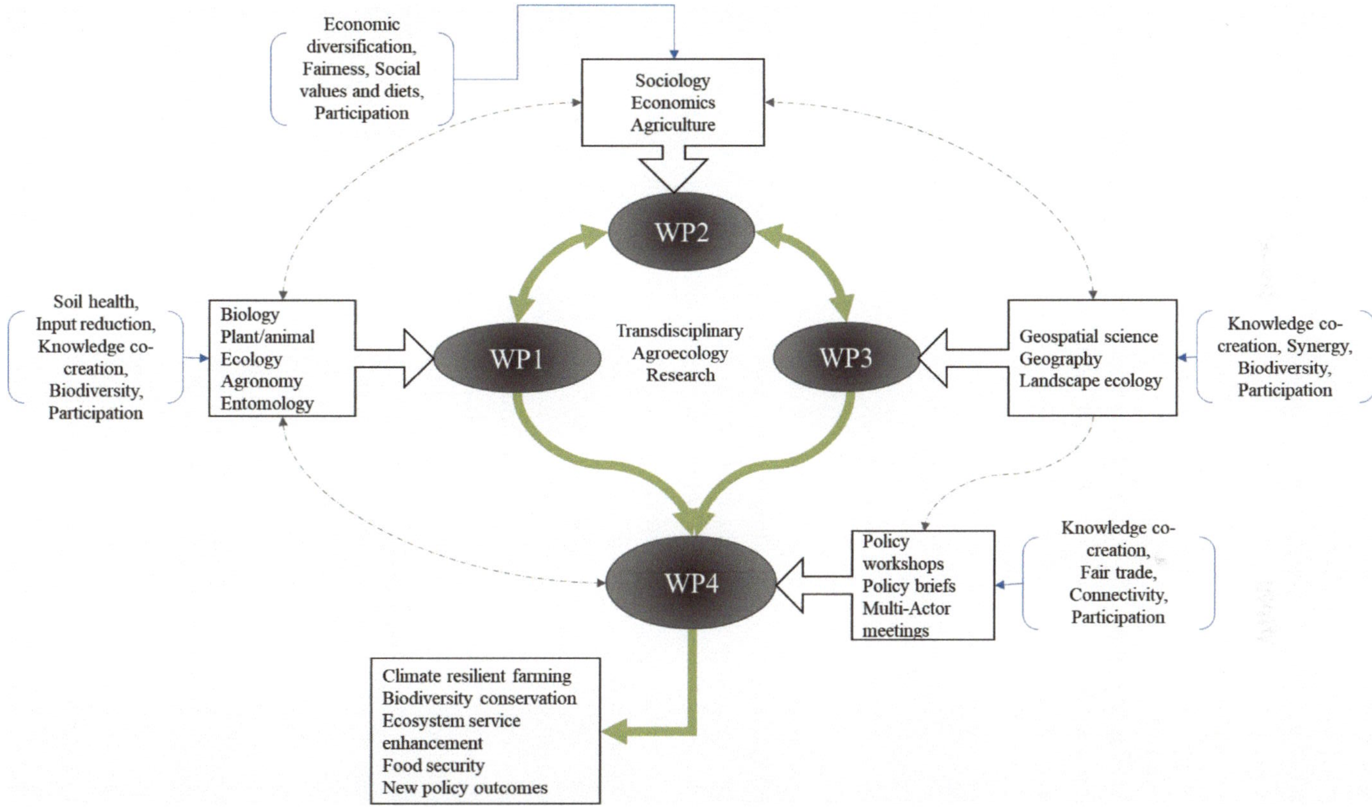

Fig. 1 Interrelationships and synergies between WPs and expected outcomes of the transdisciplinary research by various scientific disciplines. The connecting dotted lines imply the various disciplines are interrelated. The text in square brackets indicate the agroecological principles (Wezel et al., 2020) which the methods and disciplines in the white brackets are related to. WP stands for Work Page: each work page consistent of different research team members with different disciplinary background with intersecting research areas as the other researchers.

The description of the various WPs are as follows:

(i) WP1: Scenarios of Biodiversity and Ecosystem Services—farm-level research on agroecological practices' influence on land use change, biodiversity and ecosystem services.

(ii) WP2: Community Social Dynamics—large-scale surveys, in-depth interviews and document analysis to identify social factors that influence land use and biodiversity.

(iii) WP3: Participatory Scenario Planning—participatory GIS to develop maps and participatory scenario planning to examine different scenarios of land use, livelihoods and biodiversity.

(iv) WP4: Innovative Platforms and Stakeholder Engagement—Multi-Actor Platforms to discuss potential policy solutions, tools and technologies to foster viable and sustainable food security, livelihoods, land use and biodiversity in Malawi.

Farmers took the leading role in most of these WPs. For instance, in WP1 farmers were given comprehensive training on how to assess the damage caused by pests on crops in experimental and control farms. Farmers then assessed and recorded the impact of pest infestation on different crops and engaged in bird watching, identification of pollinators (bees and butterflies). In WP3, local farmers implemented different agronomic practices to examine their impacts on crop health both agroecology and non-agroecology farm. The preparation, application and assessment of the effectiveness of botanical pesticides on experimental farms were also led by the farmers. Further, the farmers played leading roles in the assessment in crop yield on experimental farms and reporting the yield to research team members. Also, the farmers were involved in the collection of spatial ecological data and field data on crop types and land cover for crop type and land cover analysis. In WP4, stakeholders led discussions on various policies and tools that can be deployed to help address biodiversity loss and ensure the sustainability of environmental resources. Farmers' leadership in these research activities increases the likelihood of decision-relevant outcomes that would improve the sustainability of biodiversity. The leadership of farmers in the research also allowed for the integration of traditional ecological knowledge in the research and with western scientific knowledge.

This farmer-led approach departs markedly from previous research (Cetrone et al., 2020; Nyantakyi-Frimpong et al., 2016) into the linkages between agroecology or agriculture more broadly, livelihoods, food security, biodiversity, ecosystem services and climate change by allowing local farmers to lead most of the research process, although it complements and adds more nuance to these previous research efforts. In this regard, the approach is innovative.

5. Pluralism of participatory and collaborative methodologies

5.1 Field surveys

Field surveys are used for two key research activities—namely ecological surveys and socio-economic surveys. Ecological surveys were used to assess on-farm biodiversity through a combination of visual bird surveys (point counts), pan traps for pollinators and parasitoid wasps, netting for butterflies, pit-fall traps for ground-dwelling arthropod predators, and malaise traps for flying insects. Farmer researchers participated in the field survey data collection (Fig. 2).

To assess general patterns and test hypotheses about relationships between community-level social factors, land use and biodiversity, panel surveys are used to collect data from 1000 smallholder farmers. The survey results will identify which social factors (e.g., income level, exposure to climatic events, access to land and agronomic practices) are relevant and where there are significant relationships between particular social factors, land use and biodiversity. Members of the non-profit farmer-led organization also participated in survey data collection as researchers.

5.2 In-depth interviews, focus group discussions and key informant interviews

In-depth interviews with smallholder farmers served several purposes in this transdisciplinary research. They presented an opportunity for examining how community-level social factors articulated with macro-level forces shape individuals' decisions on which agroecological farming practices to adopt and assess the perceptions of the farmers on the use of wild biodiversity and ecosystem services. In-depth interviews gave deeper insights into the

Fig. 2 Farmer researchers were trained on ecological data collection (e.g., birds, butterflies) and participated in field surveys. From left to right: bird counts, field survey team and pest damage assessment.

actions and decisions of farmers and served as a basis for knowledge translation. Community focus group discussions provided avenues for frank and transparent conservations about factors that shaped the use of ecosystem services and biodiversity conservation strategies and potential solutions to socio-ecological challenges.

In examining the linkages between farm practices and conservation strategies, in-depth interviews were crucial for identifying and linking individual perceptions with group perceptions solicited during focus groups. Such a holistic understanding of community perceptions formed the basis for formulating universally acceptable interventions. In-depth interviews also enabled the researchers and community members to draw the linkages between farm-level practices and processes and the adjacent ecosystems and how the two interact to shape the health of ecosystem services.

Resource persons or key informants—local chiefs and landlords, agriculture extension officers, officials from ministries and departments—were engaged through interviews to synthesize the knowledge acquired from community members and to better integrate local values and beliefs in broader conservation goals and national level development aspirations. These key informant interviews also allowed researchers to clarify information obtained from content analysis of agricultural policy documents and the media to triangulate quantitative and qualitative findings related to how macro-level social, political, and economic forces influence farmer decision-making on land use, agroecology, biodiversity, and climate change adaptation.

5.3 Participatory geographic information systems, and participatory mapping

Spatial and temporal variations emerge in the way farmers adapt to climate change and ecosystem services use over time, which need to be understood and linked to land use/cover change, climate change and changes in government policies. Spatial participation through Participatory Geographic Information Systems (PGIS) is used to empower farmers to map the spatial relations and interactions between ecological and social systems at multiple scales. Farmers are trained in clusters to use GPS devices to map farm conditions such as diseases, locations of insect and pest infestation and the area of farmlands (see Fig. 3). Farmers took charge of the process of spatial data collection based on their understanding of farm-level conditions through the application of traditional ecological knowledge.

Fig. 3 Doreen Thera, farmer researcher, reading coordinates from a GPS device to indicate the location of insect infestation on her farm while a research assistant records her observations.

PGIS activities engaged in by the farmers also included using GPS devices to walk around their farms and fallow lands to take coordinates. The coordinates were converted to polygons in a GIS environment to determine the area of the fallows and cultivated lands in agroecology and non-agroecology farming contexts. The areas were compared to determine the impact of farming methods on forest conservation in such smallholder farming systems (Kpienbaareh et al., 2022).

Participatory mapping activities were used by the farmers to delineate locations of ecosystem services in the community, elaborate on how they use these ecosystem services in the community, the impact of their use of these services and ways of conserving them (Kpienbaareh et al., 2020, 2021), using maps of the study communities printed on A3 sheets. The data generated provided information use patterns and the types and intensity of ecosystem services used. Geospatial analysis was used to identify "hot spots and 'cold spots' of use patterns which allowed community members to better visualize and appreciate the use of natural capitals in the community. The visualization of community members" use of ecosystem system services served as the basis for building consensus on appropriate conservation strategies and climate change adaptation planning.

5.4 Community engagement for collaborative learning

Effective implementation of agroecological farming methods and integration with traditional ecological knowledge requires understanding local innovations and new approaches to farming that can transform smallholder farming into climate resilient systems. Farmers are trained on agroecological principles, learned about and shared knowledge of various sustainable land management practices with each other. The training took place mainly in the farmers' communities, with farmers leading the teaching process using interactive horizontal teaching methods such as hands-on activities and small group discussions (Bezner Kerr et al., 2019). For instance, Fig. 4 shows farmers preparing bokashi compost—an organic fertilizer made from locally available resources, to be applied to their fields. The advantage of field training workshops is that they ensure continuity of the practices after the project.

In addition to onsite demonstration of various agroecological farming methods and practices, there were also training workshops at the local partner organization's Farmer Training Centre. The training centre serves as the base for researchers young academics and students with different disciplinary backgrounds to engage local stakeholders and farmers in collaboratively

Fig. 4 Farmers participating in a training workshop on the preparation of bokashi organic fertilizer.

learning ecologically friendly approaches to controlling insects and pests or learning tree grafting to create more resilient varieties that can survive the current and future climate change. For instance, farmers learned to produce botanical pesticides or plant extracts, made from locally available plants such as *Tephrosia vogelli* for use of their farms instead of synthetic agrochemicals. These plant extracts are low-cost, and more benign forms of pest control, reducing negative effects on beneficial insects (Dougoud et al., 2019; Tembo et al., 2018).

5.5 Comparative field experiments to assess the impact of agroecological practices

Various field experiments comparing agroecology and non-agroecology farms were implemented to assess the impact of agroecology on pollination, pest suppression, crop yield and crop health. For instance, to measure pest suppression, farmers recoded the damage of these pests on 30 randomly selected plants of maize, beans, pumpkin, and pigeon pea on farms in all 24 villages (Vogel et al., 2021). Record of the abundance of pest species on these selected plants was counted to determine the severity of infestation. To measure pollination deficits, we collaborative worked with the farmers to perform replicated enclosure experiments for three insect-pollinated crops namely pumpkin, cowpea, and pigeon pea and compared fruit set and yield quantity with those harvested from open control fields (Vogel et al., 2021). The experiments compared agroecology and non-agroecology farms to assess the differential impacts of farming practices on the abundance of insects and pests on the landscape. Yields on the farms were assessed by measuring seed or fruit yield for all crops within a 10×10 m transect on the farms. The yield values were used to measure the yield gap on farms by comparing each farm's actual yield to the potential yield measured across all our farms. Statistical models were used to determine the primary drivers of biodiversity, ecosystem services, and any observed yield gap.

Field experiments were also used to assess the impact of agroecological practices on the health and yield of crops. In one field experiment during the 2019/2020 growing season, farmers were given the same maize, beans, groundnuts and pigeon pea cultivars to plant on two sets of five farms—one set on which agroecological farming was practiced and another set with conventional farming methods that involved the use of synthetic fertilizers and agrochemicals. Digital Hemispherical Photographs of crop canopies were captured using a fisheye camera to compute the leaf area index (LAI) of the crops over the growing season as a proxy for assessing crop growth and health. The LAI values for the two categories of farms were compared to assess the impact of the farming practices. Generally, higher LAI values represent healthier crops. Multispectral satellite remote sensing images were used to compute vegetation indices and together with the LAI and machine learning regressions, the seasonal impact of agroecology farming methods was predicted. This approach integrates agriculture sciences, agronomy and spatial data science to identify the optimal strategies for building a resilient farming system for smallholders (Kpienbaareh et al., 2020).

The third set of field experiments was conducted over three growing seasons to explore how pest control methods are affected by farm and landscape scale processes and to examine how landscape context influences the effectiveness of plant extracts. The experiments were led by farmers living across a landscape gradient to experiment with the application of three plant extracts namely *Tephrosia vogelii, Tithonia diversifiolia,* and *Vernonia amygdalina* extracts to maize/bean intercropped farms. The farmers also maintained a no-spray control farm for comparison with farms on which the botanical pesticides were sprayed. Plant damage data were collected by a researcher and the farmers during the growing season and analysed by the research team using a generalized linear model with mixed effects. The results showed that the likelihood of maize damage decreased as the amount of forest and shrubland surrounding a farm increased, with different rates of change depending on plot type—agroecology/non-agroecology. For beans, the opposite trend was observed, though the likelihood of damage also varied by plot type. The findings revealed that the control farm of maize was 2.7 times more likely ($SE=1.33$; $P=0.04$) to be damaged for farmers less experienced with agroecology (Enloe et al., n.d.). The results demonstrate the significance of integrating traditional ecological knowledge with knowledge from other disciplines to identify effective sustainable production methods.

5.6 Crop diversity analysis using remote sensing

Biodiversity, a core principle of agroecology, is often considered a "keystone characteristic" of healthy ecosystems (FAO, 2018). Within the field of agriculture, a diversified farming system integrates biodiversity across spatial and temporal scales, including planting multiple species and varieties, intercropping, integration of livestock and planting non-crop species such as hedgerows, riparian buffers and woodlots (Kremen and Miles, 2012). Biologically diverse farming systems maintain and regenerate biological processes which support ecosystem services, including soil quality, nitrogen fixation and pollination, which are crucial for food production (Kremen et al., 2012; Snapp et al., 2010). Several reviews have shown how increasing biodiversity in farming systems can positively affect food security in the Global South (Bezner Kerr et al., 2021; Jones, 2017) and can help cushion households against climate variability and shocks (Snapp et al., 2021). Understanding crop diversity on the landscape, therefore, gives pointers to food security and provides information about the cropping patterns in a geographic location.

In this research, remote sensing analysis involving satellite data was used to map the crop types and land cover categories on the landscape for the 2019/20 and 2020/21 growing seasons (Kpienbaareh et al., 2021). Machine learning classification methods were applied on multi-source high resolution optical and radar satellite images to perform a detailed crop type mapping to identify various crop types and their distribution on the landscape and to predict the impact of agroecology on crop health in smallholder farms. Mapping crop and landscape diversity using farmers' knowledge of the landscape, crop types and cropping systems complements other studies in the area that linked agroecology, food security, livelihoods and social capital (Bezner Kerr et al., 2016, 2018; Kansanga et al., 2019). Farmers engaged in collecting reference data by identifying the crop types and land cover categories using GPS devices for use in assessing the accuracy of the crop diversity analysis.

5.7 Scenario modelling of biodiversity and ecosystems services based on land use change

To further examine and model how agroecological practices, such as legume intercropping, composting, and botanical sprays can buffer against the loss of biodiversity and ecosystem services across multiple scales, scenarios of different land use changes will be used to model the impact of land use change of crop yield. Satellite remote sensing classification of the landscape will be used as the input to assess the impact of landscape simplification on insect/pest abundance within buffers of 500 m, 1 km and 1.5 km from farmlands. GIS is used to extract land cover statistics within the buffers. We will model a range of 10 land use change scenarios reflective of realistic rates of change over the next 25 years and assess the likely impacts of that change on crop productivity, biodiversity, and ecosystem services. We will assess how these outcomes change according to the use of agroecological practices in the region, as well as the economic impacts from using those farming practices.

A second scenario planning activity in the research will apply scenarios of land use change to assess spatio-temporal distribution of the impact of anthropogenic activities on ecosystem services and biodiversity. A local professional artist will be contracted to visualize these scenarios, with narratives in local languages, to make them accessible to local communities. By combining images, remote sensing classification and narratives, we will present a richer, comprehensible version of final scenarios to local communities and policymakers for in-depth discussions.

6. Transdisciplinary knowledge co-production for farm-level and community resilience

6.1 Landscape-scale knowledge and farm-level conditions

The transdisciplinary research creates a platform for the integration of pluralistic knowledge systems—traditional ecological knowledge and multiple disciplines of formal scientific knowledge—for understanding how farm-level agroecological practices influence biodiversity, ecosystem services at a landscape level, and how this information can be used to plan for future change. Smallholder farmers understand the dynamics between farm-level practices and processes and ecosystem services surrounding the farms through various agronomic experiments and spatial data analysis.

By integrating these knowledge systems, a more detailed understanding is gathered of the factors influencing drivers and constraints for biodiversity and ecosystem services while innovative approaches to addressing challenges with the food system are unearthed. For instance, knowledge in landscape ecology and pollination biology alongside farmer practices has provided insight into the relationship between landscape composition, and the effects of the composition on both beneficial insect and pest communities on crop fields (Vogel et al., 2021) and ways to support more biodiverse-friendly pest control measures (Enloe et al., n.d.).

One of the key principles of agroecology is the facilitation of further ecological interaction between farm-level and adjacent landscapes through supporting synergistic ecological processes. Mapping spatial interactions between the farm-level and adjacent ecosystems empowered the farmers to explain the relationships between farm-level practices and processes, ecosystems services and biodiversity and to identify viable conservation strategies (Kpienbaareh et al., 2020), indicating the relevance of translational ecology in achieving environmental sustainability goals.

Research shows that ecological intensification through the adoption of agroecology practices including sustainable land management strategies can improve yield (Cassman, 1999; Kansanga et al., 2020). We used spatial data on farm sizes and fallow lands measured by farmers to compare farmlands and fallows of agroecology and non-agroecology farmers. Using participatory spatial methodologies combined with in-depth interviews, the researchers observed that farms using agroecological practices significantly increased fallow forest land compared to those not practicing agroecology, pointing to the potential for restoring degraded agroecosystems using agroecology farming strategies (Kpienbaareh et al., 2022). Table 1 presents the results of the strategies farmers outlined for conserving ecosystem services categorized according to the farming practices implemented.

6.2 Connecting micro-and macro-scale community dynamics for community resilience

The community survey data unearthed how micro-level social dynamics operating at the household and community level, involving key social roles and positions, such as gender, marital status, class, and health status, interact with macro-level social forces, such as policies and government programs (e.g., Farm Input Subsidy Program), to influence farm-level practices and land-use change. Striving to examine how community social dynamics and local land tenure institutions connect with macro social forces (e.g., agricultural policies, extension services, international agreements) to influence agroecological practices, land use and related measures of resilience are crucial for understanding progress that has and can be made in ensuring climate resilient farming and sustainable livelihoods.

6.3 Innovative multi-actor platforms for stakeholder engagement

Multi-Actor Platforms (MAPs) bring diverse stakeholders together in a formal process of interactive learning, sharing, empowerment, and collaborative governance to discuss opportunities and obstacles to the desired set of goals (Salvia and Quaranta, 2019). A MAP is a "decision-making body comprising different stakeholders who perceive the same resource management

Table 1 Strategies for conserving ecosystem services outlined by participants based on the integration of spatial ecological knowledge and agroecology.

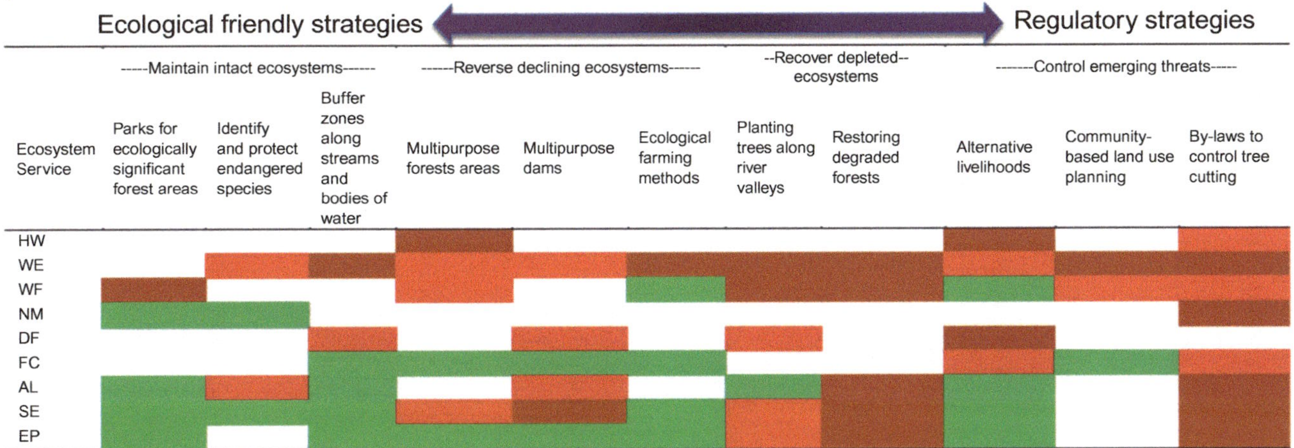

AL–aesthetic landscape; DF–dimda/dry season farming; EP–erosion prevention; FC–food (food cropss); HW–hunting/wildlife trapping; NM–natural medicines; SE–scientific/educational landscapes; WE–wild edible plants (for food); WF–wood (fuelwood and charcoal)

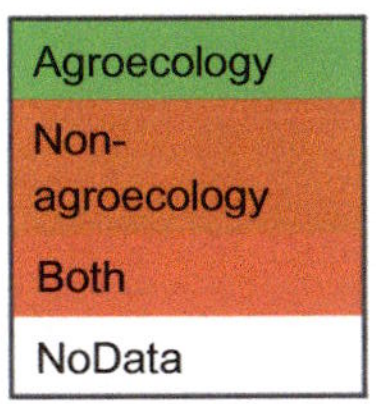

Adapted from Kpienbaareh, D., et al., 2020. Spatial and ecological farmer knowledge and decision-making about ecosystem services and biodiversity. Land 9 (10). doi:10.3390/land9100356.

problem, realize their interdependence for solving it, and come together to agree on action strategies for solving the problem" (Röling, 1994). Emerging research suggests that by building trust and dialogue among diverse stakeholders with interconnected but potentially divergent interests or viewpoints, MAPs promote resilience and innovation in the face of risk, complexity, and uncertainty (Schut et al., 2017). The objective of the MAP for this research project was to explore new, innovative institutional and policy frameworks to enable the use of agroecological practices for reducing biodiversity loss, sustaining ecosystem services and improving climate change adaptation.

The MAP in this research comprised representatives from Malawi's National Biodiversity Steering Committee, National Council on Environment, Technical Committee on the Environment, Agriculture Department Divisions, District Agriculture Committee, District Agriculture Development Office, and other relevant committees as well as local farmers and traditional leaders. Three MAP dialogue meetings were organized to share results, solicit feedback, and devise strategies for disseminating project findings. Prepared policy briefs from research findings have been shared with local partners and stakeholders to facilitate knowledge sharing on ways of strategizing for ecosystem governance. Through coordination with MAP members, several decision-making tools have been developed by farmers to provide the information needed to enable government agencies to map out strategies for assessing crop diversity (Kpienbaareh et al., 2021). There have also been community engagement workshops for knowledge translation purposes to discuss research knowledge acquired during the research process. To engage farmers and research team members for greater collaborations, field trips to monitor research processes. Engage with these diverse stakeholders and discussing research findings allows for detailed explications of complex issues that need multiple perspectives to address. These engagements align with the tenets of environmental pluralism because they allowed us to garner broad political support to outline viable solutions to address sustainability issues (Hull, 2007). By garnering broader political support for policy alternatives discussed, we were able to overcome polarization and power struggles that sometimes hinder contemporary environmental decision making, as stipulated in environmental pluralism literature (Hull, 2007). Table 2 shows the diverse disciplines and methods used in the research and some key findings that have been discussed with MAP members and the other stakeholders.

Table 2 Pluralistic transdisciplinary methods and some key findings.

Transdisciplinary and pluralistic methods	Disciplines and knowledges involved	Key findings
Farmer researchers worked with geographer to map out areas of land under fallow and cultivation	Geospatial science Local agricultural practices Forest regeneration	Farmers using agroecological practices were more likely to have more land under fallow for forest regeneration
Farmer researchers tested different botanical plant extracts	Agronomy Sociology Local agricultural knowledge	Maize damage was lower on farms managed using agroecological principles The effectiveness pesticidal extracts varied across landscape gradient
Ecologists worked with farmer researchers to collect bee and blister beetle samples	Ecology Entomology Local agricultural knowledge	Bee richness and beetle abundances were not affected by landscape characteristics, nor was flower damage or fruit set difference between bagged and open flowers
Surveys used to examine how community social dynamics and local land tenure institutions articulate with macro social forces to influence agroecological practices, land use, and related measures of resilience	Sociology Economics Local agricultural knowledge Geospatial science	Found linkages between seed security, and food insecurity among and seed-secure farmers are three times more likely to diversify crop production

7. Synthesis

Examining farming practices using multiple research instruments and experiments through a farmer-led process can enable the triangulation of knowledge across disciplinary and knowledge systems. For instance, while GIS maps pointed to the high use of specific ecosystem services in certain locations, interviews elucidated on the whys such services were highly used, surveys showed the gendered nature of these use patterns (Kpienbaareh et al., 2020), highlighting the effectiveness of methodological pluralism for detailing the intersection of knowledge systems and the powerful role such intersections play in exploring complex environmental issues. Spash

(2009) asserts that reliance on a single instrument or method to assess environmental challenges such as low productivity and biodiversity loss is misguided because no single instrument/metric is sufficiently flexible and robust enough to be able to effectively address all environmental problems in all contexts. For instance, the integration of results from landscape analysis using remote sensing (WP3) with insect abundance on farmlands and their impact on crops (WP1) and using surveys to explore community dynamics such as farming methods and socio-political contexts (WP2) led to the co-production of knowledge with local farmers that shapes the kinds of interventions that can be adopted to improve productivity while ensuring ecological integrity.

Chalmers and Fabricius (2007), however, point out that it is crucial to acknowledge and appreciate the power dynamics and inequalities involved in integrating traditional ecological knowledge and western scientific knowledge to ensure that knowledge triangulation is multidirectional. The same can be said of reliance on a single discipline to explain production shortfalls in smallholder farming systems given the multiplicity of factors that drive such shortfalls. In this regard, Spash (2009) advises against eclecticism in adopting environmental pluralism i.e., avoiding accepting everything from every discipline. This calls for the need for building knowledge through careful selections of criteria and methods for the research while appreciating the limitations of the various methods and incorporating the expertise of local collaborators. Consequently, in this farmer-led approach, we applied knowledge from multiple disciplines with carefully selected methodologies and instruments from multiple disciplines, in consistence with current thinking in (environmental) pluralism and translational ecology. For instance, cognizant of the challenges that local farmers may have with engaging with complex equipment used for the research, we integrated the ideas of participants in designing research protocols and relying on experiences and traditional knowledge in identifying the most appropriate equipment and methods for the research.

While several formal scientific disciplines shaped the basis for this farmer-led research, traditional ecological knowledge formed an integral part of the knowledge co-production process. For instance, by integrating spatial ecological knowledge of farmers collected using GPS devices and agroecological knowledge, Kpienbaareh et al. (2020) unearthed farmers' knowledge on insect/pests and how they influence yield, and how knowledge of ecosystem services among local farmers generated transparent conversations about conservation in the study communities through focus group discussions. Through this knowledge integration process, farmers identified various biodiversity conservation strategies relevant to the study context (see Table 1). Evidence from translation ecology point to the relevance of such integrated approaches to address social dimensions that underlie today's complex environmental issues (Enquist et al., 2017). We found such local knowledge to be in tandem with or complementary to the science of agroecology and local conservation strategies, in tandem with current discourses on agroecology as a transdisciplinary research area (Méndez et al., 2013). The unique perspectives from traditional knowledge systems, if not integrated into current research, would otherwise produce results that miss the world views of local communities—these communities are the most impacted by environmental challenges.

Lately, there have been efforts by various institutions, such as the Intergovernmental Science-Policy Platform on Biodiversity and Ecosystem Services, the Economics of Ecosystems and Biodiversity and the Millennium Ecosystem Assessment, to assess ecosystems within the context of a rapidly changing climate. Following these global efforts, there is now a more urgent need to develop effective locally acceptable approaches for legitimately, transparently, and constructively fostering collaborations across disciplinary backgrounds and knowledge systems for effective governance of ecosystems (Ainscough et al., 2019). Our research makes a significant contribution to these efforts by identifying, drawing linkages and synergies between agroecology, biodiversity, and ecosystem services for building resilient communities. For instance, by integrating smallholder farmers' knowledge, geospatial data science, ecology and agriculture, the research team co-produced knowledge on various biodiversity conservation strategies that are acceptable to community members for sustainable use of ecosystem services (Kpienbaareh et al., 2020) (see Table 2). Also, the use of remote sensing methods with knowledge on crop types and landscape ecology to generate crop type and cover maps (Kpienbaareh et al., 2021) demonstrate how such transdisciplinary research can generate information needed for ecosystem serviced-based governance of biodiversity by harnessing locally generated information. The experiments used in the forgoing studies and the participation of local people highlight the relevance of adaptive and collaborative research for ecosystem governance (Gunderson et al., 2016), which are tenets of translational ecology. The findings also demonstrate the significance of allowing local people who understand their terrain and with historical knowledge of their landscape to lead research work that will contribute meaningfully to current efforts to address biodiversity loss.

The assumption has always been that once research generates the information needed for improved farming and ecosystem governance, policymakers act by implementing the needed policies (De Groot et al., 2002). Though the volume of research in ecosystem services have recently increased significantly (Costanza et al., 2017), the increased rate of degradation of ecosystem services globally (IPBES, 2019) shows that there are still ecosystem service governance issues that need addressing. Additionally, ecosystems research is biased towards large-scale agriculture in temperate regions and on economically interesting crops such as coffee and cocoa, with little being studied on food crops in smallholder agriculture systems. Meanwhile, the continuous degradation of these ecosystems highlights the complexity of balancing livelihood needs, such as farming, with environmental sustainability within the context of climate change. In response to this conundrum, our approach links agroecology with ecosystem services and biodiversity with a strong collaborative participatory methodology to co-produce evidence-based outputs that are acceptable to multiple user groups. The approach used in Malawi resonates with translational ecology approaches to pluralistic governance of ecosystems, one which emphasizes the use of transdisciplinary research to address the socio-ecological, and political contexts of environmental problems (Enquist et al., 2017). An integral part of this farmer-led approach is knowledge translation which brings researchers and stakeholders in a multi-actor platform to ensure dialogue through community engagement, policy workshops and policy briefs that account for the views of all stakeholders. Such a holistic approach accounts for uncertainties and loopholes that may arise during the implementation of ecosystem service research.

8. Conclusion

With emergent challenges in a rapidly evolving human-dominated world, there is an urgent need to advance a wide range of disciplines, sources of knowledge and learning to be able to deal with global environmental change and new emerging social-ecological conditions, including low productivity and biodiversity loss. We argue, by describing an innovative farmer-led research process in Malawi, that we can achieve climate resilient farming with ecosystem service enhancement and biodiversity conservation benefits. The research demonstrates that the science-policy-practice community needs to embrace a diversity of disciplines and knowledge systems and connect to traditional ecological knowledge of communities through active participation in ways that transcend conventional models and frameworks to food production and ecosystem service governance. We show how agroecological farming can be harnessed to address food insecurity, ensure environmental sustainability, while sustaining the livelihoods of local farmers; and highlight the relevance of a pluralistic decolonized research process for addressing emerging environmental challenges, especially in the Global South.

Key challenges, however, needed to be addressed in pursuing this farmer-led transdisciplinary research. In integrating traditional ecological knowledge and western scientific knowledge systems, it is important to establish guiding frames that promote and allow equal and transparent linkages between the two knowledge systems. Setting out such frameworks closes any gaps in power relations in ways that empower local communities, and accomplish the potential of knowledge synergies for climate-resilient agriculture, food security, biodiversity conservation and ecosystem governance. Tengo et al. (2012) outline values such as respect, trust, reciprocity, and equal sharing need to characterize interactions between knowledge systems at all scales. The participatory nature of this farmer-led research enabled intercultural dialogue between scholars, traditional authorities and the farmers which add and promotes credibility to the work (Rist et al., 2011), thus making the approach and the results generalizable to other contexts in Sub-Saharan Africa.

Related to this cross-fertilization of ideas from different knowledge systems and disciplines is the challenge of developing new methods and tools for co-learning, co-producing, analysing, documenting and sharing of knowledge in such transdisciplinary research that integrates knowledge systems. We ensured close collaboration between the WPs to overcome this challenge. The fact that this research drew perspectives from disciplines in both natural science and social sciences (see Fig. 1) to achieve the research goals demonstrates that deploying methods that cut across disciplines in environmental research is relevant for facilitating collaboration between traditional ecological knowledge and western scientific knowledge, but also between different scientific knowledge systems, such as between the natural and social sciences and the humanities. Additionally, integrating methods including quantitative, qualitative and geospatial approaches within or across disciplines allows for knowledge triangulation for the effective development of policy interventions.

Glossary

Botanical sprays these are botanical insecticides produced from dried, ground plant material, crude plant extracts, or chemicals isolated from plants to kill or repel insects from croplands. For instance, *T. vogelii, Tithonia diversifiolia, and Vernonia amygdalina* are widely produced by farmers in study context for use in controlling insects on farms.

Environmental pluralism a view in which the world is seen as infinitely complex, fluid, and negotiated, frustrating any attempt to impose a master narrative as an overarching, organizing frame.

Knowledge co-production the iterative and collaborative processes involving diverse types of expertise, knowledge and actors to produce context-specific knowledge and pathways towards a sustainable future. Farmers, stakeholders and researchers co-produced relevant knowledge at the various stages of the research and the sharing of the knowledge through knowledge translation activities.

Legume intercropping the practice of planting leguminous crops with mostly cereals simultaneously during the same growing season.

Participatory agroecology an approach to practicing agroecology that involves active participation of farmers in designing, implementing and assessing agroecological farming strategies to improve yield through knowledge co-production and sharing.

Transdisciplinary research involves the integration of knowledge from different disciplines, including non-academic sources, to address complex sustainability problems. This involves processes of mutual learning between science and society.

References

Ainscough, J., et al., 2019. Navigating pluralism: understanding perceptions of the ecosystem services concept. Ecosyst. Serv. 36, 100892.

Altieri, M.A., 2000. Agroecology: principles and strategies for designing sustainable farming systems. Agroecol. Act.

Altieri, M.A., Toledo, V.M., 2011. The agroecological revolution in Latin America: rescuing nature, ensuring food sovereignty and empowering peasants. J. Peasant Stud. 38 (3), 587–612.

Arnold, M., 2013. Transdisciplinary research (transdisciplinarity). In: Encyclopedia of Creativity, Invention, Innovation and Entrepreneurship, pp. 1819–1828.

Berkes, F., 2017. Sacred Ecology. Routledge.

Bezner Kerr, R., Chirwa, M., 2004. Participatory research approaches and social dynamics that influence agricultural practices to improve child nutrition in Malawi. Ecohealth 1 (2), SU109–SU119.

Bezner Kerr, R., et al., 2016. Integrated agriculture programs to address malnutrition in northern Malawi. BMC Public Health 16 (1), 1197.

Bezner Kerr, R., et al., 2018. Knowledge politics in participatory climate change adaptation research on agroecology in Malawi. Renew. Agric. Food Syst. 33 (3), 238–251.

Bezner Kerr, R., et al., 2019. Farming for change: developing a participatory curriculum on agroecology, nutrition, climate change and social equity in Malawi and Tanzania. Agric. Human Values 36 (3), 549–566.

Bezner Kerr, R., et al., 2021. Can agroecology improve food security and nutrition? A review. Glob. Food Sec. 29 (March). https://doi.org/10.1016/j.gfs.2021.100540.

Bronen, R., Cochran, P., 2021. Decolonize Climate Adaptation Research. American Association for the Advancement of Science.
Campesina, V., 2013. De Maputo a Yakarta. In: 5 años de agroecologia em La Via Campesina. Universidade Nacional Agrária La Molina, Peru, p. 10.
Cánovas-Molina, A., García-Frapolli, E., 2021. Socio-ecological impacts of industrial aquaculture and ways forward to sustainability. Mar. Freshw. Res.
Cassman, K.G., 1999. Ecological intensification of cereal production systems: yield potential, soil quality, and precision agriculture. Proc. Natl. Acad. Sci. U. S. A. 96 (11), 5952–5959. https://doi.org/10.1073/pnas.96.11.5952.
Cetrone, H., et al., 2020. A participatory agroecological intervention reduces women's risk of probable depression through improvements in food security in Singida, Tanzania. Curr. Dev. Nutr. 4 (Suppl. 2), 819.
Chalmers, N., Fabricius, C., 2007. Expert and generalist local knowledge about land-cover change on South Africa's Wild Coast: can local ecological knowledge add value to science? Ecol. Soc. 12 (1).
Costanza, R., et al., 2017. Twenty years of ecosystem services: how far have we come and how far do we still need to go? Ecosyst. Serv. 28, 1–16.
Datta, R., 2018. Decolonizing both researcher and research and its effectiveness in Indigenous research. Res. Ethics 14 (2), 1–24.
De Groot, R.S., Wilson, M.A., Boumans, R.M.J., 2002. A typology for the classification, description and valuation of ecosystem functions, goods and services. Ecol. Econ. 41 (3), 393–408.
De Lissovoy, N., 2010. Decolonial pedagogy and the ethics of the global. Discourse 31 (3), 279–293.
Dougoud, J., et al., 2019. Efficacy of homemade botanical insecticides based on traditional knowledge. A review. Agron. Sustain. Dev. 39 (4), 1–22.
Enloe, S., Mhone, I., Moyo, P., Mkandawire, M., Shumba, L., Banda, D., Frankel, A., Iverson, A., Kpienbaareh, D., Bezner Kerr, R., and Poveda, K., 'Landscape Composition and Agroecological Farming System Influence Effectiveness of in situ Pesticidal Plant Extract Use in Malawi', in preparation.
Enquist, C.A.F., et al., 2017. Foundations of translational ecology. Front. Ecol. Environ. 15 (10), 541–550.
FAO, 2018. The 10 Elements of Agroecology: Guiding the Transition to Sustainable Food and Agricultural Systems.
FAO, et al., 2019. The State of Food Security and Nutrition in the World 2019. Safeguarding Against Economic Slowdowns and Downturns. Rome, Italy.
Francis, C., et al., 2003. Agroecology: the ecology of food systems. J. Sustain. Agric. 22 (3), 99–118. https://doi.org/10.1300/J064v22n03_10.
Francis, C.A., et al., 2008. Transdisciplinary research for a sustainable agriculture and food sector. Agron. J. 100 (3), 771–776.
Given, L.M., 2008. The Sage Encyclopedia of Qualitative Research Methods. Sage Publications.
Goldman, M.J., Nadasdy, P., Turner, M.D., 2011. Knowing Nature: Conversations at the Intersection of Political Ecology and Science Studies. University of Chicago Press.
Government of Malawi, 2016. National Forest Policy. Lilongwe, Malawi. Available at: http://www.unpei.org/sites/default/files/dmdocuments/Malawi Government National Forest Policy—June 2016.pdf.
Gumbo, D.J., et al., 2018. Sustainable Management of Miombo Woodlands: Food Security, Nutrition and Wood Energy. FAO.
Gunderson, L.H., Cosens, B., Garmestani, A.S., 2016. Adaptive governance of riverine and wetland ecosystem goods and services. J. Environ. Manage. 183, 353–360.

Guzmán, E.S., Woodgate, G., 2013. Agroecología: Fundamentos del pensamiento social agrario y teoría sociológica. Agroecología 8 (2), 27–34.
Herrero, M., et al., 2017. Farming and the geography of nutrient production for human use: a transdisciplinary analysis. Lancet Planet. Health 1 (1), e33–e42.
HLPE, 2019. Agroecological and Other Innovative Approaches for Sustainable Agriculture and Food Systems That Enhance Food Security and Nutrition, a Report by the High Level Panel of Experts on Food Security and Nutrition of the Committee on World Food Security. Rome.
Hull, B., 2007. Environmental pluralism. In: Encyclopedia of Environmental Ethics and Philosophy.
IPBES, 2019. 2019 Global Assessment Report on Biodiversity and Ecosystem Services. Bonn. Available at: https://ipbes.net/global-assessment.
IPCC, 2021. Climate change 2021: the physical science basis. In: Masson-Delmotte, V., et al. (Eds.), Contribution of Working Group I to the Sixth Assessment Report of the Intergovernmental Panel on Climate Change. Cambridge University Press. Available at: https://www.ipcc.ch/report/ar6/wg1/downloads/report/IPCC_AR6_WGI_Citation.pdf.
IPES-Food, 2016. From uniformity to diversity: a paradigm shift from industrial agriculture to diversified agroecological systems. In: International Panel of Experts on Sustainable Food systems. Brussels. IPES. Available at: www.ipes-food.org.
Jones, A.D., 2017. Food insecurity and mental health status: a global analysis of 149 countries. Am. J. Prev. Med. 53 (2), 264–273.
Kansanga, M.M., et al., 2019. Beyond ecological synergies: examining the impact of participatory agroecology on social capital in smallholder farming communities. Int. J. Sustain. Dev. World Ecol., 1–14.
Kansanga, M.M., et al., 2020. Determinants of smallholder farmers' adoption of short-term and long-term sustainable land management practices. Renew. Agric. Food Syst., 1–13.
Kpienbaareh, D., et al., 2020. Spatial and ecological farmer knowledge and decision-making about ecosystem services and biodiversity. Land 9 (10). https://doi.org/10.3390/land9100356.
Kpienbaareh, D., et al., 2021. Crop type and land cover mapping in Northern Malawi using the integration of Sentinel-1, Sentinel-2 and PlanetScope satellite data. Remote Sens. (Basel) 13 (4), 700. https://doi.org/10.3390/rs13040700.
Kpienbaareh, D., et al., 2022. Assessing local perceptions of deforestation, forest restoration, and the role of agroecology for agroecosystem restoration in northern Malawi. Land Degrad. Dev. https://doi.org/10.1002/ldr.4238.
Kremen, C., Miles, A., 2012. Ecosystem services in biologically diversified versus conventional farming systems: benefits, externalities, and trade-offs. Ecol. Soc. 17 (4).
Kremen, C., Iles, A., Bacon, C., 2012. Diversified farming systems: an agroecological, systems-based alternative to modern industrial agriculture. Ecol. Soc. 17 (4).
Levidow, L., Pimbert, M., Vanloqueren, G., 2014. Agroecological research: conforming—or transforming the dominant agro-food regime? Agroecol. Sustain. Food Syst. 38 (10), 1127–1155.
Lunduka, R., Ricker-Gilbert, J., Fisher, M., 2013. What are the farm-level impacts of Malawi's farm input subsidy program? A critical review. Agric. Econ. 44 (6), 563–579.
Mbow, C., et al., 2019. Chapter 5: food security. In: IPCC Special Report on Climate Change and Land. online at https://www.ipcc.ch/site/assets/uploads.
Méndez, V.E., Bacon, C.M., Cohen, R., 2013. Agroecology as a transdisciplinary, participatory, and action-oriented approach. Agroecol. Sustain. Food Syst. 37 (1), 3–18.
Messina, J.P., Peter, B.G., Snapp, S.S., 2017. Re-evaluating the Malawian farm input subsidy programme. Nat. Plants 3, 17013.

Mweta, N., Juma, P., 2020. Indigenous knowledge as a mitigation factor to disaster risk reduction in Malawi: a case study of Nsanje District. Int. J. Clim. Change 12 (4).
National Statistical Office, 2017. Malawi Micronutrient Survey: Key Indicators Report 2015–16. Zomba.
Ngwira, S., Watanabe, T., 2019. An analysis of the causes of deforestation in Malawi: a case of Mwazisi. Landarzt 8 (3), 48. https://doi.org/10.3390/land8030048.
Nyantakyi-Frimpong, H., et al., 2016. Agroecology and sustainable food systems: participatory research to improve food security among HIV-affected households in northern Malawi. Soc. Sci. Med. 164, 89–99.
Nyantakyi-Frimpong, H., et al., 2017. A farmer-to-farmer agroecological approach to addressing food security in Malawi. In: Wakeford, T., et al. (Eds.), Everyday Experts: How people's Knowledge Can Transform the Food System. Center for Agroecology, Water and Resilience, Coventry University, pp. 119–134.
Ribeiro, N.S., de Miranda, P.L.S., Timberlake, J., 2020. Biogeography and ecology of miombo woodlands. In: Miombo Woodlands in a Changing Environment: Securing the Resilience and Sustainability of People and Woodlands. Springer, pp. 9–53.
Rist, S., et al., 2011. Endogenous Knowledge: Implications for Sustainable Development.
Rocheleau, D., Thomas-Slayter, B., Wangari, E., 1996. A feminist political ecology perspective. In: Feminist Political Ecology: Global Issues and Local Experiences, pp. 3–26.
Röling, N., 1994. Communication support for sustainable natural resource management. IDS Bull. 25 (2), 125–133.
Rutsaert, P., Donovan, J., Kimenju, S., 2021. Demand-side challenges to increase sales of new maize hybrids in Kenya. Technol. Soc. 66, 101630.
Salvia, R., Quaranta, G., 2019. Multi-actor platform as a tool to enhance networking of sustainable socio-ecological food systems. In: Multi-Actor Platform as a Tool to Enhance Networking of Sustainable Socio-ecological Food Systems, pp. 405–427.
Schut, M., et al., 2017. Guidelines for Innovation Platforms in Agricultural Research for Development: Decision Support for Research, Development and Funding Agencies on how to Design, Budget and Implement Impactful Innovation Platforms.
Sélingué, M., 2007. Declaration of Nyéléni.
Snapp, S.S., et al., 2010. Biodiversity can support a greener revolution in Africa. Proc. Natl. Acad. Sci. U. S. A. 107 (48), 20840–20845.
Snapp, S., et al., 2021. Agroecology and Climate Change Rapid Evidence Review: Performance of Agroecological Approaches in Low-and Middle-Income Countries.
Spash, C.L., 2009. The new environmental pragmatists, pluralism and sustainability. Environ. Values 18 (3), 253–256.
Tembo, Y., et al., 2018. Pesticidal plant extracts improve yield and reduce insect pests on legume crops without harming beneficial arthropods. Front. Plant Sci. 9, 1425.
Tengo, M., Malmer, P., Borraz, P., 2012. Carin o C, Carin o J, Gonzales T, Ishizawa J, Kvarnströ m M, Masardule O, Morales A et al.: Dialogue workshop on knowledge for the 21st century: indigenous knowledge, traditional knowledge, science and connecting diverse knowledge systems. In: Workshop Report, pp. 10–13.
Tengö, M., et al., 2014. Connecting diverse knowledge systems for enhanced ecosystem governance: the multiple evidence base approach. Ambio 43 (5), 579–591.
Vogel, C., et al., 2021. Higher bee abundance, but not pest abundance, in landscapes with more agriculture on a late-flowering legume crop in tropical smallholder farms. PeerJ 9, e10732.
Wezel, A., et al., 2009. Agroecology as a science, a movement and a practice. A review. Agron. Sustain. Dev. 29 (4), 503–515.
Wezel, A., et al., 2020. Agroecological principles and elements and their implications for transitioning to sustainable food systems. A review. Agron. Sustain. Dev. 40 (6), 40. https://doi.org/10.1007/s13593-020-00646-z.

World Bank, 2019. Poverty Databank. Washington DC. Available at: http://povertydata.worldbank.org/poverty/country/MWI.
Zanotti, L., et al., 2020. Political ecology and decolonial research: co-production with the Iñupiat in Utqiaġvik. J. Polit. Ecol. 27 (1), 43–66.

Further reading

Aeberhard, A., Rist, S., 2009. Transdisciplinary co-production of knowledge in the development of organic agriculture in Switzerland. Ecol. Econ. 68 (4), 1171–1181.
Batary, P., et al., 2011. Landscape-moderated biodiversity effects of agri-environmental management: a meta-analysis. Proc. R. Soc. B Biol. Sci. 278 (1713), 1894–1902.
Bezner Kerr, R., 2012. Lessons from the old green revolution for the new: social, environmental and nutritional issues for agricultural change in Africa. Prog. Dev. Stud. 12 (2–3), 213–229.
Tscharntke, T., et al., 2012. Landscape moderation of biodiversity patterns and processes-eight hypotheses. Biol. Rev. 87 (3), 661–685.

CHAPTER TWO

Community-scientist collaboration in the creation, management and research for two National Wildlife Areas in Arctic Canada

Mark L. Mallory[a,*], Jeannie Toomasie[b], Susanne Emond[b,c], Gabby Lamarche[a], Lauren Roberts[a], Lisa Pirie-Dominix[c], and Jennifer F. Provencher[d]

[a]Biology, Acadia University, Wolfville, NS, Canada
[b]Sululiit Area Co-Management Committee (Chair, 2011–2021), Iqaluit, NU, Canada
[c]Canadian Wildlife Service, Environment and Climate Change Canada, Iqaluit, NU, Canada
[d]Science and Technology Branch, Environment and Climate Change Canada, National Wildlife Research Centre, Raven Road, Carleton University, Ottawa, ON, Canada
*Corresponding author: e-mail address: mark.mallory@acadiau.ca

Contents

Abstract

Arctic Canada hosts many remote seabird colonies supporting nationally and internationally important numbers of several species and located in geographically spectacular settings. In the spirit of co-management enshrined in the *Nunavut Agreement*, government scientists worked with the Nattivak Hunters' and Trappers' Organization in the community of Qikiqtarjuaq, Nunavut to census wildlife and document local ecological

Advances in Ecological Research, Volume 66
ISSN 0065-2504
https://doi.org/10.1016/bs.aecr.2022.04.002

knowledge to develop National Wildlife Areas (NWAs) at two of these sites: Qaqulluit and Akpait. By 2010, the sites were legally established, and the Sululiit Area Co-Management Committee (ACMC) was created with community members from Qikiqtarjuaq and a representative from the federal government. This ACMC is responsible for the management of the two NWA sites, including developing the management plans for these and annually reviewing and advising on permit applications. Since its inception, the ACMC and other research groups raised concerns about the health of the seabird colonies in Qaqulluit and Akpait given the ongoing increase in anthropogenic activity in the region. In 2018, the ACMC co-developed and implemented new research questions on environmental health of these sites, especially as they relate to contaminants and plastic pollution, two areas of concern for community members who still rely heavily on harvest of local wildlife. Collectively, this pluralism in ecosystem governance demonstrated by the ACMC model of managing these sites, and community direction of ongoing research, promotes Inuit goals of self-determination while charting a course on how to achieve local, national and international conservation and research needs in the Canadian Arctic.

1. Introduction

In the 21st Century, there has been a slow but consistent increase in scientific literature, and in many cases a shift in typical research methodology, to recognize and engage local communities in environmental research, monitoring and management (e.g. Danielsen et al., 2014; Thompson et al., 2019), and this is particularly notable for Indigenous communities (e.g. Danielsen et al., 2018; Green and Raygorodetsky, 2010; Latulippe and Klenk, 2020). In part, this has been driven by the increasing Indigenous assertion of their rights as they pursue self-determination, and recognition of their role in making environmental management systems work (e.g. Cocks, 2017; De Lissovoy, 2010; Reed et al., 2020). However, from a scientific perspective, the drive towards pluralism in ecosystem governance and the need to seek out multiple sources of data to address complex environmental issues (Anadon et al., 2009) has supported the largely global recognition of the value of incorporating local ecological knowledge (LEK); herein broadly including aspects of oral tradition, traditional ecological knowledge (TEK), aboriginal traditional knowledge, and community knowledge; (Berkes et al., 2000) into wildlife and environmental science (Gadgil et al., 2021; Gilchrist et al., 2005). This is particularly true for regions where the practice and oral tradition of Indigenous knowledge holders is present, and where long-term, community-driven scientific monitoring and management may be the best approach for tracking environmental change (Berkes et al., 2007; Danielsen et al., 2014; Thompson et al., 2019). Concomitantly, the opportunity to provide training and share scientific capacity with communities benefits all stakeholders and project outcomes related to protected areas (Healey et al., 2011; Labbé et al., 2017; Sadowsky et al., 2022). In these situations, a pluralistic model of ecosystem governance with local communities, government organizations and other stakeholders involved in sharing priorities, environmental data and decision-making seems well-suited, if not the logical, practical approach, to yield robust data for evidence-based environmental management (e.g. Wall et al., 2017).

In Canada in particular, both policy and regulation regarding the use of Indigenous perspectives, and working with Indigenous communities on almost all issues related to local activities and well-being, have become more recognized in recent years (e.g. NTI, 2006; Turcotte et al., 2021), and perhaps will accelerate following the Truth and Reconciliation Commission in 2015 (https://nctr.ca/records/reports/; see also Bozhkov et al., 2020). Of importance, for Inuit Nunangat (the Inuit homeland in Canada), Inuit Tapiriit Kanatami (ITK) developed the *National Inuit Strategy on Research* (ITK, 2018) that specifically outlines how research needs to move from an exclusion framework to an Inuit self-determination model. Among several objectives, this model includes: (1) using research as a tool for creating social equity; (2) aligning funding with Inuit research priorities; and (3) advancing Inuit governance at the centre of discussions of all things involving research within Inuit Nunangat.

In the Canadian Arctic, working collaboratively with Inuit communities, and notably considering *Inuit Qaujimajatuqangit* (IQ; specifically Inuit beliefs, laws, principles, values, skills, knowledge and attitudes) as complementary information (e.g. offering parallel worldview) to western-based scientific approaches, has been underway for several decades, especially in the natural sciences (e.g. Berkes et al., 2007; Dowsley, 2009; Ferguson et al., 1998; Gilchrist et al., 2005; Gunn et al., 1988; Mallory et al., 2003a; Nakashima and Murray, 1988; Usher, 2000). This was also prominent in negotiations of the *Nunavut Agreement* (formerly referred to as the *Nunavut Land Claims Agreement* or NLCA; INAC, 1993) for the creation of Nunavut Territory (~20% of the Canadian land mass); for example, there was a requirement for the completion of an IQ study on the bowhead whale *Balaena mysticetus* (Hay, 2000). Moreover, for all of the existing and future protected areas in territory, there is a requirement for the territorial or federal government responsible for those sites to negotiate an Inuit Impacts and Benefits

Agreement (IIBA) with the communities affected by the creation of those sites (Mallory et al., 2003b). In the case of Migratory Bird Sanctuaries (MBSs) and NWAs under the authority of the federal department Environment and Climate Change Canada (ECCC), the Umbrella IIBA (https://www.tunngavik.com/files/2017/05/6980_EC_IIBA_EN_F_av-4.pdf) included the creation of Area Co-Management Committees (ACMC) for each MBS and NWA, which are comprised mostly of community members with at least one government representative (see https://nunavutacmc.com/). These ACMCs are advisory committees and ensure that Inuit play the prominent leadership role in decision-making and governance for activities in the protected areas, including developing management plans for each site, developing research priorities with partners, reviewing permits for activities at these locations, and co-leading and actively participating in new research (consistent with ITK, 2018).

However, it has not always been this way: use of IQ, co-management, and a pluralistic model including Inuit-led development of research and management plans for federal protected areas in the North is a very recent development. In this paper, we review the history of scientific and community interest behind the creation of two NWAs, Qaqulluit (ᖃᖁᓪᓗᐃᑦ) and Akpait (ᐊᒃᐸᐃᑦ), protecting major seabird colonies on the east coast of Baffin Island near the community of Qikiqtarjuaq, Nunavut, Canada as a case study of co-management. We then describe the annual co-management activities at those NWAs, in light of known environmental concerns for the health and sustainability of the NWAs, including recent collaborative research between government, academia and the community. Through this discussion, we contend that the ACMC model for protected area management in the Canadian Arctic is an effective, mutually respectful, and beneficial approach not only for routine management issues, but promotes communication and attentiveness to new place-based concerns that otherwise scientists and environmental managers in the "south" would never experience.

2. The seabird colonies and threats in the Arctic

Two magnificent Arctic seabird colonies can be found on the east coast of Baffin Island, Nunavut, Canada, near the northern end of Davis Strait and the southern part of Baffin Island. The NWAs are located near 67°N, 62°W, in a sparsely populated part of Arctic Canada, 100 km from the nearest community (Fig. 1). Akpait and Qaqulluit represent major

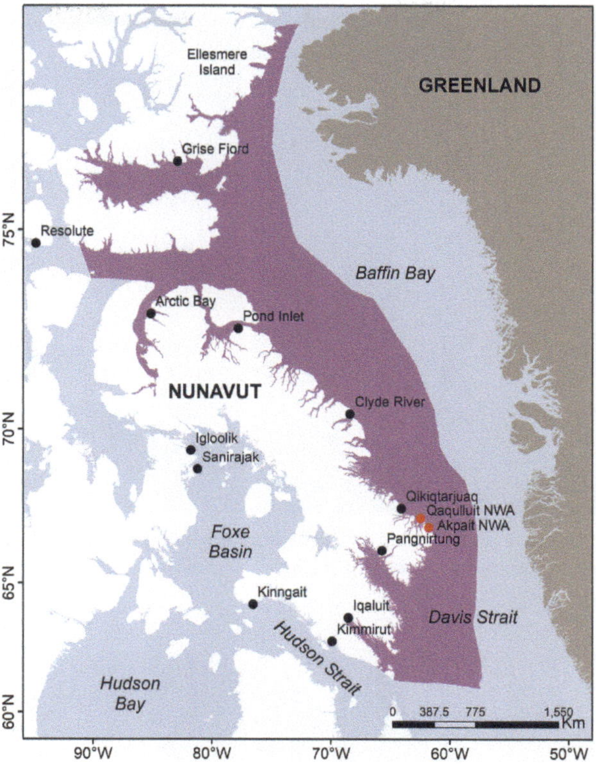

Fig. 1 Map of the study region, noting the community of Qikiqtarjuaq where the Area Co-Management Committee (ACMC) is located. Red dots represent the Akpait and Qaqulluit National Wildlife Areas. The purple region is the zone of the Baffin Bay—Davis Strait Strategic Environmental Assessment. Other communities, many of which have their own ACMCs for nearby protected areas, are also noted.

seabird colonies in Canada (Figs. 2 and 3), with each supporting >8% of the entire Canadian population of at least one seabird species (Mallory et al., 2019). Due to constraints imposed by the phenology of sea ice formation and break up, and associated effects on marine productivity in the high Arctic oceanographic zone (Nettleship and Evans, 1985), cliff-nesting seabird colonies in the Canadian Arctic tend to be large, but far apart, often near leads or polynyas in the sea ice (Mallory et al., 2019). Qaqulluit NWA protects Canada's largest colony of northern fulmars (*Fulmarus glacialis,* qaquluk), with numbers at this site once thought to be in the range of 100,000 pairs, but recent estimates suggest declines to approximately 25,500 pairs (Mallory et al., 2020). At nearby Akpait NWA, large numbers

Fig. 2 The impressive rock towers of the Qaqulluit National Wildlife Area, home to Canada's largest colony of northern fulmars (*Fulmarus glacialis*). Photograph by Michel Gendron.

Fig. 3 The pointed rock peaks of the Akpait National Wildlife Area, supporting thick-billed murres (*Uria lomvia*), northern fulmars, and black-legged kittiwakes (*Rissa tridactyla*). Photograph by Michel Gendron.

of black-legged kittiwakes (*Rissa tridactyla*, tiratira—1900 pairs), fulmars (15,000 pairs) and thick-billed murres (*Uria lomvia*, akpa,—130,000 pairs) inhabit the cliffs during the breeding season, making it one of the largest, multispecies seabird colonies in Arctic Canada (Gaston et al., 2012).

The Arctic is also distant from large, developed areas, and there is limited nearby industry, especially in the islands of the Canadian Arctic. Moreover, the number of people in the region is very small and have lived in the region for centuries; one would expect the colonies and local environments to be comparatively pristine. Despite this vastness and relatively low development density, however, a variety of threats to seabirds are rapidly increasing in Arctic Canada (e.g. Anderson et al., 2018; Wong et al., 2018). At a circumpolar scale, for example, climate change is altering the timing and extent of sea ice cover with consequent effects on the marine food webs on which these seabirds rely (e.g. Chan et al., 2013; Comiso et al., 2008; Kortsch et al., 2015), notably changing the timing of peak food supplies (Gaston et al., 2009). Within Nunavut, shipping is increasing (Dawson et al., 2020), including cruise ship expeditions that visit colonies (Dawson et al., 2016), potentially exposing the birds to greater noise and disturbance (Halliday et al., 2022), and a greater risk of contaminant or oil spill exposure (Wong et al., 2018). There remains interest, but no recent activity, in potential hydrocarbon extraction in Baffin Bay and Davis Strait (NIRB, 2018). Importantly, whether platform or ship-based, encounters with oil spilled at sea can be disastrous for seabirds (Wiese and Robertson, 2004). As well, commercial fisheries are increasing in the foraging range of Nunavut seabird colonies (Anderson et al., 2018), and many species are highly susceptible to mortality in different types of fishing gear (Ellis et al., 2013). Moreover, plastic and chemical pollution remain threats to Arctic seabirds, the sublethal effects of which continue to be evaluated (Baak et al., 2020; Braune et al., 2019; Foster et al., 2019; Provencher et al., 2020; Zahaby et al., 2021).

Additionally, millions of birds breed in the circum-Arctic region, and almost all seabirds in the region are migratory, using the Arctic for part of the year. Similar threats that are present in the Arctic affect migratory birds throughout their annual cycle, which takes them across several ecoregions and jurisdictions. For example, murres breeding in Canada travel ~3000 km away from their Arctic breeding colonies during the winter months (Gaston et al., 2011), and fulmars are recorded travelling as far as waters near Ireland (~6000 km; Mallory et al., 2008) during the non-breeding season. Consequently, Arctic seabirds are continuously exposed to numerous threats, including pollution, degradation of habitat, disturbance, and reductions in prey species across the life cycle; anthropogenic activities affect everywhere on the planet and the Arctic is not immune to these threats. Protecting these sites through national legislation was one means to exercise some control over activities near the birds, with a goal of reducing threats and conserving their populations. These threats are clearly present near the seabird colonies at Qaqulluit and Akpait; we present this below as a case study of the recent protected area creation and management process in Arctic Canada, and specifically the prominent role Inuit have in developing management, monitoring and research activities at such sites.

3. Historical perspectives on the Qaqulluit (ᖃᖅᑯᓪᓗᐃᑦ) and Akpait (ᐊᒃᐸᐃᑦ) national wildlife areas (NWAs)

Inuit and European explorers have known of the seabird colonies at Akpait and Qaqulluit for many generations (Mallory et al., 2006). Although this part of Baffin Island is relatively depauperate of terrestrial megafauna (e.g. comparatively few caribou Rangifer *tarandrus*, no large goose colonies, no muskoxen *Ovibos moschatus*), the nearby ocean is rich with marine life: whales and seals are prominent here (reviewed in Mallory et al., 2019). Inuit and their predecessors have lived near these sites for ~4000 years, with numerous archaeological sites along the coast (Graham, 1997; Kemp, 1971). Indeed, at the Akpait NWA, there is evidence of locations where people hunted the seabirds from the top of the cliffs (Gaston and Smith, 1987). Europeans came along this coast starting approximately 1000 AD with the Norse, followed many years later by whalers in the mid-1800s (Taylor, 1863). Scientists followed in the late 1800s (Boas, 1885; Kumlien, 1879), and then North American military, coast guard and transportation personnel began to appear in the 1940s, when radar stations and a community were established in the region (Mallory et al., 2006). At this time, many Inuit who had been living traditional lifestyles on the land relocated to Padloping Island (the new community). During earlier LEK interviews (Mallory et al., 2006), hunters often noted with fondness their earlier days living and hunting in this region, and especially at Akpait, if they were there when the young seabird chicks were leaping off the cliff to the water. Eventually, the community of Padloping Island was relocated to Broughton Island (renamed Qikiqtarjuaq) starting in the 1960s, where it remains today.

Although the existence of the seabird colonies had been known by western-based scientists for almost a century, the first published census of seabirds did not come until the 1950s, by Wynne-Edwards (1952a,b) and Watson (1957). In the early 1970s, the federal Canadian Wildlife Service

(CWS) conducted coastal seabird surveys across the Canadian Arctic as part of the International Biological Programme (Revel, 1981), reaffirming the 1950s information (Nettleship and Smith, 1975). Brief site visits to Qaqulluit occurred in the 1980s with little scientific product, but Gaston and Smith (1987) conducted a new census of seabirds at Akpait in 1985, adjusting the initial estimates from the 1970s. After the fervour of survey work through the 1970s and 1980s (largely supported by needs for environmental information related to oil and gas exploration (e.g. Imperial Oil Ltd., 1978; Petro-Canada Ltd., 1979) or land claim negotiations), western-based scientific work at these sites lay dormant for the next 15 years. Notably, the community of Qikiqtarjuaq continued to use the region as they had for decades (Mallory et al., 2006).

3.1 Efforts towards getting National Wildlife Areas established

Given the increasing concern over threats to Arctic seabirds, and the survey work which confirmed the importance of these colonies at a national and international scale, the CWS approached the community of Qikiqtarjuaq about protection for these sites four times between 1985 and 1995. For a variety of reasons (e.g. misconceptions on loss of hunting rights, government interest in minerals at the sites; detailed in Mallory et al., 2006), the community was not interested in pursuing protection at that time. However, in 1998, the Hamlet of Qikiqtarjuaq reopened the site protection file and sought a meeting to reinvigorate the process, largely as an economic development opportunity (for tourism). The CWS had opened an office in Iqaluit with the official creation of Nunavut Territory in 1999, and their now "local" staff initiated the process of discussions with the community. At the same time, the CWS and community began a series of site visits, guided by hunters from Qikiqtarjuaq (Fig. 4), to update information on colony size (Gaston et al., 2006), address community concerns about contaminants (Mallory et al., 2004, 2005), conduct a LEK study on important wildlife locations at the sites (Mallory et al., 2006), and discuss differences between NWAs and national parks in town hall meetings (Mallory et al., 2006). One key point that was fundamental to moving site protection ahead was that, under both NWA legislation and the *Nunavut Agreement*, all rights of Inuit access and wildlife harvest within the region were maintained even if they became NWAs. This was critically important, because many Inuit communities still rely principally on "country foods" (i.e. wildlife) to meet dietary needs (e.g. Ford, 2009; Kinloch et al., 1992).

Fig. 4 At the wharf in Qikiqtarjuaq, where Inuit hunters are preparing the boats for departure to the sites. Photograph by Michel Gendron.

While many important insights came out through this information-sharing process (Mallory et al., 2006), some key items were:

(1) Although for years the fulmar colony on Qaqulluit Island was known internationally as Cape Searle, the site name was changed to "Qaqulluit" by the ACMC, as this means "fulmars" or "place of fulmars" in Inuktitut.

(2) The second colony was known as Reid Bay or The Minarets to western-based seabird scientists, but was renamed "Akpait" which means "murres" or "place of murres" in Inuktitut.

(3) Even though Qikiqtarjuaq was 100 km north of these sites, the region was still used considerably for year-round harvesting, notably for polar bear (*Ursus maritimus*), walrus (*Odobenus rosmarus*) and ringed seal (*Pusa hispida*). The LEK study conducted as part of the formation process of the NWAs was key to helping map these data and incorporate them into boundary discussions.

(4) Because of the wildlife importance of this region, the terrestrial area supporting the colonies had been selected by the community to be retained as Inuit-owned lands during the territory-wide land claim negotiations (INAC, 1993; Riewe, 1992).

(5) While the community considered options and favoured a single, large protected area, the final site boundaries were reduced during negotiations with the Regional Inuit Association (Qikiqtani Inuit Association; QIA) and split into two NWAs, to ensure access to land in case of future needs (Mallory et al., 2006).

By 2004 the boundaries had been decided, and the Akpait and Qaqulluit NWAs were included in the development of the IIBA for CWS protected areas in Nunavut. The official designation of the sites was not completed until 2010, 25 years after the CWS had first proposed the idea of protected areas, and 12 years after the community sought to protect the sites.

The entire process of establishing these NWAs differed from the other CWS protected areas in Nunavut (and very much from the rest of Canada), in large part because Akpait and Qaqulluit were the first to be established following the implementation of the *Nunavut Agreement*. Prior to 1999, sites were recognized based on wildlife and conservation merit as determined by empirical data collected by scientists from outside of Nunavut, and then moved through the same national legislative process of creating these sites, pursuant to the *Canada Wildlife Act* (Mallory et al., 2003a), which at the time did not necessarily consider community or Indigenous interests, needs or even input. The creation of Akpait and Qaqulluit NWAs (and the nearby Ningingangiq NWA; https://www.canada.ca/en/environment-climate-change/services/national-wildlife-areas/locations/ninginganiq.html) charted a new, community-led, collaborative path to protection of key wildlife sites in Nunavut, similar to how Parks Canada has adjusted its process to respect Indigenous partners (e.g. Murray, 2011). Nonetheless, we acknowledge that history and processes differ across the Canadian Arctic, with examples where co-management has proven more challenging (e.g. Kofinas, 2005), but recent progress in Nunavut on creating larger marine protected areas speaks to the success of the modern, collaborative approach with communities (e.g. Fisheries and Oceans Canada, 2020; Parks Canada, 2022). At a national level, Canada has made commitments towards working with Indigenous communities in the collective approach to habitat protection (see Reed et al., 2021), notably for future Indigenous Protected and Conserved Areas (IPCAs) which will be Indigenous-led, and described in Government of Canada (2018), see also https://conservation-reconciliation.ca/about-ipcas.

4. Area Co-management Committees—A new model of governance for protected areas

As part of the IIBA process pursuant to the creation of NWAs in Nunavut, ACMCs were established for each site, or in the case of these two close seabird colonies, for both sites (see https://nunavutacmc.com/sululiit/). The first meeting was in May 2008, but annual meetings and development of protocols and priorities did not start until 2011. The ACMC is an advisory committee to the Minister of Environment and Climate Change Canada on all aspects related to Akpait and Qaqulluit NWAs. In this role, the ACMC is responsible for the day-to-day management of the protected Area and inform management plans, permitting and visitor access, research, and the preservation of cultural areas and resources. For example, the ACMC can make recommendations on the monitoring of wildlife, maintaining and improving wildlife habitat, periodic inspections, enforcement of regulations, and maintenance of facilities and infrastructure. The responsible Minister still retains final authority on decisions regarding the NWAs, but the main source of advice comes from the ACMC. Consequently, the ACMCs role is to ensure that the protected area is well managed. In some cases, they may also be called upon to provide advice to other government departments on issues related to the NWAs.

The ACMC is composed of six members, five of which are Inuit from Qikiqtarjuaq and the sixth member is CWS staff; by this composition, local Inuit inherently play a key role in site information-sharing and management, and a critical role in decision-making. Inuit members are usually from the Community Lands and Resources Committee (CLARC), unless there are not enough CLARC members. Appointments are done by the QIA for three members and by the federal Minister of Environment and Climate Change Canada for the other three. In general, ACMC members are nominated for appointment based on their knowledge and/or experience in conservation, wildlife, tourism, and/or Inuit culture and heritage. Members are appointed for 3 year terms that may be renewed for additional terms.

The ACMC has a Chair and Vice-chair, the latter being the CWS employee. The chair's function is to set meeting agendas and lead the meetings, while the vice-chair (currently Susanne Emond) performs the secretariat functions for the ACMC, supports the work of the chair, and works with

Environment and Climate Change Canada to acquire the annual operations budget for the committee. The ACMC also has the option of maintaining an elder advisor and a youth representative, to ensure broad representation of community views. The ACMC meets up to three times per year, pending the number of issues that may be active regarding the NWAs. These meetings are open to the public and are conducted in both English and Inuktitut, with the help of a translator. Quorum is four members, two of which must be QIA appointees.

One key aspect of the co-management regime is that all efforts are made to form decisions by consensus—agreement of all members—which is similar to how the Nunavut territorial legislature operates (Rice, 2016). However, if this is not possible, there are options to form decisions by majority vote. These operating rules vary by ACMC and are developed internally by the ACMCs, effectively adjusting rules to local community approaches and needs.

Aside from developing the management plans for the sites, key annual work includes reviewing and advising on permit applications to access the NWAs. These are typically associated with cruise ship tourism (Dawson et al., 2014), but also scientific research (principally geology, oceanography or ecology investigations). Increasingly, as the ACMC identifies work to be done in the region and reviews permit applications, the ACMC also meets with researchers to discuss recent results, future questions, and assign tasks to research teams that they would like addressed within the context of the work. Effectively, as the ACMC works more with the legislative aspects of various environmental issues, this necessitates communication between the community and scientists, which leads to new research questions and approaches, thereby better serving both the community and science. There is also an Annual Report produced in Inuktitut and English which summarizes key objectives, achievements, and ACMC membership for the year (https://nunavutacmc.com/sululiit-annual-reports-english/).

Fundamentally, the ACMC approach provides an open forum for dialogue, discussion, and idea-sharing for both community and government. Why? Because it creates a clear and approved venue for community concerns to be raised to the territorial or national level, and conversely, it serves as a regular conduit by which national or international concerns are brought to the community. A good example of this process is illustrated below (outlined in more detail in Fig. 5), on how the ACMC has been involved in a substantial environmental assessment project in Nunavut.

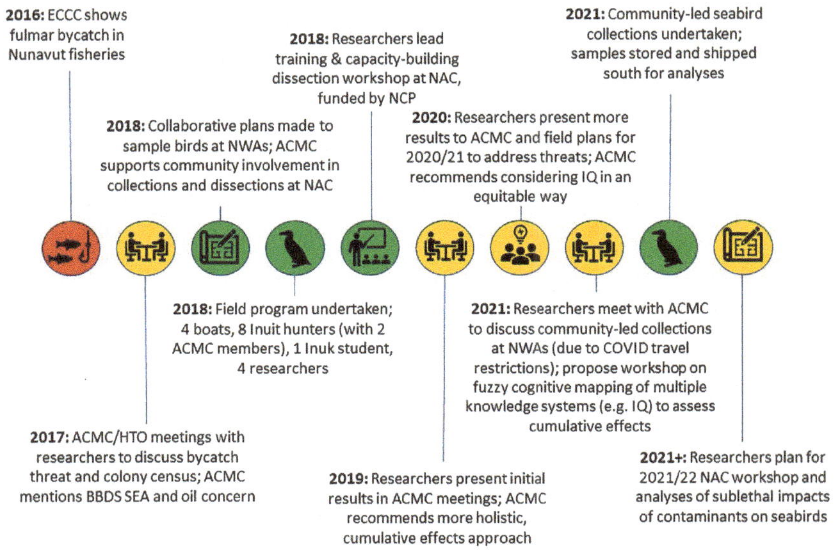

Fig. 5 Flow chart of activities from 2016 to present, focused on collaborative research directed by the ACMC and between community members and scientists from the South. Red denotes an initial concern (bycatch in fisheries), yellow denotes meetings, and green denotes "action," typically field work.

5. Co-management and shared research opportunities—An example

The creation of the ACMC and the inherent approach of co-management at the NWAs has led to new opportunities for research at these seabird colonies, which address both pressing scientific need and community concern. Given their location in the Arctic, earlier work had suggested that birds foraging from these colonies were on the cusp of marine food web changes due to global warming, observed farther south in Hudson Bay, but not detected farther north in Lancaster Sound (Provencher et al., 2012). Moreover, earlier work had suggested that the number of fulmars nesting at Qaqulluit was lower than previously estimated (Gaston et al., 2006), although the murre colony at Akpait seemed to be stable (Gaston and Smith, 1987). Furthermore, from the early work at the site (Mallory et al., 2004, 2005), the community had expressed strong concerns about contamination of local marine food webs due to known ocean dumping of wastes in the region. While these local issues were perhaps sufficient impetus for additional research, concurrently a large-scale, strategic environmental assessment had been initiated for the entire Baffin Bay—Davis Strait (BBDS) region (BBDS SEA) (https://www.nirb.ca/content/strategic-environmental-assessment; Fig. 1), most of the foraging range of the seabirds, and hunting range of people from Qikiqtarjuaq, falling within that region. The BBDS SEA process included a series of open house consultations with Qikiqtarjuaq community members on environmental issues of concern in the region.

ECCC scientists and university academics approached the Sululiit ACMC to collaborate on censusing the colonies in Akpait and Qaqulluit because of concerns that fulmars from those colonies were being deleteriously affected by offshore industrial fisheries (Anderson et al., 2018; see also Fig. 5). During these planning meetings, the committee also voiced community concerns about pollution in the region, which had also brought up in the context of the BBDS SEA that focused on the potential effects from oil and gas activities (e.g. QIA, 2019). They discussed many types of data that were of interest, but for seabird population and human health, the community was notably interested in contaminants in wildlife (including hydrocarbon-related materials) and plastic pollution (including plastic additives in birds).

Consequently, following consultations with CWS personnel, field work was planned with the Sululiit ACMC that aimed to collect a variety of samples to address multiple research questions relation to seabird population health. This was a major, collaborative field expedition that was undertaken between the ACMC, ECCC and academic researchers, with field logistics (camps, boats, fuel) and field sampling (e.g., collection of birds) run by community members, and scientific measurements (drone photography, water, air and sediment sampling) led by government scientists who joined the hunters on the trip in 2018. This work is continuing in 2021 via community collections (including Sululiit ACMC members and other Inuit hunters) which is to establish benchmark levels and sub-lethal effects on seabirds from oil-related contaminants and plastic pollution (including additives) (Fig. 5).

By any measure, the goal of updating our knowledge on birds and contamination of this region, for scientists and community members, was a success. New estimates were made for the Qaqulluit fulmar colony, showing a continued decline in numbers of breeding birds (Mallory et al., 2020). Analyses of plastic ingestion by fulmars, murres, kittiwakes and black guillemots (*Cepphus grylle*) suggested that there were some declines in number and mass of plastic pieces, but that the frequency of ingestion had not changed compared to sampling a decade earlier (Baak et al., 2020).

Environmental sampling found that the seabirds were clearly ingesting microplastics at sea and depositing them on land principally through defecation (Bourdages et al., 2021), but that air, ocean and sediment all had ample microplastics that were delivered through other means (Hamilton et al., 2021) and consequently the colonies were not predicted "hotspots" for plastic debris. All four of the seabird species had low concentrations of polycyclic aromatic compounds (PACs; Provencher et al., 2020), some of the first measures for this in the Canadian Arctic, and these were then used to develop a screening tool ("toxchip") for oil exposure and reaction in murres and guillemots (Zahaby et al., 2021). This was also followed by measures of metabolomic responses in the species (Sarma et al., 2022), which combined with the toxchip provide powerful tools for environmental assessment related to hydrocarbon exposure within the BBDS, and in fact anywhere that these species inhabit.

Of particular note, as the ACMC and researchers continue to learn about stressors to the seabirds in the NWA, the ACMC has encouraged researchers to use a more holistic approach to better understand how the stressors may interact together. In response, a project is currently underway between the Sululiit ACMC and researchers to use a fuzzy cognitive mapping approach (e.g. Hobbs et al., 2002) that will bring together both western-based science and IQ to explore how fisheries, oil pollution, plastic pollution, climate change, and other habitat changes interact and influence northern fulmar populations. This is a clear success for the ACMC, where they have directly guided research towards a community perspective on environmental change.

In the context of the SARS-CoV-2 global pandemic, it is also worth noting that the work between the Sululiit ACMC and researchers continued even though most field work globally was cancelled or limited in 2020 and 2021 due to travel restrictions (e.g. Scerri et al., 2020). Due to ongoing consultations and discussions between the ACMC and partner researchers, ACMC members and local Inuit hunters conducted sampling in the NWAs that were aimed at further exploring additional information on contaminants in the region. Thus, the successful, collaborative work between the Sululiit ACMC and researchers has been important to the resiliency of the work in the region.

6. From collaborative local research to tracking international trends

One of the concerns Nunavut communities have recognized over the last 10 years is how plastic pollution is affecting wildlife and the environment, and that this is an issue of Inuit sovereignty and wellness (Liboiron et al., 2021; J. Provencher, unpubl. data). Some of the reports that informed this listing of plastic pollution as a concern in the region were based on earlier studies from the Akpait and Qaqulluit NWAs that were conducted in 2007/08 (Provencher et al., 2009, 2010). Indeed, studies in fulmars from the Qaqulluit NWA were used to implement harmonized protocols using Canadian fulmars for the first time, and allowed researchers and communities to understand plastic pollution levels in the region in relation to other areas (e.g. northern Europe; Provencher et al., 2017). Since this time, further work has been done in Canada examining ingested plastic pollution in fulmars, which resulted in refinements of harmonized protocols for comparing plastic pollution in fulmars across their range (Provencher et al., 2019) and comparing time and spatial trends internationally (Van Franeker et al., 2021). Overall, plastic occurrence in Arctic seabirds has increased then stabilized in frequency, but mass of plastic in individual birds remains low (e.g. Baak et al., 2020).

Importantly, during ongoing consultations with communities as more plastic pollution research has been planned and developed, the Sululiit ACMC has co-developed the research questions regarding plastic occurrence, formulated the field logistics, and co-collected the field samples, to ensure the work addresses potential impacts in the NWAs (Baak et al., 2020; Bourdages et al., 2021; Hamilton et al., 2021). This has created a flow of information between the Sululiit ACMC, partner researchers, and the international plastic pollution research community that allows locally directed research to influence international collaborations, that can then inform future research objectives via the governance of the ACMCs. This has also led to the incorporation of research from the Akpait and Qaqulluit NWAs into policies related to plastic pollution, such as the Canadian Environmental Sustainability Indicators (CESI; https://www.canada.ca/en/environment-climate-change/services/environmental-indicators.html).

7. The future of collaborative research and management between Inuit communities and scientists

With the implementation of the *Nunavut Agreement*, the development of IIBAs for different protected areas, and a strong leadership role from funding organizations, regulators, government and Indigenous organizations, the move towards collaborative, community-driven environmental science and management in Arctic Canada has been slow (Brunet et al., 2014), but is clearly increasing and leading the way in Canada for Indigenous knowledge and western science based knowledge to work together in research, monitoring and management (Alexander et al., 2019). Our work on seabirds on the east coast of Baffin Island, addressing community and scientific issues, is but one example, but there are many others. The SIKU program (https://siku.org/about) has been a leader is promoting and helping these community-driven projects, supported through the Arctic Eider Project in southern Hudson Bay (https://arcticeider.com/). The eider group in particular has conducted collaborative, world-leading community research and monitoring on common eiders (*Somateria mollissima sedentaria*), now targeted at the development of a new protected area. Elsewhere, SmartICE (https://cop23.unfccc.int/climate-action/momentum-for-change/ict-solutions/smartice) links measurements of ice features in real time in a collaboration between northerners and government scientists and academics, to improve safety of travel across ice (among other products). In the western Arctic, Kutz et al. (2013) used IPY to initiate a community-led wildlife health monitoring program for terrestrial ungulates which continues. At the territory-wide level, the Nunavut Wildlife Management Board launched a community-based monitoring program (CBMP) which provides data to wildlife managers on Inuit harvest of a variety of wildlife (Ndeloh Etiendem et al., 2020), data that are exceedingly difficult to come by through other measures. In Canada and elsewhere, there has also been a rise in the number of community-led, Indigenous Guardian programs targeted for environmental monitoring and conservation (Reed et al., 2021), including in the Northwest Territories in the Łutsël K'e and the Dehcho region (Social Ventures Australia, 2016), and in British Columbia (Artelle et al., 2019).

Clearly, then, co-management led through community-driven research and monitoring, consultation, incorporation of diverse views and perspectives, and joint decision-making has a bright future in the Arctic, consistent with similar approaches promoted elsewhere (e.g. De Pourcq et al., 2015; Jaireth and Smyth, 2003). Indeed, we contend that this pluralistic model of ecosystem governance in Arctic protected areas not only aligns well with the intent of research approaches recommended by Inuit (ITK, 2018), but is well-suited to provide the information needed for evidence-based decision-making in protected area management (Wall et al., 2017). However, it is not a fast process to initiate: communities need to be engaged respectfully, with equal consideration of their concerns and culture, and appropriate training needs to be provided to ensure that ACMCs meet legislated requirements (e.g. reporting, developing measurable and achievable objectives). In our experience, this leads to much stronger community engagement, and the opportunities for additional science-based training and capacity-building. This in turn has yielded two major accomplishments: (1) enhanced, collaborative scientific research in the region; and (2) answers to community concerns over wildlife and environmental health. Thus, we suggest that the ACMC model is a mutually respectful and fulfilling approach to protected area governance within existing federal legislation, and promotes new opportunities for community and scientific exploration. However, in the future we expect that most new protected areas will be IPCAs (Finegan, 2018; Government of Canada, 2018), a new model of protected area which are fully Indigenous-led (selection, governance, management). This will take the ACMC-model further, as most of the current ACMCs (with few exceptions, such as Akpait and Qaqulluit) have developed management plans for sites that were chosen without community input, for wildlife sites deemed a priority by people not living in the Arctic.

Acknowledgements

Financial support was provided by Environment and Climate Change Canada, including an Inuit Impacts and Benefits Agreement for the sites, the Northern Contaminants Program, and Acadia University. The work of the ACMCs would not have been possible without the community involvement of numerous people, including Meeka Newkingnak, Levi Nutaralak, Stevie Aulaqiaq, Jacopie Nuqingaq, Oolana Nuqingaq, Yukipa Audlakiak, Jaypootee Aliqatuqtuq and Pasha Kooneeliusie.

References

Alexander, S.M., Provencher, J.F., Henri, D.A., Taylor, J.J., Lloren, J.I., Nanayakkara, L., Johnson, J.T., Cooke, S.J., 2019. Bridging Indigenous and science-based knowledge in coastal and marine research, monitoring, and management in Canada. Environ. Evid. 8, 1–4.

Anadon, J.D., Giménez, A., Ballestar, R., Pérez, I., 2009. Evaluation of local ecological knowledge as a method for collecting extensive data on animal abundance. Conserv. Biol. 23, 617–625.

Anderson, C.M., Iverson, S.A., Black, A., Mallory, M.L., Hedd, A., Merkel, F., Provencher, J.F., 2018. Modelling demographic impacts of a growing Arctic fishery on a seabird population in Canada and Greenland. Mar. Environ. Res. 142, 80–90.

Artelle, K.A., Zurba, M., Bhattacharyya, J., Chan, D.E., Brown, K., Housty, J., Moola, F., 2019. Supporting resurgent Indigenous-led governance: A nascent mechanism for just and effective conservation. Biol. Conserv. 240, 108284.

Baak, J.E., Provencher, J.F., Mallory, M.L., 2020. Plastic ingestion by four seabird species in the Canadian Arctic: comparisons across species and time. Mar. Pollut. Bull. 158, 111386.

Berkes, F., Colding, J., Folke, C., 2000. Rediscovery of traditional ecological knowledge as adaptive management. Ecol. Appl. 10, 1251–1262.

Berkes, F., Berkes, M.K., Fast, H., 2007. Collaborative integrated management in Canada's north: The role of local and traditional knowledge and community-based monitoring. Coast. Manag. 35, 143–162.

Boas, F., 1885. Baffin-Land. Geographische ergebnisse einer in den jahren 1883 und 1884 ausgeführten forschungsreise. Petermanns Mitteillungen. (Gotha) Erganzungsheft No. 80.

Bourdages, M., Provencher, J.F., Baak, J.E., Mallory, M.L., Vermaire, J.C., 2021. Breeding seabirds as vectors of microplastics from sea to land: evidence from colonies in Arctic Canada. Sci. Total Environ. 764, 142808.

Bozhkov, E., Walker, C., McCourt, V., Castleden, H., 2020. Are the natural sciences ready for truth, healing, and reconciliation with Indigenous peoples in Canada? Exploring 'settler readiness' at a world-class freshwater research station. J. Environ. Stud. Sci. 10, 226–241.

Braune, B.M., Gaston, A.J., Mallory, M.L., 2019. Temporal trends of legacy organochlorines in eggs of Canadian Arctic seabirds monitored over four decades. Sci. Total Environ. 646, 551–563.

Brunet, N.D., Hickey, G.M., Humphries, M.M., 2014. The evolution of local participation and the mode of knowledge production in Arctic research. Ecol. Soc. 19, 2.

Chan, F.T., Bailey, S.A., Wiley, C.J., MacIsaac, H.J., 2013. Relative risk assessment for ballast-mediated invasions at Canadian Arctic ports. Biol. Invasions 15, 295–308.

Cocks, S., 2017. Environmental pluralism, polar harmonies and resolution. Dialog. Univers. 4, 9–21.

Comiso, J.C., Parkinson, C.L., Gersten, R., Stock, L., 2008. Accelerated decline in the Arctic sea ice cover. Geophys. Res. Lett. 35, 1.

Danielsen, F., Topp-Jørgensen, E., Levermann, N., Løvstrøm, P., Schiøtz, M., Enghoff, M., Jakobsen, P., 2014. Counting what counts: using local knowledge to improve Arctic resource management. Polar Geogr. 37, 69–91.

Danielsen, F., Burgess, N.D., Coronado, I., Enghoff, M., Holt, S., Jensen, P.M., Poulsen, M.K., Rueda, R.M., 2018. The value of Indigenous and local knowledge as citizen science. In: Hecker, S., Haklay, M., Bowser, A., Makuch, Z., Vogel, J., Bonn, A. (Eds.), Citizen Science: Innovation in Open Science, Society and Policy. UCL Press, London, pp. 110–123.

Dawson, J., Johnston, M.E., Stewart, E.J., 2014. Governance of Arctic expedition cruise ships in a time of rapid environmental and economic change. Ocean Coast. Manag. 89, 88–99.

Dawson, J., Stewart, E.J., Johnston, M.E., Lemieux, C.J., 2016. Identifying and evaluating adaptation strategies for cruise tourism in Arctic Canada. J. Sustain. Tour. 24, 1425–1441.

Dawson, J., Carter, N., Van Luijk, N., Parker, C., Weber, M., Cook, A., Grey, K., Provencher, J., 2020. Infusing Inuit and local knowledge into the low impact shipping corridors: an adaptation to increased shipping activity and climate change in Arctic Canada. Environ. Sci. Policy 105, 19–36.

De Lissovoy, N., 2010. Decolonial pedagogy and the ethics of the global. Discourse: Stud. Cult. Polit. Educ. 31, 279–293.

De Pourcq, K., Thomas, E., Arts, B., Vranckx, A., Léon-Sicard, T., Van Damme, P., 2015. Conflict in protected areas: who says co-management does not work? PLoS One 10, e0144943.

Dowsley, M., 2009. Community clusters in wildlife and environmental management: using TEK and community involvement to improve co-management in an era of rapid environmental change. Polar Res. 28, 43–59.

Ellis, J., Wilhelm, S., Hedd, A., Fraser, G., Robertson, G., Rail, J.F., Fowler, M., Morgan, K., 2013. Mortality of migratory birds from marine commercial fisheries and offshore oil and gas production in Canada. Avian Conserv. Ecol. 8, 2.

Ferguson, M.A.D., Williamson, R.G., Messier, F., 1998. Inuit knowledge of long-term changes in a population of arctic tundra caribou. Arctic 51, 201–219.

Finegan, C., 2018. Reflection, acknowledgement, and justice: A framework for indigenous-protected area reconciliation. Internat. Indigen. Policy J. 9, 3.

Fisheries and Oceans Canada, 2020. Tuvaijuittuq Marine Protected Area (MPA). https://www.dfo-mpo.gc.ca/oceans/mpa-zpm/tuvaijuittuq/index-eng.html. accessed 24 March 2022.

Ford, J.D., 2009. Vulnerability of Inuit food systems to food insecurity as a consequence of climate change: a case study from Igloolik, Nunavut. Reg. Environ. Change 9, 83–100.

Foster, K.L., Braune, B.M., Gaston, A.J., Mallory, M.L., 2019. Climate influence on legacy organochlorine pollutants in Arctic seabirds. Environ. Sci. Technol. 53, 2518–2528.

Gadgil, M., Berkes, F., Folke, C., 2021. Indigenous knowledge: From local to global. Ambio 50, 967–969.

Gaston, A.J., Smith, S.A., 1987. Seabirds in the Cape Dyer—Reid Bay area of Cumberland Peninsula, Baffin Island, Northwest Territories. Can. Field-Nat. 101, 49–55.

Gaston, A.J., Mallory, M.L., Gilchrist, H.G., O'Donovan, K., 2006. Status, trends and attendance patterns of the northern fulmar *Fulmarus glacialis* in Nunavut, Canada. Arctic 59, 165–178.

Gaston, A.J., Gilchrist, H.G., Mallory, M.L., Smith, P.A., 2009. Changes in seasonal events, peak food availability and consequent breeding adjustment in a marine bird: a case of progressive mis-matching. Condor 111, 111–119.

Gaston, A.J., Smith, P.A., Tranquilla, L.M., Montevecchi, W.A., Fifield, D.A., Gilchrist, H.G., Hedd, A., Mallory, M.L., Robertson, G.J., Phillips, R.A., 2011. Movements and wintering areas of breeding age thick-billed Murre *Uria lomvia* from two colonies in Nunavut, Canada. Mar. Biol. 158, 1929–1941.

Gaston, A.J., Mallory, M.L., Gilchrist, H.G., 2012. Populations and trends of Canadian Arctic seabirds. Polar Biol. 35, 1221–1232.

Gilchrist, G., Mallory, M., Merkel, F., 2005. Can local ecological knowledge contribute to 2ildlife management case studies of migratory birds. Ecol. Soc. 10, 1.

Government of Canada, 2018. We rise together. Achieving pathway to Canada Target 1 through the creation of Indigenous Protected and Conserved Areas in the spirit and practice of reconciliation. Indigenous Circle of Experts Report and Recommendations, Indigenous Affairs, Ottawa, Canada, ISBN: 978-0-660-25571-2.

Graham, M., 1997. A preliminary study of representative marine areas for the Baffin Island Shelf Region. Vancouver Aquarium (for Parks Canada), Vancouver.

Green, D., Raygorodetsky, G., 2010. Indigenous knowledge of a changing climate. Clim. Change 100, 239.

Gunn, A., Arlooktoo, G., Kaomayok, D., 1988. The contribution of ecological knowledge of Inuit to wildlife management in the Northwest Territories. In: Freeman, M.M.R., Carbyn, L.N. (Eds.), Traditional knowledge and renewable resource management in northern regions. Edmonton: IUCN Commission on ecology and the Boreal Institute for Northern Studies Occ. Publ. 23, pp. 22–30.

Halliday, W.D., Dawson, J., Yurkowski, D.J., Doniol-Valcroze, T., Ferguson, S.H., Gjerdrum, C., Hussey, N.E., Kochanowicz, Z., Mallory, M.L., Marcoux, M., Watt, C.A., Wong, S.N.P., 2022. Vessel risks to marine wildlife in the Tallurutiup Imanga National Marine Conservation Area and eastern entrance to the Northwest Passage. Environ. Sci. Policy. in press.

Hamilton, B.M., Bourdages, M.P.T., Geoffroy, C., Vermaire, J.C., Mallory, M.L., Rochman, C.M., Provencher, J.F., 2021. Microplastics around an Arctic seabird colony: particle community composition varies across environmental matrices. Sci. Total Environ. 773, 145536.
Hay, K., 2000. Final report of the Inuit bowhead knowledge study. Nunavut Wildlife Management Board, Iqaluit, Nunavut.
Healey, G.K., Magner, K.M., Ritter, R., Kamookak, R., Aningmiuq, A., Issaluk, B., Mackenzie, K., Allardyce, L., Stockdale, A., Moffit, P., 2011. Community perspectives on the impact of climate change on health in Nunavut, Canada. Arctic 64, 89–97.
Hobbs, B.F., Ludsin, S.A., Knight, R.L., Ryan, P.A., Biberhofer, J., Ciborowski, J.J., 2002. Fuzzy cognitive mapping as a tool to define management objectives for complex ecosystems. Ecol. Appl. 12, 1548–1565.
Imperial Oil Ltd., 1978. Environmental impact statement for exploratory drilling in Davis Strait region. Unpublished Report, Imperial Oil Ltd., Aquitane Co. Canada Ltd., Canada Cites Service Ltd.
INAC, 1993. Agreement between the Inuit of the Nunavut settlement area and Her Majesty the Queen in right of Canada. Indian and Northern Affairs Canada, Ottawa.
ITK, 2018. National Inuit Strategy on Research. Inuit Tapiriit Kanatami, Ottawa, Canada, p. 48.
Jaireth, H., Smyth, D., 2003. Innovative governance—indigenous peoples, local communities, and protected areas. Ane Books, New Delhi, p. 338.
Kemp, D., 1971. The flow of energy in a hunting society. Sci. Am. 225, 105–115.
Kinloch, D., Kuhnlein, H., Muir, D., 1992. Inuit foods and diet: a preliminary assessment of benefits and risks. Sci. Total Environ. 122, 247–278.
Kofinas, G.P., 2005. Caribou hunters and researchers at the co-management interface: emergent dilemmas and the dynamics of legitimacy in power sharing. Anthropologica 47, 179–196.
Kortsch, S., Primicerio, R., Fossheim, M., Dolgov, A.V., Aschan, M., 2015. Climate change alters the structure of arctic marine food webs due to poleward shifts of boreal generalists. Proc. R. Soc. B 282, 20151546.
Kumlien, L., 1879. Birds—Contributions to the natural history of Arctic America, made in connection with the Howgate Expedition, 1877–78. U.S. Natl. Mus. Bull. 15, 69–105.
Kutz, S., Ducrocq, J., Cuyler, C., Elkin, B., Gunn, A., Kolpashikov, L., Russell, D., White, R.G., 2013. Standardized monitoring of *Rangifer* health during International Polar Year. Rangifer 33, 91–114.
Labbé, J., Ford, J.D., Araos, M., Flynn, M., 2017. The government-led climate change adaptation landscape in Nunavut, Canada. Environ. Rev. 25, 12–25.
Latulippe, N., Klenk, N., 2020. Making room and moving over: knowledge co-production, Indigenous knowledge sovereignty and the politics of global environmental change decision-making. Curr. Opin. Environ. Sustain. 42, 7–14.
Liboiron, M., Zahara, A., Hawkins, K., Crespo, C., de Moura Neves, B., Wareham-Hayes, V., Edinger, E., Muise, C., Walzak, M.J., Sarazen, R., Chidley, J., 2021. Abundance and types of plastic pollution in surface waters in the Eastern Arctic (Inuit Nunangat) and the case for reconciliation science. Sci. Total Environ. 782, 146809.
Mallory, M.L., Gilchrist, H.G., Fontaine, A.J., Akearok, J.A., 2003a. Local ecological knowledge of ivory gull declines in Arctic Canada. Arctic 56, 293–298.
Mallory, M.L., Johnston, V.H., Latour, P.B., 2003b. Striking a new balance: co-management of federal protected areas in the Northwest Territories and Nunavut, Canada. In: Jaireth, H., Smyth, D. (Eds.), Innovative governance—indigenous peoples, local communities, and protected areas. Ane Books, New Delhi, pp. 295–313.
Mallory, M.L., Wayland, M., Braune, B.M., Drouillard, K.G., 2004. Trace elements in marine birds, arctic hare and ringed seals breeding near Qikiqtarjuaq, Nunavut, Canada. Mar. Pollut. Bull. 49, 136–141.

Mallory, M.L., Braune, B.M., Wayland, M., Drouillard, K.G., 2005. Persistent organic pollutants in marine birds, arctic hare and ringed seals near Qikiqtarjuaq, Nunavut, Canada. Mar. Pollut. Bull. 50, 95–104.

Mallory, M.L., Fontaine, A.J., Akearok, J.A., Johnston, V.H., 2006. Synergy of local ecological knowledge, community involvement and scientific study to develop marine wildlife areas in eastern Arctic Canada. Polar Rec. 42, 205–216.

Mallory, M.L., Akearok, J.A., Edwards, D.B., O'Donovan, K., Gilbert, C.D., 2008. Autumn migration and wintering of northern fulmars (*Fulmarus glacialis*) from the Canadian high Arctic. Polar Biol. 31, 745–750.

Mallory, M.L., Provencher, J.F., Wong, S.N.P., Anderson, C.A., Elliott, K.H., Gaston, A.J., Gilchrist, H.G., Janssen, M., Lazarus, T., Pirie-Dominix, L., Spencer, N.C., 2019. Identifying key marine habitat sites for seabirds and sea ducks in the Canadian Arctic. Environ. Rev. 27, 215–240.

Mallory, M.L., Dey, C.J., McIntyre, J., Pratte, I., Mallory, C.L., Francis, C.J., Black, A.L., Provencher, J.F., 2020. Long-term declines in the size of northern fulmar (*Fulmarus glacialis*) colonies on eastern Baffin Island, Canada. Arctic 73, 187–196.

Murray, D., 2011. Parks Canada: working with aboriginal peoples, establishing new National Parks. In: Human dimensions of ecological conservation in the Tibetan Plateau Region. Xining, Qinghai, China, 21-26 August, 2011, pp. 82–89.

Nakashima, D.J., Murray, D.J., 1988. The common eider (*Somateria mollissima sedentaria*) of eastern Hudson Bay: a survey of nest colonies and Inuit ecological knowledge. Environmental Studies Revolving Fund Report No. 102, Ottawa.

Ndeloh Etiendem, D., Jeppesen, R., Hoffman, J., Ritchie, K., Keats, B., Evans, P., Quinn, D.E., 2020. The Nunavut Wildlife Management Board's community-based monitoring network: documenting Inuit harvesting experience using modern technology. Arctic Sci. 6, 307–325.

Nettleship, D.N., Evans, P.J., 1985. Distribution and status of the Atlantic Alcidae. In: Nettleship, D.N., Birkhead, T.R. (Eds.), The Atlantic Alcidae. Academic Press, London, pp. 53–154.

Nettleship, D.N., Smith, P.A., 1975. Ecological sites in northern Canada. In: Ottawa: Canadian Committee for the International Biological Programme, Conservation Terrestrial—Panel 9.

NIRB, 2018. Baffin Bay Davis Strait Strategic Environmental Assessment. URL https://www.nirb.ca/content/strategic-environmental-assessment.

NTI, 2006. Inuit impact and benefit agreement-in-principle for national wildlife areas and migratory bird sanctuaries in the Nunavut Settlement Area. Nunavut Tunngavik Incorporated, Iqaluit, Nunavut, p. 96.

Parks Canada. 2022. Tallurutiup Imange National Marine Conservation Area. https://www.pc.gc.ca/en/amnc-nmca/cnamnc-cnnmca/tallurutiup-imanga; accessed 24 March 2022.

Petro-Canada Ltd, 1979. Initial environmental assessment. Proposed Baffin Bay exploratory drilling program. Petro-Canada Ltd. (Unpublished report), Calgary.

Provencher, J.F., Gaston, A.J., Mallory, M.L., 2009. Evidence for increased ingestion of plastics by Northern Fulmars (Fulmarus glacialis) in the Canadian Arctic. Mar. Pollut. Bull. 58, 1092–1095.

Provencher, J.F., Gaston, A.J., Mallory, M.L., O'Hara, P.D., Gilchrist, H.G., 2010. Ingested plastic in a diving seabird, the thick-billed murre (Uria lomvia), in the Eastern Canadian Arctic. Mar. Pollut. Bull. 60, 1406–1411.

Provencher, J.F., Gaston, A.J., Hara, P.O., Gilchrist, H.G., 2012. Seabird diet indicates changing Arctic marine communities in eastern Canada. Mar. Ecol. Prog. Ser. 454, 171–182.

Provencher, J.F., Bond, A.L., Avery-Gomm, S., Borrelle, S.B., Rebolledo, E.L.B., Hammer, S., Kühn, S., Lavers, J.L., Mallory, M.L., Trevail, A., Van Franeker, J.A., 2017. Quantifying ingested debris in marine megafauna: a review and recommendations for standardization. Anal. Meth. 9, 1454–1469.

Provencher, J.F., Borrelle, S.B., Bond, A.L., Lavers, J.L., Van Franeker, J.A., Kühn, S., Hammer, S., Avery-Gomm, S., Mallory, M.L., 2019. Recommended best practices for plastic and litter ingestion studies in marine birds: collection, processing, and reporting. FACETS 4, 111–130.

Provencher, J.F., Thomas, P., Pauli, B., Braune, B.M., Franckowiak, R., Gendron, M., Savard, G., Sarma, S.N., Crump, D., Zahaby, Y., O'Brien, J., Mallory, M.L., 2020. Polycyclic aromatic compounds (PACs) and trace elements in four marine species from northern Canada in a region of natural marine oil and gas seeps. Sci. Total Environ. 744, 140959.

QIA, 2019. UQAUSIRISIMAJAVUT: What we have said. The Inuit view of how oil and gas development could impact our lives. Summary document. Qikiqtani Inuit Association, Iqaluit, Nunavut. https://www.qia.ca//wp-content/uploads/2019/03/QIA-SEA-Summary.pdf.

Reed, G., Brunet, N.D., Natcher, D.C., 2020. Can indigenous community-based monitoring act as a tool for sustainable self-determination? Extract. Indust. Soc. 7, 1283–1291.

Reed, G., Brunet, N.D., Longboat, S., Natcher, D.C., 2021. Indigenous guardians as an emerging approach to indigenous environmental governance. Conserv. Biol. 35, 179–189.

Revel, R.D., 1981. Conservation in northern Canada: International Biological Program conservation sites revisited. Biol. Conserv. 21, 263–287.

Rice, R., 2016. How to decolonize democracy: Indigenous governance innovation in Bolivia and Nunavut, Canada. Boliv. Stud. J. 22, 220–242.

Riewe, R., 1992. Nunavut atlas. Edmonton: Canadian Circumpolar Institute and Tunngavik Federation of Nunavut.

Sadowsky, H., Brunet, N.D., Anaviapik, A., Kublu, A., Killiktee, C., Henri, D.A., 2022. Inuit youth and environmental research: exploring engagement barriers, strategies, and impacts. FACETS 7, 45–70.

Sarma, S.N., Thomas, P.J., Naz, S., Pauli, B., Crump, D., Zahaby, Y., O'Brien, J.M., Mallory, M.L., Franckowiak, R.P., Gendron, M., Provencher, J.F., 2022. Metabolomic profiles in relation to benchmark polycyclic aromatic compounds (PACs) and trace elements in two seabird species from Arctic Canada. Environ. Res. 204, 112022.

Scerri, E.M., Kühnert, D., Blinkhorn, J., Groucutt, H.S., Roberts, P., Nicoll, K., Zerboni, A., Orijemie, E.A., Barton, H., Candy, I., Goldstein, S.T., 2020. Field-based sciences must transform in response to COVID-19. Nat. Ecol. Evol. 4, 1571–1574.

Social Ventures Australia, 2016. Analysis of the current and future value of Indigenous Guardian work in Canada's Northwest Territories. Dehcho First Nations, Lutsel K'e Dene First Nation, Indigenous Leadership Initiative, Tides Canada.

Taylor, J., 1863. Notice of flowering plants and ferns collected on both sides of Davis Straits and Baffin's Bay. Trans. Botan. Soc. Edinburgh 7, 323–324.

Thompson, K.L., Reece, N., Robinson, N., Fisher, H.J., Ban, N.C., Picard, C.R., 2019. "We monitor by living here": community-driven actualization of a social-ecological monitoring program based in the knowledge of Indigenous harvesters. FACETS 4, 293–314.

Turcotte, A., Kermany, N., Foster, S., Proctor, C.A., Gilmour, S.M., Doria, M., Sebes, J., Whitton, J., Cooke, S.J., Bennett, J.R., 2021. Fixing the Canadian Species at Risk Act: identifying major issues and recommendations for increasing accountability and efficiency. FACETS 6, 1474–1494.

Usher, P.J., 2000. Traditional ecological knowledge in environmental assessment and management. Arctic 53, 183–193.

Van Franeker, J.A., Kühn, S., Anker-Nilssen, T., Edwards, E.W., Gallien, F., Guse, N., Kakkonen, J.E., Mallory, M.L., Miles, W., Olsen, K.O., Pedersen, J., 2021. New tools to evaluate plastic ingestion by northern fulmars applied to North Sea monitoring data 2002–2018. Mar. Pollut. Bull. 166, 112246.

Wall, T.U., McNie, E., Garfin, G.M., 2017. Use-inspired science: making science usable by and useful to decision makers. Front. Ecol. Environ. 15, 551–559.

Watson, A., 1957. Birds in Cumberland Peninsula, Baffin Island. Can. Field-Nat. 71, 87–109.

Wiese, F.K., Robertson, G.J., 2004. Assessing seabird mortality from chronic oil discharges at sea. J. Wildl. Manage. 68, 627–638.

Wong, S.N., Gjerdrum, C., Gilchrist, H.G., Mallory, M.L., 2018. Seasonal vessel activity risk to seabirds in waters off Baffin Island, Canada. Ocean Coast. Manag. 163, 339–351.

Wynne-Edwards, V.C., 1952a. The fulmars of Cape Searle. Arctic 5, 105–117.

Wynne-Edwards, V.C., 1952b. Zoology of the Baird expedition (1950). I. The birds observed in central and southeast Baffin Island. Auk 69, 353–391.

Zahaby, Y., Xia, P., Crump, D., Provencher, J., Thomas, P., Pauli, B., Braune, B., Franckowiak, R., Gendron, M., Savard, G., Sarma, S., Mallory, M.L., O'Brien, J., 2021. ToxChip PCR arrays for two arctic-breeding seabirds: applications for regional environmental assessments. Environ. Sci. Technol. 55, 7521–7530.

CHAPTER THREE

Enhancing collaboration across the knowledge system boundaries of ecosystem governance

Nathan A. Badry* and Gordon M. Hickey
Department of Natural Resource Sciences, McGill University, Montreal, QC, Canada
*Corresponding author: e-mail address: nathan.badry@mail.mcgill.ca

Contents

Abstract

In ecosystem governance, due to the ecological, social, and jurisdictional complexity of these systems, pluralities of knowledge are increasingly necessary for informed decision-making. However, there are frictions between different kinds of knowledges, whether scientific, bureaucratic, or local. The bringing together of Indigenous and Western knowledge systems is an especially intractable challenge. Indigenous knowledge is frequently incommensurable with dominant scientific frameworks, particularly in how many Indigenous Peoples conceptualize the social roles of nonhumans. Nonetheless, Indigenous and scientific knowledges are regularly brought together in contexts ranging from wildlife co-management to global environmental assessments, and this can involve Indigenous knowledge being selectively adopted, integrated, translated, or just ignored to fit within those frameworks. In this literature review we suggest that Actor-Network Theory—a way of accounting for the webs of relations constituting and generating social and natural worlds—has the potential to help researchers and practitioners challenge the ontological boundaries that make this work difficult. ANT, as well as other approaches under the wider umbrella of posthumanism, could fit within existing pluralist frameworks like Two-Eyed Seeing which strive to address entrenched power dynamics in work involving Indigenous and non-Indigenous Peoples, and help inspire new participatory methods for understanding and enhancing knowledge pluralism in governance.

Advances in Ecological Research, Volume 66
ISSN 0065-2504
https://doi.org/10.1016/bs.aecr.2022.04.010

1. Introduction

"Animals, it would seem, can serve as food or as food for thought, but they can never interact with humans as intelligent actors in their own right"

(Nadasdy, 2007, p. 30).

At the ecosystem level, the governance of natural resources is typically conceived of in social terms. It is determined by social structures and processes, people, and organizations. But natural resource governance is about more than just the social. Microbes, animals, plants, laws, policy documents, new technologies, rivers, and weather (among many other things) can affect governance processes in their own ways. However, these nonhuman things, and the roles they play, are often obscured or left out of governance accounts. In the epigraph above, Paul Nadasdy, drawing on their research on Dall sheep co-management in northern Canada with the Kluane First Nation, observes that animals are always viewed in one of two ways: as natural resources ("food") by biologists and ecologists, or symbols ("food for thought") by anthropologists. Nadasdy contrasts these conceptions of animals, as resources or symbols, with how the Kluane conceptualize animals, as social actors with agency. This understanding of sheep as agential actors is prevented from influencing co-management decisions, limiting Kluane participation in governance processes, and leading to poor decision-making based on insufficient knowledge of the system. If this obstacle to collaboration was overcome, what would that kind of governance look like, and what could it achieve?

Indigenous knowledges often have separate ontological assumptions than the scientific, bureaucratic, and local knowledges of other governance actors, and this can challenge communication and collaboration (Diver, 2017; Verran, 2002; Walsh et al., 2013; Watts, 2013). Is a river, like the Whanganui, famously granted legal personhood by the New Zealand Parliament, a body of flowing water, or an honoured ancestor of the Māori (Salmond, 2018)? Among some Northern Peoples of North America, nonhumans have relationships with humans defined by respect, reciprocity, and agency (Miller and Davidson-Hunt, 2013). Are nonhumans like wildlife, for example, resources waiting to be harvested by humans, or do they actively offer themselves to the hunter (Watson and Huntington, 2008)? Boundaries—between human and nonhuman, social and natural, subject and object, and scientific and Indigenous—all contribute to challenging collaborative governance (Norman, 2015). Nevertheless, Indigenous knowledge and Western science are still commonly brought together in governance structures and processes, from wildlife co-management boards (Nadasdy, 1999), to environmental impact assessments (Huntington, 2000), to bodies of global environmental governance like the Intergovernmental Science-Policy Platform on Biodiversity and Ecosystem Services (IPBES; Diaz et al., 2019).

A key objective behind this knowledge synthesis is pluralism in governance, and in particular, epistemic pluralism. Pluralism has long been a topic of study in environmental governance, particularly in how it influences the processes and outcomes of decision-making (Davidson and Frickel, 2004). Pluralism in environmental governance generally describes a diversity of approaches and perspectives that can be applied in problem-solving. In the case of epistemic pluralism, it describes a diversity in theories of knowledge and ways of knowing. Different knowledge systems, like Indigenous and Western scientific ones, are often brought together in natural resource governance to advance this epistemic pluralism, improving understandings of complex governance systems (Apetrei et al., 2021), and hopefully leading to more informed and responsive action (Bennett and Satterfield, 2018). However, compared to epistemic pluralism, ontological pluralism within natural resource governance is not as well understood. Ontological pluralism refers to a diversity of deeply held understandings of what exists in the world and how those things interact (Theriault, 2017). There are few tools to help governance scholars and practitioners navigate different ontological commitments, like those regarding the social roles of Dall sheep or rivers. These diverse understandings are intertwined with other forms of pluralism like epistemic pluralism, and hamper efforts to improve collaboration in pluralistic governance contexts (Brennan, 2022).

For more pluralistic natural resource governance structures and processes to be advanced, new approaches for addressing ontological pluralism may be needed. In this literature review, we explore this need, as well as the potential for new methods to address it. We specifically consider what actor-network theory (ANT)—an analytical and empirically driven strand of posthuman thought—can offer a new methodological approach for collaborative spaces (Callon, 1984; Latour, 2005; Law and Joks, 2019). ANT has a long history of transdisciplinary applications (Donaldson et al., 2002; Watts and Scales, 2015), and synergies with network methods already well established in the field of natural resource policy and governance (Bodin and Crona, 2009). We review the literatures on governance in social-ecological systems (SESs), traditional ecological knowledge (TEK), and ANT, with the aim of identifying opportunities for supporting collaborative work across knowledge system boundaries.

2. Governing transboundary social-ecological systems

SES theory has been called a new ontological approach to doing science; an approach that integrates the ecological and social, embraces transdisciplinary approaches, and refutes equilibrium-based understandings of systems (Schoon and van der Leeuw, 2015). SESs are a type of complex adaptive system made up of interdependent social and ecological components (Biggs et al., 2021). The interactions of these components lead to emergent, system-wide patterns, which in turn influence these components and their interactions. In SESs, the problems that arise are complex (Brehony et al., 2020; Scarlett and McKinney, 2016), spanning political, ecological, social, biophysical, and cultural boundaries; and require the collective action of diverse actors to address them (Cash et al., 2006; Plummer et al., 2013). Because natural resources are often entangled within complex social and ecological relationships, when it comes to these problems, it can be productive to consider natural resources as embedded within SESs (Abrams et al., 2021; Hotte et al., 2019; Kobluk et al., 2021; Miller et al., 2020).

Natural resource governance[a] scholarship has increasingly adopted systems-thinking perspectives to better account for the complex social-ecological processes driving management problems (Ban et al., 2013). This has resulted in a shift from state-centric government approaches, to new governance approaches (DePuy et al., 2021; Turnhout et al., 2014) that recognise the governance of natural resources as being "characterized by complexity and contestation originating from multiple problem causes, divergent problem perspectives and solution strategies, and fragmented institutional settings" (Lockwood et al., 2010, p. 986), as well as numerous social and ecological interdependencies (Bodin et al., 2005; Steins, 2001). Faced with complexity, the need for collaborative governance has become clear; defined here as the "processes and structures of public policy decision making and management that engage people constructively across the boundaries of public agencies, levels of government, and/or the public, private and civic spheres in order to carry out a public purpose that could not otherwise be accomplished" (Emerson et al., 2012, p. 2). In particular, collaborative forms of networked governance can allow for diverse actors to collectively address complex issues which they may struggle to address in

[a] Natural resource governance can be defined as: "the interactions among structures, processes and traditions that determine how power and responsibilities are exercised, how decisions are taken, and how citizens or other stakeholders have their say" (Graham et al., 2003, p. ii).

isolation (Bodin and Crona, 2009; Emerson et al., 2012; Malekpour et al., 2021), and which may require agreement on rules and practices, coordination of resource use, the building of common knowledge, or other forms of cooperation (Bodin and Crona, 2009). In natural resource systems, collaborative strategies like co-management help varied governance actors—holding scientific, bureaucratic, and local knowledges—to work across boundaries through "knowledge co-production, mediation, translation, and negotiation" (Cash et al., 2006, p. 1), and "to foster synergies across different knowledge systems to enrich understanding of, and solutions for, environmental challenges" (Pyke et al., 2021, p. 27). For collaborative governance strategies, knowledge co-production and sharing is a key determinant of successful outcomes.

However, bringing together diverse actors and knowledges creates significant challenges for transboundary governance networks in SESs (Folke, 2004; Zetina-Rejón et al., 2020). Different actors understand systems in different ways and have different priorities. Theories, frameworks, models, and methods of collaborative governance promote particular ontological perspectives of how SESs work (DePuy et al., 2021). For example, Ostrom's frequently cited framework (Ostrom, 2009) divides all SESs into subsystems of resource units, resource systems, users, and governance systems, all situated within a wider social, political, and ecological setting. What a resource is, and the ways in which it can interact with other variables to result in outcomes, is prescribed. The conceptual framework adopted by IPBES, used to explain linkages between human and natural worlds, includes Indigenous-inspired concepts like Mother Earth derived from Bolivia's Mother Earth Law, but fits these within broader categories and relationships determined by the ontological underpinnings of ecology and ecosystem services (Díaz et al., 2015). While valuable, these governance and management perspectives do not always resonate with, and can even be exclusive of, the diverse actors and knowledge systems within SESs (Howitt and Suchet-Pearson, 2006). In particular, differences in values, priorities, knowledge systems, and worldviews, particularly between Indigenous and non-Indigenous actors, create barriers to communication and collaboration on transboundary issues, limiting the potential for collective action (Adade Williams et al., 2020).

3. Traditional ecological knowledge and natural resource governance

TEK has been described in SES literatures as "a cumulative body of knowledge, practice, and belief, evolving by adaptive processes and handed down through generations by cultural transmission, about the relationship of living beings (including humans) with one another and with their environment" (Berkes, 2018, p. 8). Alternatively, TEK has been described as "the culturally and spiritually based way in which indigenous peoples relate to their ecosystems" (LaDuke, 1994, p. 127). The term TEK has been widely criticized from a variety of angles (see Berkes, 2018, pp. 8–10 for a brief overview), among them that speaking in terms of TEK and Western science leads to dichotomous thinking, which lacks nuance and the diverse perspectives to be found under these umbrella terms (Henriksen et al., 2021). Combining all the knowledges and relationships within TEK into a single category is as much an oversimplification as distilling all of science into a single category (Agrawal, 1995). TEK is unique to people and place, and can often be characterized more by difference than similarity. Further, making a clear distinction between TEK and Western science is not always useful, as one influences the other whenever they come into contact (Law and Joks, 2019). However, TEK is above all an academic construct, and the use of these terms is often necessary in more general discussions of the role of Indigenous knowledge in governance. We refer to TEK in this paper, while acknowledging that some of the above complexity will invariably be lost, despite our intentions to do justice to the nuance and diversity of perspectives. And while we focus on TEK, we also draw on concepts like Indigenous and local knowledge (ILK; Rarai et al., 2022) and local ecological knowledge (LEK; Shaffer et al., 2022) when their use overlaps with TEK.

The term TEK was first used by anthropologists in the field of ethnoecology during the 1980s (Berkes, 2018), but concepts like Indigenous knowledge and traditional knowledge have been in use in the development literature since the 1950s, and were used similarly (Agrawal, 1995). Until relatively recently, the prevailing academic attitude was that this knowledge was unscientific, imprecise, and too embedded within social and cultural practice to be of substantial management use (Berkes, 2018). This attitude changed as recognition grew among a wide variety of scholars and practitioners that TEK held relevance to the achievement of management and sustainability goals (Armitage et al., 2011). The potential for TEK to aid in these goals, as a complement to Western science, is now well established (Berkes, 2004; Bohensky and Maru, 2011; Davis, 2006; Folke, 2004), particularly in natural resource management and climate change adaptation contexts (Rarai et al., 2022). The possible applications of TEK in conservation science and restoration ecology have also been noted, especially for advancing understandings of biodiversity and traditional management practices (Molnár and Babai, 2021; Pyke et al., 2021). More specific ecological applications have included scientific research into multi-species management, resource rotation, succession management, landscape patchiness management, and responses to ecological pulses and surprises (Berkes, 2004). Beyond the value TEK could bring as a complement to Western science, it is further thought that the inclusion of TEK will empower Indigenous Peoples and communities (Agrawal, 1995; Bohensky and Maru, 2011; Nadasdy, 1999, 2005), and could even promote Reconciliation and epistemic justice through the advancement of full and equal participatory processes (Castleden et al., 2017; Toncheva and Fletcher, 2021).

In light of these advancements, questions regarding the relationship between Western science and TEK have been raised by both Indigenous (Reo, 2019) and non-Indigenous scholars (Agrawal, 1995; A. Miller and Davidson-Hunt, 2013; Verran, 2002). In academic discourses, the combining of TEK and Western science approaches has gone by many names, from the much criticized "integration" to the more recently popular "bridging" (Bohensky and Maru, 2011). And while progress has been made, and terms and approaches have evolved, problems remain. Many studies have recognized that true integration of knowledge systems is difficult (Ban et al., 2018; Scott, 1996; Watts, 2013), or even impossible (Watts, 2013). Complementary strengths and weaknesses of TEK and Western science are usually presented in a simplified manner, often in the form of a table or Venn diagram (Albuquerque et al., 2021), which can lead to integration being treated like a box-ticking exercise (Bohensky and Maru, 2011) and risks tokenizing TEK in governance processes (Huntington, 2000). Combined with the historically rooted and uneven power dynamics often present in these systems, working across knowledges can be fraught with challenge and risk.

Berkes (2018) described TEK as a nested knowledge-practice-belief complex, comprising knowledge of plants and animals, management systems, social institutions, and a worldview. However, when TEK is integrated with Western science, it is often only in the form of specialised knowledge of plant and animal species, or climate change, or the medicinal qualities of plants, while everything else is excluded (Shaffer et al., 2022). Similarly, Nadasdy (1999) has described how TEK is distilled and compartmentalised when it is made to fit within scientific paradigms in order to be incorporated into management and policy. As Berkes puts it: "it is clear that the positivist–reductionist paradigm holds little promise as a framework for understanding Indigenous knowledge or for integrating Western science and other kinds of knowledge" (Berkes, 2018, p. 287). Indigenous scholars like Vanessa Watts (2013) have argued that Western science is based on a fundamentally different ontological-epistemological frame than place-based Indigenous cosmologies. For Indigenous peoples, "TEK is not just knowledge about the relationships with the environment, but it is the relationship itself; it is the way that one relates" (Ramos, 2022, p. 2). Integrating TEK into governance processes—for example, distilling rights and customs into Valued Components for Environmental Impact Assessment—can advance sovereignty and pluralism (Butler et al., 2021), but is still antithetical to many Indigenous cultures, and represents only incremental progress.

One aspect of TEK that is often excluded from formal natural resource governance processes is relationships between humans and nonhumans. For many Indigenous Peoples, wildlife and rivers have a much more active role in social relations than the passive role typically conceptualised in the mainstream natural and social sciences (Chaplier and Scott, 2018; Watts, 2013). This understanding of nonhuman social roles falls well outside of the Cartesian duality between social and natural that dominates Western science–based approaches to natural resource governance. Nadasdy (2007) describes how the relationships between the hunter and hunted are defined by reciprocity for many Indigenous Peoples, and that this ontological commitment is antithetical to many approaches in natural resource sciences. Other contestations over the relational roles of nonhumans—like supernatural beings among the Palawan in the Philippines (Theriault, 2017), or the Orca *Tsu-xiit* among the Mowachaht/Muchalaht in Canada (Norman, 2015)—are common, and this poses significant challenges for collaborative, transboundary natural resource management. As Shaffer et al. describe in their work with spider monkeys in Guyana:

> *"it is much easier for scientists to integrate scientific and Indigenous knowledge on spider monkey ecology when they both agree that a spider monkey is an animal and not a person. However, finding commensurability (the extent to which these knowledges overlap and can be integrated) becomes more difficult if one ontological system identifies both humans and monkeys as persons" (Shaffer et al., 2022, p. 4).*

This kind of ontological friction makes efforts at knowledge integration very difficult. Indigenous knowledge of species and ecosystems cannot be understood if divorced from its relational and place-based context, let alone Indigenous knowledge of management systems, social institutions, or worldviews (Molnár and Babai, 2021). With increasing attention being given to Indigenous sovereignty and epistemic justice in Canada (Castleden et al., 2017), and internationally (Harcourt, 2021; Rarai et al., 2022; Sundberg, 2014; Verran, 2002), Indigenous knowledges and ways of knowing, and how these can help achieve broadly shared goals of sustainability and conservation are increasingly part of the discourse. Models of governance in Indigenous contexts often acknowledge the importance of reconciling different worldviews to successful management, but rarely provide tools for addressing conflicts between different worldviews (Maxwell et al., 2020). Indigenous frameworks and approaches such as Two-eyed Seeing (Popp et al., 2020), Indigenous Métissage (Donald, 2012), and the Herringbone Stitch Model (Andrews, 2021) are increasingly being developed and applied to make work at the interface of Indigenous and non-Indigenous Peoples more equitable (Hessami et al., 2021). However, there are limited tools available within these frameworks to support researchers and practitioners to, for example, reconcile knowledge system contestations like the roles nonhuman actors play in SESs to inform collaborative governance.

This paper arose from observations by the authors of TEK's role in transboundary natural resource governance, and the gaps between aspirations and practice identified by both Indigenous and non-Indigenous peoples. In what follows, we aim to prompt a discussion that moves beyond the integration of TEK and towards practical steps to achieving more collaborative, boundary-spanning natural resource governance, and suggest ANT as one possible way forward. Within this pluralistic space, the question of who can speak and write on matters of Indigenous knowledges and integration is increasingly, and rightly, being raised. This is particularly true in academia, in which Indigenous voices have historically been marginalised. As scholars of white, settler descent, our goal is not to explain Indigenous worldviews through Western posthumanist frameworks, but rather to explore how ANT-inspired methods might help Indigenous and non-Indigenous researchers and practitioners challenge the ontological boundaries that make transboundary collaboration difficult, thereby supporting the implementation of pluralist knowledge frameworks (Hessami et al., 2021).

4. Actor-network theory and other ways of accounting for complexity and nonhuman agency

ANT emerged from the field of science and technology studies during the late 20th century, largely through the works of Bruno Latour, Michel Callon, and John Law. ANT is a family of tools, sensibilities, and analytical methods used to account for the webs of relations which constitute the social (Law, 2009). In the context of ANT, 'social' includes heterogenous, interwoven human and more-than-human actors (or actants in ANT parlance;

Latour, 2005). These actors, whether human or not, all have the potential for demonstrating agency, which, rather than conscious thought and intentionality, can be understood simply as the ability of actors to affect change on other associated actors (Toncheva and Fletcher, 2021). In this way ANT differentiates itself from mainstream sociological perspectives, seeing many more human and more-than-human actors as possessing this ability (mediators) than not (intermediaries). Classic examples include the structural characteristics of Portuguese caracks enabling the control of colonial empires (Law, 1984), and French scallop larvae refusing to behave as their biologists expect and dooming fledgling aquaculture management plans (Callon, 1984).

ANT allows for the tracing of associations among heterogenous human and nonhuman actors whose actions were driven by multiple entangled agencies (Law, 1992). ANT was described by Latour not as a "sociology of the social", but a "sociology of associations", an explicit recognition of the wide and diverse array of things that come together to do 'social' work (2005). While ANT has 'theory' in its name, it is better thought of as a way of doing things, or a methodological approach. The role of the ANT analyst is to make no *a priori* assumptions about the actors in an actor-network. Whether an actor is human or not, no differences are assumed until they are observed by the analyst. According to ANT, collective action is complex; it cannot be fully understood without considering everything that makes that action possible; Networks are more than the sum of their (human) parts. ANT can be considered an analytically focussed part of a wider 'posthuman' movement in academic discourse that seeks to decentre the human subject (Coffey, 2021). The legal field has seen several prominent examples of this movement, including the previously mentioned Whanganui River, granted legal personhood in New Zealand. Posthuman ideas have also been taken up by and overlap with a variety of other academic fields, including in geography (Watson and Huntington, 2008), political ecology (Gesing, 2021), critical posthumanism (Braidotti, 2016), new materialisms (Alaimo, 2012), queer and feminist theory (Hayward, 2010), and multispecies ethnography (Kirksey and Helmreich, 2010).

Some recent examples of posthuman scholarship include the work of Eduardo Kohn, who used more-than-human ethnography and semiotics to move anthropology beyond the human to include human-nonhuman interaction, with dogs in the Amazon (Kohn, 2007), and later with forests (Kohn, 2013). The role of oysters in coastal infrastructure in New York, and how that role requires not only work by humans but also the oysters themselves, has been studied through the lens of biopolitics (Wakefield, 2020). Oysters have also been studied for their role in the gentrification of coastal communities in the UK through the lens of posthumanism and ANT (Brooks and Hubbard, 2021). Paul Robbins, in *Lawn People* (2007), drew and built upon apolitical ecology to examine lawns as environmental actors, with agential relations towards everything from humans to earthworms. While Theriault (2017), grounding their work in ontological multiplicity, showed how social relationships between supernatural beings and Indigenous Peoples relates to and contests environmental governance in the Philippines. Cambodian irrigation infrastructure has been studied through a science and technology studies perspective (Venot and Jensen, 2021), which described how different related practices and enactments result in a socionatural mosaic landscape of human and more-than-human. And Gesing (2021) applied a more-than-human approach to political ecology to explain how coastal restoration in New Zealand could be made more sustainable by acknowledging the entanglements of plants, animals, humans, artefacts, and abiotic elements that do restoration work.

Much of the work has also focussed on human-animal interactions. In a recent study, Ampumuza and Driessen (2021) applied tools from ANT and more-than-human geography to help understand how the agency of gorillas in Uganda shaped conservation and tourism development. Li and von Essen (2020) adopted a new materialist perspective in their work to explore interactions between macaques and the farmers whose crops they damage in China, and how these interactions were mediated by other nonhumans: crops, topography, and management tools. Human-shark encounters in Australia were contextualised by analysing how agency is distributed among ocean-users and sharks and their relations with the ocean (Gibbs, 2021). Bear (2021) employed more-than-human-geography to show how UK insect farmers' care for their charges was shaped by the insects themselves. And Toncheva and Fletcher (2021) used a case study of human-bear relationships in the mountains of Bulgaria to add to discussions of how nonhumans fit within conservation and policy-making.

Outside of academia, the rise of posthuman ideas and concepts can also be observed in popular discourses. At the 2020 Sundance Film Festival, there were multiple works that grappled with environmental concerns and which were experienced from the perspective of the more-than-human (Mendez, 2020). One example, Natalie Cabrera's virtual reality film Hypha (Cabrera, 2020), imagined the viewer as a fungal spore, which over the course of the film grows into a hyphae, forms relationships with other fungi and plants, and eventually realizes its great potential for bioremediation and healing (Harms et al., 2011). This trend comes at a time of worsening environmental and sustainability challenges, during which academia and society at large is struggling to come to terms with how to understand the relationships and divisions among humans and nonhumans.

Crucially however, this widespread shift towards the posthuman overlaps with, and draws from, the lived experience of many Indigenous Peoples, and the great number of Indigenous scholars within and without academia who work on ideas of ontology and more-than-human agency (Hunt, 2014; Reo, 2019; Todd, 2016; Watts, 2013). Some have gone as far as to say that "Indigeneity can be positioned as ontologically prior to posthumanism" (Henriksen et al., 2021, p. 1). We take seriously Todd's claim that: "when we cite European thinkers who discuss the 'more-than-human' but do not discuss their Indigenous contemporaries who are writing on the exact same topics, we perpetuate the white supremacy of the academy" (Todd, 2016, p. 18). Posthumanism (Sundberg, 2014; Watts, 2013) and Latour's ANT more specifically (Todd, 2016), has been criticized for privileging Euro-Western thought, and not doing enough to work against the colonial systems they are embedded in. However, there is considerable overlap between these bodies of work. Collaborative projects incorporating elements of posthuman and Indigenous perspectives are beginning to emerge (Engman and Hermes, 2021), and posthumanist philosophies, and their more-than-human assemblages, which could include everything from spiritual stories, to animals, to tools, to landscapes, have been recognised for their flexibility in facilitating work across knowledge systems in Indigenous spaces (Watson and Huntington, 2008).

Among this cross-disciplinary posthuman work, ANT is relatively distinct for its grounding in empirical and applied approaches, focussed on tracing the associations among heterogenous actors, often through case studies. ANT and other similar approaches have now been applied in a wide variety of disciplinary and research contexts, from livestock regulatory policy (Donaldson et al., 2002) to financial risk in Australian housing markets (Palmer, 2014), to low-carbon commercial development (Rydin, 2013). Over the past several decades, it has become a particularly popular theoretical foundation in the fields of geography, health studies, anthropology, and tourism (Deason et al., 2022). This may be partly due to that, while ANT is often linked to the ideas of individuals scholars like Bruno Latour, ANT is unique to each researcher, and contemporary ANT approaches draw on many of the other influences discussed within this section. Some researchers divide ANT work into two camps: a "true" ANT, grounded in the social studies of science, and often applying early ANT concepts like moments of translation, and ANT-adjacent, which borrows elements of ANT and applies them in a variety of transdisciplinary approaches (Williams, 2020). Regardless of this distinction's accuracy or usefulness, the dichotomy demonstrates ANT's increasingly widespread relevance outside of its field of origin. And among those suggesting the discipline-spanning value of ANT and ANT-inspired ideas are natural resource governance scholars.

In the 1990s, Freudenberg et al. wrote on the complexity of natural resources, and how they are not actually so natural after all. Rather, they are "a complex mix of social, technological, and biophysical conditions through which a given element of the natural environment, at a particular point in time, comes to be socially defined as valuable" (1995; p. 387). Later, Steins (2001), drawing on ANT, would propose that a conceptualisation of relational agency that emerges from the interactions of more-than-humans could help understand the complexity, uncertainty, and dynamism of natural resource management processes. Trosper and Parrotta (2012) have argued that ANT could be a useful perspective for understanding traditional forest knowledge, and a new ANT-inspired understanding of agency was recently recognized by both Resilience (Dwiartama and Rosin, 2014) and SES scholars (Nabavi and Daniell, 2017), for having the potential to provide new perspectives on natural resource governance. Some of these ideas are now being taken up by natural resource scholars in applied contexts. This has included using ANT to study the networks related to salmon runs in British Columbia, with the aim of improving conservation practice (Massey et al., 2021), as well as to study an Indigenous, community-based ecotourism organization in southern Mexico and its resilience to climate change (Deason et al., 2022).

An ANT-inspired methodological approach to natural resource governance, by addressing ontological pluralism in natural resource governance, appears to have the potential to help displace the central role of the human in SESs scholarship, and advance transboundary governance that is more informed, effective, and equitable. However, there are still very few examples of ANT being applied within local natural resource management and governance contexts. The studies that have been done rarely dig deeply into the ANT and posthuman methodological toolbox, and often do not stray too far from human governance actors, including only a select few nonhumans. This is despite increasing calls for the democratic inclusion of nonhumans in governance processes (Deason et al., 2022), and the clear potential for improving governance outcomes by understanding how SESs are shaped in agential ways by nonhumans. There have also been no explorations of how ANT and related approaches could help span the ontological boundaries between knowledge systems, potentially supporting the implementation of pluralist frameworks such as Two-Eyed-Seeing.

5. New directions for scholarship and practice in ecosystem governance

ANT offers under-explored opportunities to facilitate transboundary collaboration and tools for advancing ontological pluralism in natural resource governance. Through relational tracing, visualization, and description, ANT has the potential to help diverse policy actors recognize that Dall sheep and rivers are not just resources, but can actively affect change on the humans who interact with them, sometimes directly, and sometimes through mediated relationships including other humans and nonhumans, thereby facilitating pluralistic understandings of governance in complex SESs. Collaborative governance arrangements like co-management are already working across spatial and jurisdictional boundaries (Cash et al., 2006), and methods like the multiple-evidence-based approach are meant to help integrate knowledge systems (Pyke et al., 2021). However, despite recognition of the need to address nonhuman agencies and the spaces between worldviews in governance (Norman, 2015), there are still few practical tools for practitioners to draw on when seeking to span ontological boundaries. ANT could help fill this gap, facilitating the shared learning and knowledge co-production crucial to transboundary networks (Agranoff and McGuire, 2001), and drawing in actors and perspectives from across the diverse disciplinary and bureaucratic spheres involved in natural resource governance. One of the manners in which ANT could be applied is as a complement to research on network governance.

ANT offers inspiration for a new analytical network approach because of its synergies with existing network governance research methods, and its deep grounding in empiricism, focussed on the careful tracing of associations among human and nonhuman actors using case studies or ethnography (Law, 2009). Most ANT studies use methods like interviews, participant observation, or document analysis to understand how networks function (Deason et al., 2022). Employing a wide variety of methods both directly with human participants (interviewing, focus groups) and without (document analysis, participant observation) can help illuminate these associations.

For example, in a recent study of lobster fishing areas in Atlantic Canada, researchers utilized archival research, participant observation, interviews, and focus groups to understand how the boundaries of fishing areas interacted with changing lobster species' distributions, new technologies, licensing procedures, and competition with other marine industries (Wiber and Barnett, 2021). The researchers found that the fishing areas acted almost as social actors, controlling where and when fishing could take place. These kinds of insights into the impact of nonhumans could benefit governance within complex SESs, helping policy actors to untangle the relationships between interwoven social and natural components, as well as how these lead to emergent patterns and feedbacks.

Common network governance research methods, such as social network analysis, also have the potential to be used with ANT, helping to qualify the links between governance network structures and outcomes, and identifying obstacles and opportunities within the network for enhancing collective action (Palmer, 2014). Network analysis in SESs is a growing body of work, but a lack of methodological pluralism, especially in analysing social and ecological data together, has been noted as a barrier to the advancement of this approach (Eider et al., 2021). While there are theoretical obstacles to combining ANT with social network analysis (Latour, 2005)—differences in understandings of the relationship between network structure and agency, for instance (Stevenson and Greenberg, 2000)—there could also be some benefit to applying social network analysis to a posthuman network elaborated through ANT. In a study of Australian urban development, Palmer (2014) used ANT to explore the role of human actors like "property developers", as well as nonhuman actors like "market value" and "tax legislation", in determining what kinds of buildings get built. They then applied social network analysis measures like betweenness and centrality to their network to help identify important components like focal actors, mediators, and obligatory passage points, or features of the network which force actors to conform to certain goals or perspectives (Callon, 1984; Fig. 1). These were then used to help identify obstacles and opportunities for more sustainable development, for instance, properties of the network which prevented the building of higher-density housing developments. This is an area that would benefit from future participatory research in different governance contexts.

However, maybe the application of ANT with the most potential for facilitating collaborative governance outcomes is as a boundary object. Boundary objects are tools which facilitate collaboration in the absence of

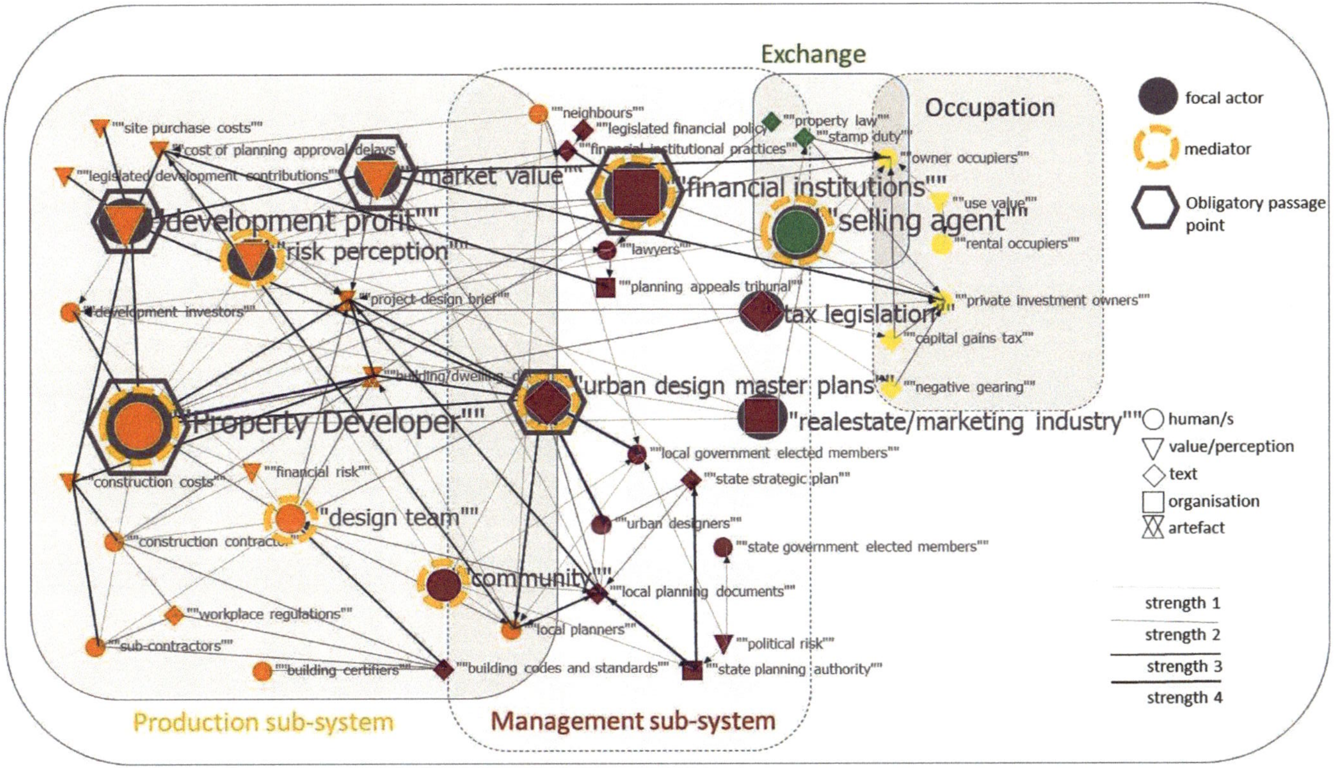

Fig. 1 The development of medium-density housing in Australia depicted as an actor-network, with key actors and relationships identified using social network analysis tools (Palmer, 2014).

consensus (Star, 2010). They exist in the space between social groups (Star and Griesemer, 1989), creating common ground for collaboration (Enqvist et al., 2018), while still being flexible enough for each group to interpret and understand the object in their own way (Amundsen and Hermansen, 2020). These objects are not necessarily physical things, but can be abstract concepts like Resilience (Brand and Jax, 2007), Ecosystem Services (Abson et al., 2014), Stewardship (Enqvist et al., 2018), or Green Transformations (Amundsen and Hermansen, 2020). Boundary objects make it easier for actors to act collectively across boundaries, for example, by translating co-learning into evidence-based decision making. In natural resource governance, pluralities of actors, knowledges systems, and ontologies make collaboration challenging, but actors often have broadly shared conservation or sustainability goals, despite not necessarily agreeing on how to meet those goals. In this context, boundary-spanning objects that can act as intermediaries are extremely valuable (Cash et al., 2006).

Applying ANT as a boundary object may help researchers and practitioners to incorporate ontological pluralism when working across knowledge systems (Adade Williams et al., 2020; Molnár and Babai, 2021), and better address existing power imbalances within governance networks (DePuy et al., 2021). By allowing space for multiple understandings of the social roles of Dall sheep, or rivers, or even Mother Earth, ontological frictions in governance networks could perhaps be navigated, and trust and knowledge sharing/co-production enhanced, without necessarily requiring agreement on, or the imposition of, a single worldview. This could also help to move beyond critiques of participatory governance practices, which see some of these processes as being tokenistic and serving only to reproduce existing power imbalances within governance networks (Guibrunet et al., 2021).

An example of these possible applications is offered by the case of the orca *Tsu-xiit* (Norman, 2015). A lone juvenile orca in British Columbia, Canada, famous for approaching boats, he became known as Luna to the general public, L98 to scientists, and *Tsu-xiit* to the Mowachaht/Muchalaht First Nation, who believed he embodied the spirit of their recently deceased chief. Due to concerns about *Tsu-xiit's* behaviour leading to unsafe conditions, and driven by political pressures, a plan to relocate him was developed. This plan was opposed by the Mowachaht/Muchalaht First Nation who did not want him removed from the traditional waters where he had chosen to reside. *Tsu-xiit*'s multiple identities "as an animal and a human, a whale and a chief, and a victim and a victor" (Norman, 2015, p. 171), created ontological frictions which prevented this transboundary governance challenge from being resolved quickly and equitably. *Tsu-xiit* would eventually be struck by a tugboat and killed.

Although there are similarities in how more-than-humans are conceptualised according to posthumanism and TEK, ANT has yet to be regularly applied to the transboundary governance of natural resources in Indigenous contexts. That may be partly due to the well-founded concern that translating TEK through ANT, itself a Western academic construct, would only exacerbate the existing problems with knowledge integration (Todd, 2016). Conception of nonhuman agency can differ between ANT (agency within webs of moderating relationships) and Indigenous ontologies (agency that is often tied to spirituality). ANT, and especially Latour's ANT, has also faced criticism for ignoring inequalities and power relations at the expense of maintaining the status quo, and privileging scientists and bureaucrats over marginalised peoples (Holifield, 2009). Two-Eyed Seeing offers insights for weaving Indigenous and non-Indigenous knowledges and guiding co-learning, but it lacks specific tools for addressing ontological frictions and communicating across ontological spaces in practice. By adding a boundary object like ANT to the methodological toolbox of frameworks like Two-Eyed Seeing, these criticisms of ANT might be better addressed, and the pluralist goals of these frameworks advanced. This is an area that requires further empirical exploration. ANT has always been "a diaspora that overlaps with other intellectual traditions" (Law, 2009, p. 142) and future work developing methodological discourses in which the voices of Indigenous scholars play a much more prominent role is needed if ANT is to meaningfully inform new approaches in natural resource governance. This would further ANT's goal of understanding the messy and multiple interwoven realities at play in society, and perhaps better realise the benefits of collaboration in transboundary natural resource governance.

It should be noted that the term natural resources is itself problematic. It does not evoke agency, but instead the exact boundaries between natural and social that posthumanism seeks to confront. Natural resources are generally treated as inert biophysical units, perceived to hold economic value. They are acted upon or acted with, but are not actors themselves. The term is also indelibly linked to extractivism, dispossession, and colonization. In that sense, a turn towards the posthuman in natural resource governance is perhaps contradictory. While new language is likely needed to replace that of natural resource discourses, we leave that important work to others for now. Governance methods and discourses have, however, evolved in other ways: from primarily state-led and hierarchical models to ones which are more inclusive, decentralised, networked, and collaborative. Drawing on insights from Indigenous scholars, ANT, and posthumanism scholarship more broadly, these governance perspectives have the potential to be productively decentralised further still, to include the nonhuman, and in ways which can further realise pluralist objectives within ecosystem governance.

Acknowledgements

Thank you to Professors Murray Humphries and Monica Mulrennan for their help refining the ideas within this chapter. We are also grateful for the valuable feedback provided by the anonymous reviewers. This work was financially supported by the Social Sciences and Humanities Research Council (SSHRC).

References

Abrams, J., Huber-Stearns, H., Steen-Adams, M., Davis, E.J., Bone, C., Nelson, M.F., Moseley, C., 2021. Adaptive governance in a complex social-ecological context: emergent responses to a native forest insect outbreak. Sustain. Sci. 16 (1), 53–68. https://doi.org/10.1007/s11625-020-00843-5.

Abson, D.J., von Wehrden, H., Baumgärtner, S., Fischer, J., Hanspach, J., Härdtle, W., Heinrichs, H., Klein, A.M., Lang, D.J., Martens, P., Walmsley, D., 2014. Ecosystem services as a boundary object for sustainability. Ecol. Econ. 103, 29–37. https://doi.org/10.1016/j.ecolecon.2014.04.012.

Adade Williams, P., Sikutshwa, L., Shackleton, S., 2020. Acknowledging indigenous and local knowledge to facilitate collaboration in landscape approaches—lessons from a systematic review. Land 9 (9), 331. https://doi.org/10.3390/land9090331.

Agranoff, R., McGuire, M., 2001. Big questions in public network management research. J. Public. Adm. Res. Theory 11 (3), 295–326. https://doi.org/10.1093/oxfordjournals.jpart.a003504.

Agrawal, A., 1995. Dismantling the divide between indigenous and scientific knowledge. Dev. Change 26 (3), 413–439. https://doi.org/10.1111/j.1467-7660.1995.tb00560.x.

Alaimo, S., 2012. Sustainable this, sustainable that: new materialisms, posthmanism, and unknown futures. Mod. Lang. Assoc. 127 (3), 558–564.

Albuquerque, U.P., Ludwig, D., Feitosa, I.S., de Moura, J.M.B., Gonçalves, P.H.S., da Silva, R.H., da Silva, T.C., Gonçalves-Souza, T., Ferreira Júnior, W.S., 2021. Integrating traditional ecological knowledge into academic research at local and global scales. Reg. Environ. Change 21 (2), 45. https://doi.org/10.1007/s10113-021-01774-2.

Ampumuza, C., Driessen, C., 2021. Gorilla habituation and the role of animal agency in conservation and tourism development at Bwindi, South Western Uganda. Environ. Plan A: Economy Space 4 (4), 1601–1621. https://doi.org/10.1177/2514848620966502.

Amundsen, H., Hermansen, E.A., 2020. Green transformation is a boundary object: An analysis of conceptualisation of transformation in Norwegian primary industries. Environ. Plan A: Economy Space, 22. https://doi.org/10.1177/2514848620934337.

Andrews, S., 2021. Qualitative analysis at the interface of indigenous and Western knowledge systems: the herringbone stitch model. Qual. Res. 21 (6), 939–956. https://doi.org/10.1177/1468794120965365.

Apetrei, C.I., Caniglia, G., von Wehrden, H., Lang, D.J., 2021. Just another buzzword? A systematic literature review of knowledge-related concepts in sustainability science. Glob. Environ. Chang. 68, 102222. https://doi.org/10.1016/j.gloenvcha.2021.102222.

Armitage, D., Berkes, F., Dale, A., Kocho-Schellenberg, E., Patton, E., 2011. Co-management and the co-production of knowledge: learning to adapt in Canada's Arctic. Glob. Environ. Chang. 21 (3), 995–1004. https://doi.org/10.1016/j.gloenvcha.2011.04.006.

Ban, N.C., Frid, A., Reid, M., Edgar, B., Shaw, D., Siwallace, P., 2018. Incorporate indigenous perspectives for impactful research and effective management. Nat. Ecol. Evol 2 (11), 1680. https://doi.org/10.1038/s41559-018-0706-0.

Ban, N.C., Mills, M., Tam, J., Hicks, C.C., Klain, S., Stoeckl, N., Bottrill, M.C., Levine, J., Pressey, R.L., Satterfield, T., Chan, K.M., 2013. A social–ecological approach to conservation planning: embedding social considerations. Front. Ecol. Environ. 11 (4), 194–202. https://doi.org/10.1890/110205.

Bear, C., 2021. Making insects tick: responsibility, attentiveness and care in edible insect farming. Environ. Plan A: Economy Space 4 (3), 1010–1030. https://doi.org/10.1177/2514848620945321.

Bennett, N.J., Satterfield, T., 2018. Environmental governance: a practical framework to guide design, evaluation, and analysis. Conserv. Lett. 11 (6), e12600. https://doi.org/10.1111/conl.12600.

Berkes, F., 2004. Rethinking community-based conservation. Conserv. Biol. 18 (3), 621–630. https://doi.org/10.1111/j.1523-1739.2004.00077.x.

Berkes, F., 2018. Sacred Ecology, fourth ed. Routledge.

Biggs, R., Clements, H., de Vos, A., Folke, C., Manyani, A., Maciejewski, K., Martín-López, B., Prieser, R., Selomane, O., Maja, S., 2021. What are social-ecological systems and social-ecological systems research? In: *The Routledge Handbook of Research Methods for Social-Ecological* Systems. Routledge.

Bodin, Ö., Crona, B.I., 2009. The role of social networks in natural resource governance: what relational patterns make a difference? Glob. Environ. Chang. 19 (3), 366–374. https://doi.org/10.1016/j.gloenvcha.2009.05.002.

Bodin, Ö., Crona, B., Ernstson, H., 2005. Social Networks in Natural Resource Management: What Is There to Learn from a Structural Perspective? https://www.ecologyandsociety.org/vol11/iss2/resp2/.

Bohensky, E.L., Maru, Y., 2011. Indigenous Knowledge, Science, and Resilience: What Have we Learned from a Decade of International Literature on "Integration"? Ecol. Soc. *16* (4). art6 https://doi.org/10.5751/ES-04342-160406.

Braidotti, R., 2016. Posthuman critical theory. In: Banerji, D., Paranjape, M.R. (Eds.), Critical Posthumanism and Planetary Futures. Springer India, pp. 13–32. https://doi.org/10.1007/978-81-322-3637-5_2.

Brand, F.S., Jax, K., 2007. Focusing the meaning(s) of resilience: resilience as a descriptive concept and a boundary object. Ecol. Soc. 12 (1), 23. https://doi.org/10.5751/ES-02029-120123.

Brehony, P., Tyrrell, P., Kamanga, J., Waruingi, L., Kaelo, D., 2020. Incorporating social-ecological complexities into conservation policy. Biol. Conserv. 248, 108697. https://doi.org/10.1016/j.biocon.2020.108697.

Brennan, R., 2022. Making space for plural ontologies in fisheries governance: Ireland's disobedient offshore islands. Marit. Stud. 21 (1), 35–51. https://doi.org/10.1007/s40152-021-00257-8.

Brooks, A., Hubbard, P., 2021. Oysteropolis: Animals in Coastal Gentrification. Nature and Space, Environment and Planning E. https://doi.org/10.1177/25148486211024901.

Butler, C., Watkinson, B., Witzke, J., 2021. The immovable object: mitigation as indigenous conservation. Collaborative Anthropologies 13 (2), 1–28. https://doi.org/10.1353/cla.2021.0001.

Cabrera, N., 2020. Hypha. Maltrato Films.
Callon, M., 1984. Some elements of a sociology of translation: domestication of the scallops and the fishermen of St Brieuc Bay. Sociol. Rev. *32* (1_suppl), 196–233. https://doi.org/10.1111/j.1467-954X.1984.tb00113.x.
Cash, D.W., Adger, W.N., Berkes, F., Garden, P., Lebel, L., Olsson, P., Pritchard, L., Young, O., 2006. Scale and cross-scale dynamics: governance and information in a multilevel world. Ecol. Soc. *11* (2), art8. https://doi.org/10.5751/ES-01759-110208.
Castleden, H., Hart, C., Cunsolo, A., Harper, S., Martin, D., 2017. Reconciliation and relationality in water research and Management in Canada: implementing indigenous ontologies, epistemologies, and methodologies. In: Renzetti, S., Dupont, D.P. (Eds.), Water Policy and Governance in Canada. Vol. 17. Springer International Publishing, pp. 69–95. https://doi.org/10.1007/978-3-319-42806-2_5.
Chaplier, M., Scott, C., 2018. Introduction: from beavers to land: building on past debates to unpack the contemporary entanglements of algonquian family hunting territories. Anthropologica. https://doi.org/10.3138/anth.60.1.t06.
Coffey, J., 2021. Images as 'potentials': feminist new materialist orientations to photovoice. Qual. Res. https://doi.org/10.1177/14687941211049334.
Davidson, D.J., Frickel, S., 2004. Understanding environmental governance: a critical review. Organ. Environ. 17 (4), 471–492. https://doi.org/10.1177/1086026603259086.
Davis, M., 2006. Bridging the gap or crossing a bridge? Indigenous knowledge and the language of Law and policy. In: Bridging Scales And Knowledge Systems: Concepts and Applications in Ecosystem Assessment: A Contribution to The Millennium Ecosystem Assessment. Island Press, pp. 145–163.
Deason, G., Seekamp, E., Barbieri, C., 2022. Actor-network theory and organizational resilience to climate change in community-based tourism. J. Outdoor Recreat. Tour. https://doi.org/10.1016/j.jort.2021.100483.
DePuy, W., Weger, J., Foster, K., Bonanno, A.M., Kumar, S., Lear, K., Basilio, R., German, L., 2021. Environmental governance: Broadening ontological spaces for a more livable world. Environ. Plan A: Economy Space. https://doi.org/10.1177/25148486211018565.
Díaz, S., Demissew, S., Carabias, J., Joly, C., Lonsdale, M., Ash, N., Larigauderie, A., Adhikari, J.R., Arico, S., Báldi, A., Bartuska, A., Baste, I.A., Bilgin, A., Brondizio, E., Chan, K.M., Figueroa, V.E., Duraiappah, A., Fischer, M., Hill, R., Zlatanova, D., 2015. The IPBES conceptual framework—connecting nature and people. Curr. Opin. Environ. Sustain. 14, 1–16. https://doi.org/10.1016/j.cosust.2014.11.002.
Diaz, S., Settele, J., Brondizio, E., 2019. Summary for Policymakers of the Global Assessment Report on Biodiversity and Ecosystem Services of THE INTERGOVERNMENTAL SCIENCE-Policy Platform on Biodiversity and Ecosystem Services. (Advance Version). https://www.ipbes.net/document-library-categories/meeting-documents.
Diver, S., 2017. Negotiating indigenous knowledge at the science-policy interface: insights from the Xáxli'p community Forest. Environ. Sci. Policy 73, 1–11. https://doi.org/10.1016/j.envsci.2017.03.001.
Donald, D., 2012. Indigenous Métissage: a decolonizing research sensibility. Int. J. Qual. Stud. Educ. 25 (5), 533–555. https://doi.org/10.1080/09518398.2011.554449.
Donaldson, A., Lowe, P., Ward, N., 2002. Virus-crisis-institutional change: the foot and mouth actor network and the governance of rural affairs in the UK. Sociol. Rural. 42 (3), 201–214. https://doi.org/10.1111/1467-9523.00211.
Dwiartama, A., Rosin, C., 2014. Exploring agency beyond humans: the compatibility of actor-network theory (ANT) and resilience thinking. Ecol. Soc. *19* (3). art28 https://doi.org/10.5751/ES-06805-190328.
Eider, D., Partelow, S., Albrecht, S., Adrianto, L., Kluger, L.C., 2021. SCUBA tourism and coral reefs: a social-ecological network analysis of governance challenges in Indonesia. Curr. Issue Tour., 1–20. https://doi.org/10.1080/13683500.2021.2006612.

Emerson, K., Nabatchi, T., Balogh, S., 2012. An integrative framework for collaborative governance. J. Public. Adm. Res. Theory 22 (1), 1–29. https://doi.org/10.1093/jopart/mur011.

Engman, M.M., Hermes, M., 2021. Land as interlocutor: a study of ojibwe learner language in interaction on and with naturally occurring 'materials.'. Mod. Lang. J. 105 (S1), 86–105. https://doi.org/10.1111/modl.12685.

Enqvist, J.P., West, S., Masterson, V.A., Haider, L.J., Svedin, U., Tengö, M., 2018. Stewardship as a boundary object for sustainability research: linking care, knowledge and agency. Landsc. Urban Plan. 179, 17–37. https://doi.org/10.1016/j.landurbplan.2018.07.005.

Folke, C., 2004. Traditional knowledge in social–ecological systems. Ecol. Soc. 9 (3). https://doi.org/10.5751/ES-01237-090307.

Freudenburg, W.R., Frickel, S., Gramling, R., 1995. Beyond the nature/society divide: learning to think about a mountain. Sociol. Forum 10 (3), 361–392. https://doi.org/10.1007/BF02095827.

Gesing, F., 2021. Towards a more-than-human political ecology of coastal protection: coast care practices in Aotearoa New Zealand. Environ. Plan A: Economy Space 4 (2), 208–229. https://doi.org/10.1177/2514848619860751.

Gibbs, L., 2021. Agency in human–shark encounter. Environ. Plan A: Economy Space 4 (2), 645–666. https://doi.org/10.1177/2514848620929942.

Graham, J., Amos, B., Plumptre, T., 2003. Governance Principles for Protected Areas in the 21st Century. https://iog.ca/research-publications/publications/governance-principles-for-protected-areas-in-the-21st-century/.

Guibrunet, L., Gerritsen, P.R.W., Sierra-Huelsz, J.A., Flores-Díaz, A.C., García-Frapolli, E., García-Serrano, E., Pascual, U., Balvanera, P., 2021. Beyond participation: how to achieve the recognition of local communities' value-systems in conservation? Some insights from Mexico. People Nat. 3 (3), 528–541. https://doi.org/10.1002/pan3.10203.

Harcourt, W., 2021. Rethinking life-in-common in the Australian landscape. Environ. Plan A: Economy Space 4 (4), 1330–1345. https://doi.org/10.1177/2514848621989602.

Harms, H., Schlosser, D., Wick, L.Y., 2011. Untapped potential: exploiting fungi in bioremediation of hazardous chemicals. Nat. Rev. Microbiol. 9 (3), 177–192. https://doi.org/10.1038/nrmicro2519.

Hayward, E., 2010. FINGERYEYES: impressions of cup corals. Cult. Anthropol. 25 (4), 577–599. https://doi.org/10.1111/j.1548-1360.2010.01070.x.

Henriksen, D., Creely, E., Mehta, R., 2021. Rethinking the politics of creativity: Posthumanism, indigeneity, and creativity beyond the Western Anthropocene. Qual. Inq. https://doi.org/10.1177/10778004211065813.

Hessami, M.A., Bowles, E., Popp, J.N., Ford, A.T., 2021. Indigenizing the north American model of wildlife conservation. FACETS 6, 1285–1306. https://doi.org/10.1139/facets-2020-0088.

Holifield, R., 2009. Actor-network theory as a critical approach to environmental justice: a case against synthesis with urban political ecology. Antipode 41 (4), 637–658. https://doi.org/10.1111/j.1467-8330.2009.00692.x.

Hotte, N., Wyatt, S., Kozak, R., 2019. Influences on trust during collaborative forest governance: a case study from Haida Gwaii. Can. J. For. Res. 49 (4), 361–374. https://doi.org/10.1139/cjfr-2018-0222.

Howitt, R., Suchet-Pearson, S., 2006. Rethinking the building blocks: Ontological pluralism and the idea of 'management.'. Geogr. Ann. B: Hum. Geogr. 88 (3), 323–335. https://doi.org/10.1111/j.1468-0459.2006.00225.x.

Hunt, S., 2014. Ontologies of indigeneity: the politics of embodying a concept. Cult. Geogr. 21 (1), 27–32. https://doi.org/10.1177/1474474013500226.

Huntington, H.P., 2000. Using traditonal ecological knowledge in science: methods and applications. Ecol. Appl. 10 (5), 1270–1274. https://doi.org/10.1890/1051-0761(2000)010[1270:UTEKIS]2.0.CO;2.

Kirksey, S.E., Helmreich, S., 2010. The emergence of multispecies ethnography. Cult. Anthropol. 25 (4), 545–576. https://doi.org/10.1111/j.1548-1360.2010.01069.x.

Kobluk, H.M., Gladstone, K., Reid, M., Brown, K., Krumhansl, K.A., Salomon, A.K., 2021. Indigenous knowledge of key ecological processes confers resilience to a small-scale kelp fishery. People Nat. 3 (3), 723–739. https://doi.org/10.1002/pan3.10211.

Kohn, E., 2007. How dogs dream: Amazonian natures and the politics of transspecies engagement. Am. Ethnol. 34 (1), 3–24. https://doi.org/10.1525/ae.2007.34.1.3.

Kohn, E., 2013. How Forests Think: Toward an Anthropology Beyond the Human (Vol. 1–1 online resource (xiii, 267 pages) : illustrations). University of California Press. WorldCat.org. http://www.vlebooks.com/vleweb/product/openreader?id=none&isbn=9780520956865.

LaDuke, W., 1994. Traditional ecological knowledge and enviromental futures. Colo. J. Int'l Envtl. L. & Pol'y 5 (27).

Latour, B., 2005. Reassembling the Social: An Introduction to Actor-Network-Theory. Oxford University Press.

Law, J., 1984. On the methods of long-distance control: vessels, navigation and the Portuguese route to India. Sociol. Rev. *32* (1_suppl), 234–263. https://doi.org/10.1111/j.1467-954X.1984.tb00114.x.

Law, J., 1992. Notes on the theory of the actor-network: ordering, strategy, and heterogeneity. Syst. Pract. 5 (4), 379–393. https://doi.org/10.1007/BF01059830.

Law, J., 2009. Actor-network theory and material semiotics. In: The New Blackwell Companion to Social Theory. Wiley-Blackwell, pp. 141–158.

Law, J., Joks, S., 2019. Indigeneity, science, and difference: notes on the politics of how. Sci. Technol. Hum. Values 44 (3), 424–447. https://doi.org/10.1177/0162243918793942.

Li, W., von Essen, E., 2020. Guarding crops from monkey troops: farmer-monkey interaction near a nature reserve in Guangxi, China. Environ. Sociol. 7 (1), 12–24. https://doi.org/10.1080/23251042.2020.1811004.

Lockwood, M., Davidson, J., Curtis, A., Stratford, E., Griffith, R., 2010. Governance principles for natural resource management. Soc. Nat. Resour. 23 (10), 986–1001. https://doi.org/10.1080/08941920802178214.

Malekpour, S., Tawfik, S., Chesterfield, C., 2021. Designing collaborative governance for nature-based solutions. Urban For. Urban Green. 62, 127177. https://doi.org/10.1016/j.ufug.2021.127177.

Massey, C.D., Vayro, J.V., Mason, C.W., 2021. Conservation values and actor networks that shape the Adams River Salmon run in Tsútswecw Provincial Park, British Columbia. Soc. Nat. Resour. 34 (9), 1174–1193. https://doi.org/10.1080/08941920.2021.1946225.

Maxwell, K., Awatere, S., Ratana, K., Davies, K., Taiapa, C., 2020. He waka eke noa/we are all in the same boat: a framework for co-governance from aotearoa New Zealand. Mar. Policy 121, 104213. https://doi.org/10.1016/j.marpol.2020.104213.

Mendez, S.R., 2020. Sundance VR Tackled Climate Change With Groundbreaking New Projects. IndieWire. https://www.indiewire.com/2020/02/sundance-2020-vr-climate-change-1202208386/.

Miller, A., Davidson-Hunt, I., 2013. Agency and resilience: teachings of Pikangikum first nation elders, northwestern Ontario. Ecol. Soc. 18 (3). https://doi.org/10.5751/ES-05665-180309.

Miller, R., Marsh, H., Benham, C., Hamann, M., 2020. Stakeholder engagement in the governance of marine migratory species: barriers and building blocks. Endanger. Species Res. 43, 1–19. https://doi.org/10.3354/esr01049.

Molnár, Z., Babai, D., 2021. Inviting ecologists to delve deeper into traditional ecological knowledge. Trends Ecol. Evol. https://doi.org/10.1016/j.tree.2021.04.006.

Nabavi, E., Daniell, K.A., 2017. Rediscovering social–ecological systems: taking inspiration from actor-networks. Sustain. Sci. 12 (4), 621–629. https://doi.org/10.1007/s11625-016-0386-0.

Nadasdy, P., 1999. The politics of Tek: power and the "integration" of knowledge. Arct. Anthropol. 36 (1/2), 1–18.

Nadasdy, P., 2005. The anti-politics of TEK: the institutionalization of co-management discourse and practice. Anthropol. Forum 47 (2), 215–232.

Nadasdy, P., 2007. The gift in the animal: the ontology of hunting and human–animal sociality. Am. Ethnol. 34 (1), 25–43. https://doi.org/10.1525/ae.2007.34.1.25.

Norman, E.S., 2015. What Boundary? What Whale? Whose Responsibility? The Blurring of Political and Cultural Boundaries in Marine Governance. In: Governing Transboundary Waters: Canada, the United States, and Indigenous communities. Routledge.

Ostrom, E., 2009. A general framework for analyzing sustainability of social-ecological systems. Science 325 (5939), 419–422. https://doi.org/10.1126/science.1172133.

Palmer, J.S., 2014. (Actor-) network mapping of housing systems employing social network analysis tools: the case of medium-density dwelling design in Australia., p. 19.

Plummer, R., Armitage, D.R., de Loë, R.C., 2013. Adaptive Comanagement and its relationship to environmental governance. Ecol. Soc. *18* (1). art21 https://doi.org/10.5751/ES-05383-180121.

Popp, J.N., Priadka, P., Young, M., Koch, K., 2020. Indigenous guardianship and moose monitoring: weaving indigenous and Western ways of knowing. Hum.-Wildl. Interact. 14 (2), 296–308.

Pyke, M.L., Close, P.G., Dobbs, R.J., Toussaint, S., Smith, B., Cox, Z., Cox, D., George, K., McCarthy, P., Angus, B., Riley, E., Clifton, J., 2021. 'Clean him up…make him look like he was before': Australian aboriginal Management of Wetlands with implications for conservation, restoration and multiple evidence base negotiations. Wetlands 41 (2), 28. https://doi.org/10.1007/s13157-021-01410-z.

Ramos, S.C., 2022. Understanding Yurok traditional ecological knowledge and wildlife management. J. Wildl. Manag. 86 (1). https://doi.org/10.1002/jwmg.22140.

Rarai, A., Parsons, M., Nursey-Bray, M., Crease, R., 2022. Situating climate change adaptation within plural worlds: The role of Indigenous and local knowledge in Pentecost Island, Vanuatu. Environ. Plan A: Economy Space. https://doi.org/10.1177/25148486211047739.

Reo, N.J., 2019. Inawendiwin and relational accountability in Anishnaabeg studies: the crux of the biscuit. J. Ethnobiol. 39 (1), 65. https://doi.org/10.2993/0278-0771-39.1.65.

Robbins, P., 2007. Lawn People: how grasses, weeds, and chemicals make us who we are. Temple University Press.

Rydin, Y., 2013. Using actor–network theory to understand planning practice: exploring relationships between actants in regulating low-carbon commercial development. Plan. Theory 12 (1), 23–45. https://doi.org/10.1177/1473095212455494.

Salmond, A., 2018. Rivers as ancestors and other realities: governance of waterways in Aotearoa/ New Zealand. In: Martin, B., Te Aho, L., Humphries-Kil, M. (Eds.), ResponsAbility: Law and Governance for Living Well with the Earth, second ed. Routledge. https://doi.org/10.4324/9780429467622.

Scarlett, L., McKinney, M., 2016. Connecting people and places: the emerging role of network governance in large landscape conservation. Front. Ecol. Environ. 14 (3), 116–125. https://doi.org/10.1002/fee.1247.

Schoon, M., van der Leeuw, S., 2015. The shift toward social-ecological systems perspectives: insights into the human-nature relationship. Nat. Sci. Soc. 23 (2), 166–174. https://doi.org/10.1051/nss/2015034.

Scott, C., 1996. Science for the West, myth for the rest? The case of James Bay Cree knowledge production. In: Nader, L. (Ed.), Naked Science: Anthropological Inquiry into Boundaries, Power, and Knowledge, first ed. Routledge, pp. 69–86.
Shaffer, C.A., Milstein, M.S., Lindsey, L.L., Wolf, T.M., Suse, P., Marawanaru, E., Kipp, E.J., Garwood, T., Travis, D.A., Terio, K.A., Larsen, P.A., 2022. "Spider monkey cotton": bridging Waiwai and scientific ontologies to characterize spider monkey (Ateles paniscus) Filariasis in the Konashen community owned conservation area, Guyana. Int. J. Primatol. https://doi.org/10.1007/s10764-021-00272-w.
Star, S.L., 2010. This is not a boundary object: reflections on the origin of a concept. Sci. Technol. Hum. Values 35 (5), 601–617. https://doi.org/10.1177/0162243910377624.
Star, S.L., Griesemer, J.R., 1989. Institutional ecology, "translations" and boundary objects: amateurs and professionals in Berkely's museum of vertebrate zoology, 1907–39. Soc. Stud. Sci. 19, 387–420.
Steins, N., 2001. New directions in natural resource management: the offer of actor-network theory. ResearchGate, IDS Bull. (4), 18–25. https://doi.org/10.1111/j.1759-5436.2001.mp32004003.x.
Stevenson, W.B., Greenberg, D., 2000. Agency and social networks: strategies of action in a social structure of position, opposition, and opportunity. Adm. Sci. Q. 45 (4), 651–678. https://doi.org/10.2307/2667015.
Sundberg, J., 2014. Decolonizing posthumanist geographies. Cult. Geogr. *21* (1), 33–47. WorldCat.org.
Theriault, N., 2017. A forest of dreams: ontological multiplicity and the fantasies of environmental government in the Philippines. Polit. Geogr. 58, 114–127. https://doi.org/10.1016/j.polgeo.2015.09.004.
Todd, Z., 2016. An indigenous Feminist's take on the ontological turn: 'ontology' is just another word for colonialism: an indigenous Feminist's take on the ontological turn. J. Hist. Sociol. 29 (1), 4–22. https://doi.org/10.1111/johs.12124.
Toncheva, S., Fletcher, R., 2021. Knowing bears: An ethnographic study of knowledge and agency in human–bear cohabitation. Environ. Plan A: Economy Space. https://doi.org/10.1177/25148486211015037.
Trosper, R.L., Parrotta, J.A., 2012. Introduction: the growing importance of traditional Forest-related knowledge. In: Parrotta, J.A., Trosper, R.L. (Eds.), Traditional Forest-Related Knowledge. Vol. 12. Springer, Netherlands, pp. 1–36. https://doi.org/10.1007/978-94-007-2144-9_1.
Turnhout, E., Neves, K., de Lijster, E., 2014. 'Measurementality' in biodiversity governance: knowledge, transparency, and the intergovernmental science-policy platform on biodiversity and ecosystem services (IPBES). Environ. Plan A: Economy Space 46 (3), 581–597. https://doi.org/10.1068/a4629.
Venot, J.-P., Jensen, C.B., 2021. A multiplicity of *prek(s)*: Enacting a socionatural mosaic in the Cambodian upper Mekong delta. Environ. Plan A: Economy Space. https://doi.org/10.1177/25148486211026835.
Verran, H., 2002. A postcolonial moment in science studies: alternative firing regimes of environmental scientists and aboriginal landowners. Soc. Stud. Sci. 32 (5–6), 729–762.
Wakefield, S., 2020. Making nature into infrastructure: the construction of oysters as a risk management solution in new York City. Environ. Plan A: Economy Space 3 (3), 761–785. https://doi.org/10.1177/2514848619887461.
Walsh, F.J., Dobson, P.V., Douglas, J.C., 2013. Anpernirrentye: a framework for enhanced application of indigenous ecological knowledge in natural resource management. Ecol. Soc. *18* (3). art18 https://doi.org/10.5751/ES-05501-180318.
Watson, A., Huntington, O.H., 2008. They're here—I can feel them: the epistemic spaces of indigenous and Western knowledges. Soc. Cult. Geogr. 9 (3), 257–281. https://doi.org/10.1080/14649360801990488.

Watts, V., 2013. Indigenous place-thought & agency amongst humans and non-humans (first woman and sky woman go on a European world tour!). Decolonization: Indig., Educ. Soc. 2 (1), 20–34.

Watts, N., Scales, I.R., 2015. Seeds, agricultural systems and socio-natures: towards an actor–network theory informed political ecology of agriculture. Geogr. Compass 9 (5), 225–236. https://doi.org/10.1111/gec3.12212.

Wiber, M.G., Barnett, A., 2021. (Re)assembling marine space: lobster fishing areas under conditions of technological and legal change in Atlantic Canada. Sci. Technol. Hum. Values. https://doi.org/10.1177/01622439211042416.

Williams, I., 2020. Contemporary Applications of Actor Network Theory (Vol. 1–1 online resource). Palgrave Macmillan. WorldCat.org. https://eui.idm.oclc.org/login?url=https://doi.org/10.1007/978-981-15-7066-7.

Zetina-Rejón, M.J., Zepeda-Domínguez, J.A., Rodríguez-Fuentes, M., Fumero-Andreu, C.M., 2020. Stakeholder diversity correlates with governance network performance in two artisanal fisheries in Northwest Mexico. Ocean Coast. Manag. 196, 105313. https://doi.org/10.1016/j.ocecoaman.2020.105313.

Amplifying muted voices

CHAPTER FOUR

Of green spaces and grey areas: An ethnography of everyday forest management in peri-urban Bengaluru, India

Lingaraj G. Jayaprakash and Gordon M. Hickey*
Department of Natural Resource Sciences, McGill University, Montreal, QC, Canada
*Corresponding author: e-mail address: gordon.hickey@mcgill.ca

Contents

Abstract

India's forest landscapes are reeling under severe demographic and developmental pressures. These tensions are particularly pronounced in peri-urban regions that are marked by rapid land use change, weak and overlapping institutions and competing claims on land. This paper examines how rapid urbanization influences place-making and the practice of forest management in Bannerghatta National Park (BNP), India. BNP is an elephant corridor traversing through a human-dominated landscape bordering the megacity of Bangalore. Straying of elephants and leopards into village settlements has led to significant destruction of crops and life. Unregulated real-estate development driven by a boom in the information technology (IT) industry, combined with granite quarries that operate illegally around BNP have altered the characteristics of the surrounding villages. These processes have revealed or created newer forms of inequity that challenge forest institutions to respond ingeniously. Applying concepts from critical institutionalism and ethnography, our research aims to inform policy efforts to

Advances in Ecological Research, Volume 66
ISSN 0065-2504
https://doi.org/10.1016/bs.aecr.2022.04.003

incorporate pluralism in sustainable forest management and rural development. Through data collected over 16 months via observation, interviews, and oral histories we examine the workings of forest management institutions in everyday social life. We map specific modes through which four marginalized agents navigate institutional plurality and intersect with each other in the peri-urban forest landscape. The results illustrate the potency of 'institutional bricolage', wherein agents renegotiate their space in the landscape by actively drawing from available resources to ensure workable institutional arrangements, challenging government efforts to conserve and manage ecosystems.

1. Introduction

It is mid-September, close to the end of another suboptimal monsoon. The dust hangs thick in the dry air as you turn towards the Bannerghatta National Park (BNP henceforth) off the Bengaluru-Jigani highway, not far from the Indian Institute of Management and a cluster of the information technology (IT) companies that Bengaluru in South India is famous for. BNP is an elephant corridor traversing through a human-dominated landscape bordering the megacity of Bangalore. The park forms the eastern-most extent of India's largest elephant reserve, the 6724 sq. km Mysore reserve. With a population of 150 resident elephants, BNP occupies an important place in the distribution of the Asian elephant (*Elephas maximus*) in South India. The Government of India recognizes the park as a priority area for conservation under the Project Elephant program (Varma et al., 2009). In addition to the Asian Elephant, the park houses several important faunal species such as Leopard (*Panthera pardus*), Sloth bear (*Melurus ursinus*), Spotted deer (*Axis axis*), Wild pig (*Sus scrofa*) and Pangolin (*Manis crassicaudata*) among others. Around 180 species of avian fauna, including migratory birds such as the Indian Pitta and Orange Headed Thrush are present in BNP (Singh, 2008).

We are headed towards Ragihalli - a village named for the millet that is now grown less and less around here. The pothole-ridden road has a steady upward incline and can barely accommodate two cars side by side. The landscape is dotted with the occasional open-air brick industry, and signs of the beginnings of several new suburban residential layouts. Multilingual hoardings encourage you to invest in these yet-to-materialize real estate projects, with idyllic names such as "Hill-view", "Forest View" and "Green Nest", and other billboards helpfully offer loans with 'easy' mortgage plans to help you buy them. But what is most striking about this path are forbidding monster haul-trucks that have unquestioned right of way, even as they rumble downhill weighed down by granite quarried from the gneissic hills. This is one of the three motorable roads that intercept BNP. By the time you cross the seven-kilometre section of forest to reach Ragihalli village, you are likely to have encountered over 20 trucks either going or coming from the quarries that lie on the other side of the forest. A gateway with a display-board that reads "Bannerghatta National Park" in Kannada and English reminds people that they are entering a protected area. The road gets strikingly better within forest limits. Thorn and scrub forests alternate with thicker/ greener patches and open stretches with granitic boulders from one of the oldest rock formations in the world. The flora is diverse and is characterized by patches of deciduous and thorny scrub-type vegetation, interspersed with stretches of grassland. Bamboo (*Dendrocalamus strictus*), the deciduous Jaalari (*Shorea talura*) and the thorny *Acacia chundra* dot the landscape. Boards installed by the forest department announcing platitudes in Kannada and English, such as 'Protecting our woods is our duty' and 'we are, if the forest is', are frequent. Soon, the only reminders of Bengaluru city are the tell-tale beer bottles and overstuffed garbage bags that are sometimes dumped at night, surreptitiously. Fig. 1 shows the location of BNP (D) with respect to its relative position in the region (A), in India (B) and in the state of Karnataka (C).

BNP contributes significantly towards the wellbeing of local communities by providing income, employment, food and vital ingredients for their farming systems (Rajeev, 2002). While the park is exclusively controlled and managed by state and federal authorities, there are valuable common property resource systems such as, *gomala* grazing lands, community wetlands and sacred groves outside the park that are of immense ecological value. Around 112 villages are located within the one-kilometre buffer boundary of the national park (Gopalakrishna et al., 2015) and five villages are inside the park. Most of these settlements are agrarian and have traditionally been dependent on the surrounding forests for wild fruits, meat, manure, firewood, lac, and honey (Varma et al., 2009). There is a sizable tribal population, as well as inhabitants belonging to the lower castes (Singh, 2008). Local people largely depend on natural resource-intensive activities such as agriculture and livestock rearing for their subsistence.

Ragihalli and 15 other hamlets lie in the north-eastern arm of the BNP. They are a peninsula surrounded by forest on three sides, and small, arid agricultural fields on the other. Agriculture is practiced during monsoon and winter seasons, acting as the primary source of income for most of the

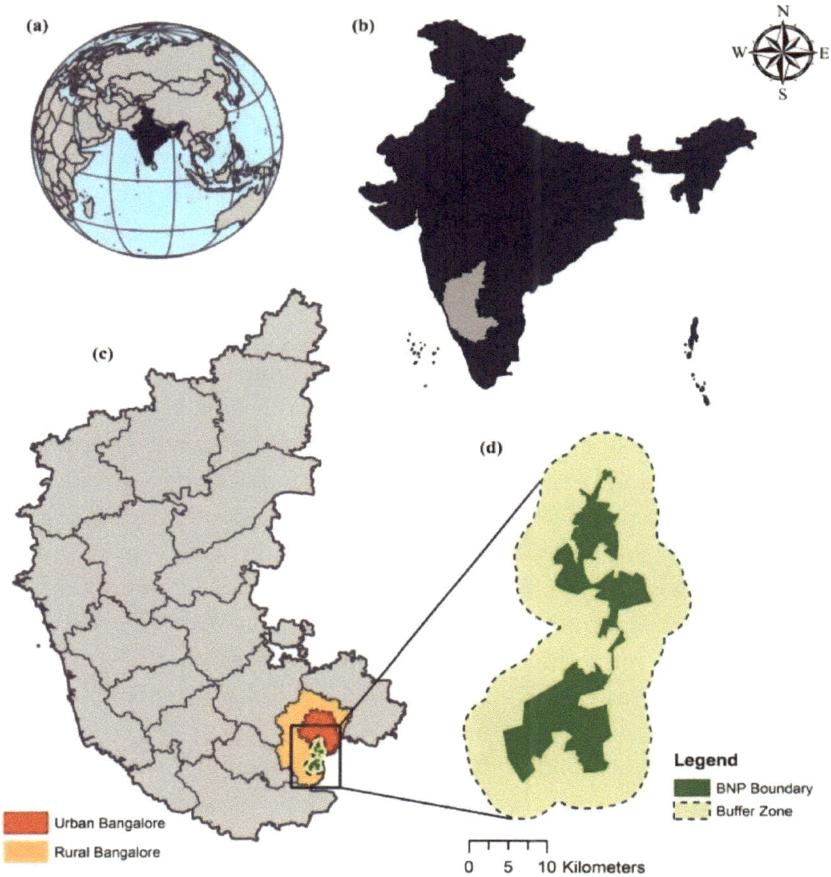

Fig. 1 Map of Bannerghatta National Park.

people. Landholdings are invariably small, agriculture is rainfed and farmers primarily grow finger millet (*Cynoasurus corocanaus*). Farmer's supplement this with crops such as banana, mulberry (for silk-worm rearing), along with seasonal flowers and seasonal vegetables. Given the proximity of the settlements to the national park, human-wildlife conflicts are observed regularly. Loss of livelihoods, crops and life have led to significant conflict between conservation efforts and the developmental needs of the people.

Ragihalli and surrounding villages are at the crossroads not just of routes cutting through the forest, but also of diverse aspirations, policies, institutions, and attitudes. Bangalore city has undergone rapid urban expansion over the last three decades and is a dominant presence in the BNP landscape, observed in the form of granite quarries, brick kilns, real estate projects, stone crushing units, tourist resorts – many of which are operating here illegally (Varma et al., 2009). Cumulatively, structures and processes of urban expansion have revealed or created newer forms of inequity that challenge a wide plurality of forest-related institutional arrangements (formal, informal, traditional, modern, social, economic) to respond ingeniously (Cleaver, 2002).

According to Vira et al. (1998) institutional pluralism in forestry describes the interplay of "a multiplicity of ideologies, interests, actors and organizations, which have combined to create the current scenario, and which can contribute to future development". Importantly, institutional plurality "creates opportunity, processes of bricolage and institutional improvization that can open spaces for different voices to be heard" in forest resource management (Cleaver, 2002). Wollenberg et al. (2001) consider pluralism to offer a foundation for understanding how less powerful groups' interests can be accommodated in forest management, and how different actors deal with multiple systems of rules (Wollenberg et al., 2005). Adopting a pluralist approach to forest management therefore requires a careful examination of the competing values and interests of different actors, particularly those that have been historically excluded or marginalized, to expand the range of possibilities for sustainable management (Anderson et al., 1998; Wollenberg et al., 2001).

This chapter aims to "hear the voices" of marginalized forest agents operating at the periphery of BNP to better understand the complexity and practice of everyday forest management and inform efforts to incorporate pluralism in sustainable forest management and rural development (Anderson et al., 1998; Lynch, 1998; Vira et al., 1998; Wollenberg et al., 2001). More specifically, we explore the values and experiences of different forest dependent actors in their everyday social life to identify processes of bricolage and institutional improvization affecting forest ecosystem management. By adopting an emic perspective (Kottak, 2006), and embedding ourselves in the study area, we piece together a fuller picture of the multi-level governance system by refracting divergent viewpoints against each other, to give a "sense of the place" (Convery et al., 2014) and uncover the lived complexity of forest ecosystem management in a human dominated landscape. Furthermore, through this ethnographic narrative we underline the different political spaces available to forest-dependent agents by mapping their strategies employed to navigate complexity.

2. Methods

Applying concepts from critical institutionalism (Cleaver and De Koning, 2015) and environmental ethnography (Kohn, 2007), we collected qualitative data over 16 months. Data collection was spread across the monsoon, winter and the summer seasons in 20 villages on the north-eastern part of BNP. We selected these villages based on the following criteria: proximity to BNP, natural resource dependency, community diversity (e.g. tribal, caste, livelihoods), urbanization pressures and prevalence of human-wildlife conflicts. We collected data through interviews (semi-structured and unstructured interviews in Kannada language spoken locally) and observation of "daily life" (Gupta and Ferguson, 1997) at various locations such as homes, farms, the forest department and in transit. Conversely, we participated in a range of activities such as: afforestation drives, a forest department outreach program for school children (*chinnara vanadarshana* in Kannada), public meetings to "resolve man-animal conflicts", self-help group meetings of milk cooperatives, anti-snare drills with forest staff, night vigils with farmers to prevent elephant depredation, among others. Data were collected with the informed consent of all participants.

Data were coded and manually analysed using NVIVO (Bazeley and Jackson, 2013). We framed our analysis using the standpoints of four agents that operate at the margins of the BNP forest landscape viz.: a traditional forest dweller from the Iruliga Adivasi tribe, a "daily-wage" forest department "watcher", a smallholder farmer whose husband was trampled by an elephant, and a bull elephant known locally as "Rowdy Ranga" that has a reputation for raiding crops habitually. We incorporate a multispecies lens that has been used previously to analyse the contact zone between humans and elephants (e.g.: Kirksey and Helmreich, 2010; Locke, 2017). Our analysis is guided by the work of Cleaver (2012), using 'institutional bricolage' as an analytical frame for studying the messiness of practice in natural resource governance. She builds on the work of Claude Lévi-Strauss, who first introduced the terminology of bricolage and bricoleurs in social sciences, and Mary Douglas, who advanced it to criticize rational choice views on institutions (Cleaver, 2012; Douglas, 1986; Johnson, 2012). "Bricolage" denotes the creative practices by which actors draw on, shape and adapt a range of new and pre-existing formal and informal institutions in particular situations. Institutional bricolage (Cleaver, 2012) or "doing institutions" (Etzold et al., 2012) describes the mechanisms by which institutions are rearranged.

Bricolage is seen as the product of "untamed thinking" that has been celebrated as resourcefulness and as resistance actions and spaces against adversity (Real, 2008). A bricolage approach to institutional analysis is of relevance in the Indian context where boundaries between the state, traditional authorities, private and informal sectors are particularly messy (Benjaminsen and Lund, 2002).

Our study involved the collection of information regarding human-environment interaction in general and the human-elephant relationship in particular, as well as collecting information about the implementation of different forest governance mechanisms around BNP. These issues are politically and socially sensitive. Therefore, we have used pseudonyms and have not listed any identifiers or personal information.

3. Results

This section maps the specific modes through which four forest-dependent agents navigate complexity and intersect with each other and forest institutions.

3.1 Anna—The Iruliga Adivasi

We are standing at the foot of one of the many granitic outcrops dotting the region. Almost continuous with the forest land is a farm of a businessperson from Bangalore. The holding has small haphazard tracts of teak and coconut plantations interspersed with larger patches of scrub vegetation that are indistinguishable from the surrounding forest. At the centre of the farm is a modest lake, frequented by elephants and other fauna. This lake is surprisingly full despite a bad monsoon and has been created by intercepting a seasonal stream. The lake and the farm were added to the landscape no more than 30 years ago. We are here to meet Anna, the custodian of this farm. A tall, wiry man in his early fifties, he is of the adivasi (Indigenous) community called Iruliga (literally people of the darkness). From his job on the farm, Shivanna earns less than 3 dollars a day.

Iruliga are traditional forest dwellers scattered across the forests of Karnataka, Kerala and Tamil Nadu states (Guruprasad et al., 2013). They hold a deep cultural connection with the forests and go by the name of *Kaadu Poojararu* (forest priests) around here. According to Iruliga accounts, the very purpose of their existence is tied with the forest. Anna recounted this story that he had first heard as a child:

"Long long ago, there were no Iruligas. And there was something unusual about our forests. Honeycombs hung from every branch and rock. But the bees were so fierce that no one could even go near them. The honey just lay in the hives, untasted. People prayed to goddess *Beeramma* for help. *Beeramma* taking pity on her children, created a man from her body's sweat. She gave him a hatchet and a staff and blessed him. He could approach the honeycombs and take some, without being stung by the bees. The very scent of his sweat would drive them away. We Iruligas are his descendants. Even now when I go to collect honey, if bees buzz around me, all I have to do is run my hand over my sweaty brow—they fly away. I have heard that my grandfather's sweat had even greater powers—larger beasts—tigers bears, and elephants would change their course upon smelling his presence".[a]

Iruligas originally subsisted by harvesting minor forest produce like honey, bee-wax, and bamboo (Manjunatha et al., 2014). Just like his ancestors, Anna is a great connoisseur of honey. He claims to be able to recognize 22 different kinds of honey and identify bee species based on the taste of the honey. With the coming of the Government of India's Wildlife Protection Act (WPA) of 1972, Iruligas were evicted from the forests of Bannerghatta and others. The WPA took a landscape approach towards the conservation of dwindling wildlife species—primarily the elephant and tiger, and large tracts of land (about 4% of India's total area) were demarcated as no-go protected areas. Despite restrictions on entering the national park, we often see Anna seamlessly walking in and out of the national park in search of honey.

In spite of their proximity to a megacity like Bangalore, Iruliga in BNP lived in relative isolation until the late 1980s and rank severely low on many human development indicators (RKM 2013). The Indian constitution makes special provisions for safeguarding the rights of adivasi communities; however, they often do not translate effectively on the ground. Regardless of the rehabilitation schemes of the government, Iruligas live in abject socio-economic conditions. They now work primarily as *coolies* (labourers) and are often taken advantage of by the upper caste communities (Manjunatha et al., 2014). Many of Anna's relatives have left for Bangalore and do a variety of odd jobs there.

As a corrective, the Government of India introduced the "Scheduled Tribes and other Traditional Forest Dwellers Act" (also known as the Forest Rights Act or FRA) in 2006. The Act was co-written by tribal

[a] A variant of this story has also been recorded in the 1930s Ethnographical Survey of Mysore.

activists through the union ministry of tribal affairs; and was ostensibly meant to correct the historical injustices done to the adivasis through various forest regulations across colonial and post-colonial administrations that dispossessed them of their customary rights to land and livelihoods. Now Anna and other adivasis like him can claim individual and community forest land rights (CFRs) in all categories of forest areas such as PAs, reserve forests and state forests. However, Anna acts hassled at the very mention of the FRA. "It's all an eyewash" he says. "This is not the first such law. Early 1990s—the department said we are valuable partners in managing forests. They organized many events to commemorate the launch of *grama aranya samithi.*[b] What came out of it? Nothing! We are still working as *coolies*. In our name everybody else benefited." Anna is brooding over the Joint Forest Management scheme that was launched in 1988 with much fanfare with claims of "democratizing forest governance".

Anna is uneducated but knows the laws and procedures like the back of his hand. "We have been trying hard to own arable land since my grandfather's time. We have gone from pillar to post seeking help. Officers come and go, legislations come and go, but nothing has changed for us. Just a few months ago we did a semi-nude *satyagraha.*[c] Our pictures were all over the media. Some social workers approached us on their own. I have seen it time and again, they are so full of words we don't understand; and promises that never materialize". The main problem, as Anna himself acknowledges, pertains to documentation. "Government only understands what is on paper", He adds. To claim land titles and community forest rights, adivasis must show documentary evidence that they have been occupying the land from before December 13, 2005. The basis of Anna's land claims it turns out are entirely based on an Iruliga cultural practice called *Kalluseve* (stone service). Traditionally, when an Iruliga dies, their body is buried under concentric stones laid similar to megalithic dolmens. Everyone that attends the funeral, prays, and keeps a stone on the dead body. Prehistoric dolmens are found scattered all across BNP and nobody practices *Kalluseve* anymore. Anna points to yellow flowers growing amidst one such dolmen inside the forest and says it has healing properties, and that it will cure a child of illness if tied around the neck. For Anna and his kinsmen, the very presence of *Kalluseve* remains is good enough proof to lay claim to the forestland. Anna contends that the FRA fails to recognize the intricate association between Adivasi and

[b] Village forest committees established as part of Joint Forest Management.

[c] Gandhian mode of protest that simply means "polite insistence" or "non-violent resistance".

the forest. While FRA, and JFM before that, have been celebrated for their claims of empowering local communities in forest conservation and management, as Kashwan (2017) observed, implementation has often failed due to bureaucratic quagmires and "non-cooperation from entrenched forest administration".

Anna, as we found over time, is a man of many parts—based on his judgement of you he will effortlessly juggle between—a honey loving Iruliga, forest guide, political aspirant, and more actively, a real estate agent who can point to the nearest *gomaala* land[d] or agriculturist in need of money (who is likely to sell land). He now feels an uneasy sense of resentment, as he and his kinsfolk are increasingly limited from visiting their sacred sites, or collecting forest produce or even grazing their cattle and sheep in lands that they know and believe their folk have been traversing and safeguarding, in a manner of speaking, for centuries, long before the Indian state even existed. At the same time, he sees no contradiction between viewing forest as their spiritual home and pieces of real-estate.

3.2 Ayya—the watcher

It is the week before *Sankranti,*[e] the nights are chilly, and it gets colder (by South Indian standards) towards the morning. It is 4 AM, the air heavy with dew, and it is still dark all around. Our ears and nose are assaulted by the thunder of firecrackers and their distinct phosphorus stink; as we see several people running about shouting "*aane aane*". A herd of elephants (*aane*) has wandered into the nearby fields in Gollarahatti village. The sweet ragi millet, heavy on stalk and ready for harvest, is a great favourite of the elephants. Even when a crop is not part of their diet, the elephant herds cause serious damage to farmlands by just strolling through them. Desperate to contain the damage, people are trying their best to drive the elephants back into the national park using firecrackers and torches. The only representative from the forest department present there, the "watcher" Ayya, is handing out firecrackers to the others.

Ayya has spent much of his working life as a *hangami* (temporary) watcher in BNP. His wrinkled face has been tanned a deeper shade of brown by constant rambles in the sun. Dressed in mismatched and threadbare khakis and cheap "*hawaai*" flip-flops, he carries a small hatchet to clear the undergrowth. He has no other equipment—neither flashlight nor rifle but has

[d] *Gomaala* lands are village commons that often encroached upon by land grabbers using dubious means.

[e] Sankranti is a harvest festival celebrated in the month of January.

traversed all 20 km^2 of his forest "beat" countless times. He knows the landscape like the back of his hand. All our conversations with Ayya have been on the go, never sitting down. I am now accompanying him in one of his many routine walks around the forest. Ayya is monitoring the recently initiated excavation of an elephant-proof trench (EPT) around Gollarahatti.

L: Wasn't there an elephant raid recently in your village? It must be harder when it strikes close to home. What do you watchers usually do when you hear about a raid?

A: It used to be simpler. Earlier, elephants would steer shy of any signs of humans or human activity. All one had to do was switch on a flashlight to turn them away. They are getting bolder and bolder now. They actually charge when they see us now. You must have heard of this Rowdy Ranga. No elephant has given us as much trouble as him. For people here, I "am" the department, and the elephants are "my" responsibility. The villagers direct much of their anger and frustration at us watchers when their crops are raided. It is worse if someone is injured or killed by an elephant. Emotions run high. Even my own kinsfolk has abused me in such situations.

Then there are other mischief-mongers who stoke tensions further. I had recently caught two sand miners sneaking out a tractor load of riverine sand collected from the forest.[f] When Ramappa from Hudi village was killed by an elephant, the villagers held a demonstration. These sand miners incited the protest and used it as an opportunity to settle scores. They assaulted me and other forest officials who had come to assess the situation. In the pandemonium of the protests, it is difficult to pinpoint who is doing what.

L: What sort of training do you get to handle all of this?

A: *Ayyo,*[g] forget training—we aren't even paid regularly. Sometimes I go for up to 8 months without getting any salary. Still, I have to pay for the firecrackers from my own pocket sometimes. For the villagers, we are the forest-*avaru* (forest department people), but our own department neglects us. Last year another watcher fell into a ditch while trying to escape a charging elephant. He had to drag himself out with a broken leg for nearly 3 hours till he could get help. Nowadays, no matter where I am, I head towards the main road as it begins to get dark. As

[f] Riverine sand is a priced commodity and a key input fueling the "construction boom" in Bangalore. With limited reserves, illegal sand extraction has given rise to organized crime across India.

[g] A common refrain in south Indian languages.

you saw, we are called even in the middle of the night if there is fire or elephants, Sunday or festival, day or night, there is no respite for us. And what do I get for rushing there to help? Abuse and accusations. The ranger once ordered me to clear an encroachment in my village. The encroacher is a cousin of mine. I went to his house and told him to clear out. An argument ensued, and I gave him a gentle shove. He tripped and fell to the ground. Two weeks later, I get an arrest warrant for assault and "attempt to murder". I have been fighting the case in the *Kanakapura* sessions court for over 6 months now with zero assistance from the forest department. Tell me sir, if my own department forsakes me like this, where do I go?

L: If things are so bad, why do you continue as a watcher?

A: I was born around here. We do not have any lands to cultivate. I could not complete school also. All I know is this forest—every tree and stone. Either I had to migrate to the city and do menial jobs or stay here. This job made me stay back. Recently, an NGO honoured us all for our service.

Two years ago, I came up with a way to make sure the farmers that go into the forest to graze livestock are more careful in keeping the gates closed while returning. I proposed that whosoever leaves the gates open - will have to pay a fine of 400 rupees as donation to the village temple. It has been 2 years now and there are fewer instances of the gates being left open. When it does happen, people voluntarily make the donation.

Ayya has no officially issued flashlight or weapon, no health coverage or retirement benefits, no legal support from his department, and the working hours extend well beyond his official shifts. His situation is not unique. He is one of the many daily wage workers hired by the forest department, usually picked from the local community, who assume the role of institutional intermediaries (Mosse and Lewis, 2006). Even with advancements in wildlife monitoring technology and vehicles, "foot soldiers" such as the watchers are extremely critical to realize conservation goals. As Gubbi (2011) notes "there is no alternative to the old system of foot patrolling as most illegal activities such as poaching, or timber smuggling can only be detected through this mode of perambulation".

The typical forest watcher is constantly putting out fires, both literally and figuratively. He has to keep the fire-lines clear of undergrowth, he is expected to be a constable constantly patrolling the forests keeping an eye for poachers and other trespassers, he has to counsel the locals when there are crop raids or deaths, and has to appraise the situation on behalf of the department to determine "compensation", he has to monitor construction and "development" activities within the forest, and is also expected to be an authority on all aspects of the surrounding landscape. All this for the equivalent of 6 dollars a day, paid irregularly, and often in arrears. However, talk to people around here, and there is no sympathy for the forest watcher. The watcher is usually seen as corrupt, self-serving, and inefficient.

Over the last few years, the department has numerous projects going on inside the forest purportedly to reduce "man-animal conflicts". These projects usually concern structural measures to conserve water and contain elephants. Check dams are being built, ponds are being lined with concrete, new EPT are being dug and existing barriers are being reinforced with rubble walls. The watcher must monitor and document the quality of work and inputs. Each of this presents an opportunity for various rent-seeking actions. Further, in exchange for allowing shepherds to graze their sheep inside the forest, Ayya exacts one sheep from each shepherd just before the annual festival of "Bakr-Id[h]". Despite it being illegal to enter the national park under WPA, Ayya occasionally leads birdwatchers and photographers from the city on nature walks for a fee.

For Ayya, the forest is a source of ambivalence. It is a space he has grown up knowing intimately, he knows every pebble and leaf, but his relationship is hardly one of romantic idealism. Underpaid and overworked by his employers, abused, and mistrusted by the villagers who see him as an emblem of an ineffectual authority; he yearns for a stabler job, better hours, better pay, health benefits, better equipment, most of all to regain the trust of his kinsfolk and villagers. And he feels no compunctions resorting to his own common sense in determining right and wrong. He will let farmers graze their livestock in the forest, as long as they shut the gate on their way out. He will impose a fine, payable at the temple when they break this rule. He will guide urban trekkies through the forest to augment his meagre income and will take an occasional "gift" from the contractors working in the forest. Left to hold the fort with no resources, the guard must employ some assortment of techniques to survive in the face of the rapid changes around him.

[h] Bakrid is an Islamic festival celebrated usually in the month of August when there is high demand for mutton.

3.3 Ranga—The elephant

Rowdy Ranga is a celebrity in the hamlets around Hallikal-betta in the northern part of BNP. Talk to anybody here, they are bound to have a personal anecdote (real or imagined) to recount about Ranga. If you begin to google "Rowdy R", the search engine auto- suggests "Rowdy Ranga Elephant"—such is the reputation of this 35-year-old wild asian elephant (*Elephant maximus*). Like Anna and Ayya, Ranga has spent most of his life in and around BNP. It is unusual for wild elephants to be given names, but Ranga too, is also quite unusual.

It is 2 pm now. Ranga and his comrades are splashing around in a murky lake skirting the National Highway 948, about five kilometres radially from Akka's village and the BNP boundary. Unmindful of the sweltering sun, villagers, students, and factory workers are all jostling to catch a glimpse of the giant beasts while a couple of forest guards with no more than bamboo staffs for protection wait for reinforcements to come. The onlookers are busy shouting, shooting videos on their cell phones, and hurling stones towards Ranga and his posse. Twenty years ago, this locale would have been unrecognizable. Small village settlements each with one or more ponds, expanses of uncultivated land with sparse vegetation, and rain-fed farms interspersed with coconut groves made up this landscape that was cheek by jowl with BNP. Today, the unmistakable imprint of the city is visible everywhere. The villages are no longer agrarian, and the ponds have fallen into disuse. Farmlands have been converted to the Harohalli Industrial area, one of three industrial hubs that surround BNP. Engineering colleges of indifferent quality have mushroomed all over, handing out promises of remunerative jobs and social status.

After nearly 2 hours, a green jeep arrives carrying officials and guards from the range office. They light burst firecrackers to scare them off towards Bannerghatta. Only 2 days ago this group faced a similar ordeal on the other side of BNP. Ranga, until he was captured in 2017, was the leader of an all-male herd of 11 elephants that had gone "rogue". Mature males usually lead solitary lives, as elephant herds are matriarchal—made up of females and calves (Menon et al., 2017). Ranga and his band are an exception. Owing to their exploits outside BNP, be it habitual raiding of farmlands and maiming humans, they have been categorized as "rogue" in official records.

In BNP, Elephants have come to be identified as a keystone species as they help propagate 70 species of plants and are crucial to the stability of the national park (Bantalpad et al., 2017). These mega-herbivores forage and feed for 12 to 18 hours a day (Gray and Phan, 2011) and consume over 100 species of plants (Daniel, 1998). Conversely, they are nomadic and move long distances (Menon et al., 2017; Sukumar, 2006). While the average home range of an elephant herd in South India is 651 km^2, elephants in BNP have to make do with only 200 km^2 of forested land. Therefore, they come into intense contact and competition with humans (Lenin and Sukumar, 2008). There are two main elephant depredation seasons around BNP. Between October and January, when *Kharif* crops come for harvest and in the summer months between March and June when the forest is dry. Summer months also see trees such as jackfruit (*Artocarpus heterophyllus*), tamarind (*Tamarindus indica*) and mango (*Mangifera indica*) bear fruit causing elephants to move out of BNP (Bantalpad et al., 2017).

BNP today is claimed to be a site along one of the oldest migratory routes of the Asian Elephant in South India (Varma et al., 2009). While the picture one gathers from this information is one of an ancient association– local memories have a different story to tell. Most respondents interviewed, including forest department personnel, speak about the rarity of elephants in the BNP region till the 1990s. Even if they did maraud fields, the infrequency of the phenomenon made people more understanding. They would let the elephants feed undisturbed, and later worship their footprints, as auspicious omens, and harbingers of good agricultural yield. Peri-urban Bangalore has changed starkly since the 1980s, and if changing economic values are one factor, the increased frequency of the elephant raids and associated economic impact must have acted in tandem to change the locals' attitude towards the animal—what was once worshipped is now treated as vermin and kept away with electric fences and firecrackers. Lowe and Munster (2016) observed in their work on elephants in human dominated landscapes of South India, that as part of an array of coping mechanisms, elephants "cooperate, feel compassion, and care for each other to survive alongside humans. Others, often elephant bulls, have taken advantage of their strength, size, and body power to force their way through and across human dominated landscapes". Ranga and his band circumnavigate the city of Bangalore to Savanadurga and beyond, areas with no prior history of elephant distribution.

We ask a watcher if barriers like EPT and electric fences are effective in keeping away elephants, he says:

"Who can stop the elephants from going wherever they want? The electric fences seem to work for a while, but the elephants soon learnt to deal with them. They push down the stone posts of the electrified fence, column and cross safely walking over them. The EPTs pose a barrier when they are newly dug. But during the rains, they invariably get filled with mud, and then are clogged with quick growing weeds like lantana. The elephants can easily walk over them now. Even if the trenches and fences are in good repair, elephants manage to "escape" through other means. Villagers who go to the forest to graze their livestock are sometimes negligent and forget to shut the gate properly when they are returning. That is practically an open invitation to the elephants to feast on the crops. There is no single foolproof solution, and the elephants are not easy to contain. We do what we can and they will do what they will do".

Nobody here quite knows who first started addressing the rogue *makhna* (tuskless male elephant) as Rowdy Ranga.[i] Through various accounts of Ranga, what we are presented now is an anthropomorphized being endowed with personality, voice and agency. As Lowe and Munster (2016) point, elephants in "anthropogenic landscapes" of South India "learn to live their complex psychological, social, cultural, and gendered lives in close proximity to humans". They transcend the binaries of nature and culture where the animal is often seen as natural and the human-cultural (Locke and Buckingham, 2016). Ranga has had to deal with barriers of several kinds obstructing his earlier route and has been forced to muster a posse of other bull elephants. The dreaded poacher Veerappan[j] may be no more, but there are countless other sources that threaten Ranga and his kin's home and their way of life. He has broken into fields, destroyed electric fences, even killed a few humans, hoping that they will let him alone, but things continue as before, rather they worsen. Ranga was captured last year and sent to an elephant kraal (corral). The forest department is now in the process of taming him to induct into their fold for capturing other "rogue" elephants and carrying timber. The elephants are as much casualties of the changing statuses and land-use patterns of the peri-urban space, as are the farmer and the watcher.

3.4 Akka—A smallholder farmer

"All I know is to grow *ragi*. I married off my daughter from money saved from previous ragi yields. From the last five years, I am not even able to

[i] The Rowdy Ranga name is presumably based on a popular Kannada language movie character from the 1960s

[j] Koose Munisamy Veerappan was a notorious forest brigand who is believed to have poached 2000 elephants and killed over **120** people (including forest department staff) in forests over 100 km south of BNP. He died in a police encounter in 2004.

recover my investment on seeds and manure. Once, two years ago, an elephant herd came through and devastated my crop. They did not even eat them, just mowed through the field. It was all over in ten minutes. For fear of these vermin, I have stopped growing ragi and have planted *Nilagiri* (*Eucalyptus Spp*). My only sustenance now is through selling milk from my five cows to the dairy cooperative. We are not allowed to graze them in the forest, but what choice do we have? Where else can we go?"

Akka, a smallholder farmer from Sampigedoddi village is exasperated. She is voicing her grievance at a public consultation organized by the forest department in "high conflict villages". An ecologist from a prominent Bangalore based conservation NGO had just finished his ten-minute presentation on conserving BNP. Employing flowery jargon, his Kannada language talk had more English than Kannada, where the ecologist had "schooled" the locals on the need to adopt "eco-friendly agricultural practices". Among other things, the ecologist enthusiastically pointed that grazing inside BNP and, creating Eucalyptus (known locally as *Nilagiri*) groves have been outlawed by the state. Fast growing Australian Eucalyptus were first introduced in 1860s by the British colonial administration and received huge impetus in World Bank supported social forestry schemes of the 1980s to meet growing timber needs. Owing to mounting scientific evidence on the adverse impact of Eucalyptus cultivation on ground water, the Government has tried to regulate the creation of Eucalyptus plantations since 2011 culminating in the ban of 2017. While there is a consistent demand for timber from Bangalore's construction sector, farmers here grow Eucalyptus as Elephants do not consume them.

In that gathering of 150 people, Akka was the only woman to speak up. Many villagers requested officials to repair, maintain and reinforce forest barriers such as EPTs and electric fences, some also made requests to clear their long-pending dues that the department owed them as compensation following depredation events. Some events of the day resonated strongly with what Münster and Münster (2012) had observed in South India:

"With the growing neo-liberalization of agriculture, the economic stakes have increased for farmers. Agriculture has become more capital-intensive and speculative. Farmers take higher loans and higher risk today. For capitalist smallholders, a partial loss due to wildlife raids has come to represent a real threat to livelihood. Farmers respond to this situation by demanding a modernization of surveillance, defence and fencing technologies. Fortification of the forest boundary is thus more than a state led top-down process: farmers at the forest edge don't demand democratization of the forest but improved fortification".

Akka owns three acres of land and is from the peasant caste. Akka came to Sampige doddi 40 years ago, as a 16-year-old bride of Bhimanna, a smallholder farmer. In those days, it used to be a trek of 4 km through the forest trail connecting her natal village Ramapura to Sampigedoddi. It is now forbidden to take the route that traverses through the National Park. There is a much more circuitous route that involves changing three public-buses and more than 2 hours of travelling time. Many people, either to save time, or out of habit, prefer to sneak through the old trail. It was on one such surreptitious hike through the BNP (in 2011) that Bhimanna was killed by a wild elephant, leaving Akka to fend for herself and their three children.

Akka's small tin-roofed two room house is very close to Kardikal-Madeshwara elephant corridor, where BNP is at its narrowest width with less than 600 m. This makes the Sampigedoddi particularly vulnerable to depredation by elephants. Two people died each year between 1997 and 2010, due to elephant attacks. "What can we do? The elephants have been on this earth since long before we have. We are the real interlopers here. What sins did we commit that the very *Ganapati*[k] we worship took away my husband? An elephant dies, scores of officials and important people rush to the spot. But when a man dies because of an elephant attack, nobody turns up. Is man not even as important as that beast?" she wondered, as Bhimanna's black and white portrait stares pointedly from the wall. The villagers (like Akka) who live in the hamlets surrounding the national park are victims of the swirl of changing policies and the idiosyncratic implementation of those, ordered by far-away government officers, and deployed by the villagers own—watchers chosen from among the villagers in what is seen as a shocking defection.

Around Sampigedoddi, notwithstanding the proximity to the elephant corridor, two granite quarries, a stone crushing unit and a resort operate illegally. There is a conspicuous silence around them as they are operated by "big people" as per Akka. She adds "the department impose all kinds of unreasonable rules on farmers, and they cannot even keep their elephants under control. That day in the meeting, not a single person spoke about 'stone blasting'. Not even the officials!". She and other smallholders like her understand the Forest department's remit and want to be able to continue their agricultural and pastoral practices unharassed. They feel that

[k] Ganapati is an elephant-headed Hindu deity revered as "remover of obstacles".

everyone else seems to have power, or importance, or a say in matters related to the forest—the Forest department, "quarry people", politicians, real estate folk, environmentalists from the city, even the elephants. However, the very people who have lived in and around these forests, tending to them after a fashion, are now feeling disenfranchised.

"Before our forests became BNP, I would wake up to the chirping of birds from the forests. Now all you hear are the noises of man and—trucks thundering-rumbling by, dynamite exploding every now and then. It is just man's noise everywhere. As a result, wild animals are getting scattered. It brings tears to my eyes to think of days gone past".

These observations of Akka neither represent an uninformed villager's nostalgia for a simpler past, nor is sentimental about the vagaries of the present. Akka is from a demographic that is truly a victim of its location and circumstance. To add insult to injury, their own children are no longer interested in their land or their way of life. Many are selling land and moving to Bangalore city. They simply have no patience for the struggle to retain rights to live, farm and graze in these lands that Akka and her compeers feel so passionately about. As Akka says "If we have to persist here, we have to break one or the other law on a daily basis. There is no other option. That is the way it is".

4. Discussion

In the context of managing pluralism in forest ecosystem management, Babin and Bertrand (1998) ask: "How should the wide range of actors, uses and often conflicting and incompatible interests in the same area be managed?". This is an essential (potentially existential) question facing the governance of BNP, where a plurality of forest uses and forest institutions require diverse actors to operate through multi-layered processes of institutional bricolage. Without adequate inclusion, these processes can present significant challenges to government-led forest management and conservation efforts, which require an understanding of the existing plurality of institutions, social relationships and processes of local decision-making.

In this paper, we identify how some of the most marginalized and disempowered forest actors in BNP have become "bricoleurs"—making the best of their situation by creatively drawing on available capitals to "construct new governance arrangements and/or remould existing ones to meet their current objectives, circumstances and livelihoods" (Ingram et al., 2015) to uncover opportunities for more pluralistic forest conservation and management approaches. According to Vira et al. (1998), pluralism management efforts can include: recognizing and drawing on local knowledge, skills, and experience; understanding the motives of forest dependent actors; identifying and strengthening local institutional capacities; relating general environmental concerns to local contexts; and challenging assumptions about the nature and causes of local environmental problems. Babin and Bertrand (1998) identify the importance of shared decision-making and decentralized management strategies to facilitate joint decisions on objectives and the "rules of the game", however these objectives are very difficult to realize for disempowered actors in practice (Wollenberg et al., 2005). Further, as noted by Cleaver and De Koning (2015), the numerous dimensions of plurality in forest resource management mean that the institutional arrangements "are multi-stranded, overlapping and imbued with a variety of meanings and interests", raising questions concerning the extent to which processes of bricolage can be managed.

Key to forest ecosystem management outcomes is the need to better address situationality and complexity in multi-level governance processes. Situationality is defined as a "temporal and spatial condition in which actors find themselves, 'to the extent that they get challenged by it to act upon it" (Freire, 1970/2018). In the case of BNP, an ever-growing metropolis inches closer and closer to the peri-urban villages, threatening to engulf it in space and spirit. As the urban slowly percolates into the peri-urban, patterns of land use change, and the hamlets rather than being somewhat self-sufficient units that needed little truck with the big city, now become entirely oriented towards servicing the tastes and fancies of the city. The villages around BNP, as a unit, slowly turns into a shell of its former self, soon to become an attenuated appendage, supplying labour and other resources to the city. It is particularly bad for marginal actors inhabiting the BNP, since they are stuck between a rock and a hard place, as it were. Old-timers like Akka find themselves wedged between a city that they don't want to go into and a forest that they can't get into. They feel beleaguered on all sides and their children are no longer interested in their way of life, looking for easy employment in the city. For a farmer caught between the shrinking margins of a forest and the growing/ all-engulfing borders of a megacity, it is no surprise that Akka and other farmers in these villages feel beleaguered and have a sense that all elements are conspiring to disinherit them of the forests that they have been living around for a very long time. All of which act in tandem to drive these forest agents to resort to a assorted set of bricolage practices, or "what local actors actually do in practice" (De Koning, 2011).

Each of the actors in our ethnographic narrative are not passive instruments, they are fully engaged in exercising their agency to cope with the pressures, prescriptions and proscriptions that bind them—be it Akka, Ayya, the guard or Ranga the elephant, we can see their actions as transgressions of prevalent norms. Transgressions not as willful acts of thumbing the nose at authority, but as coping strategies to maintain their own survival. As Akka said, "If we are to survive here, we are breaking one or the other rule everyday". This statement captures the disconnect between the formal institutional mechanisms that declares a forest as a National Park, and the consequences on the life of a smallholder farmer that lives around it.

Bricolage practices are mediators between introduced institutions and the existing logic of practice, organically structured by "fragments of norms, regulations, practices, path-dependencies, artifacts, interventions, and incidental factors derived from history" (Jayaprakash and Hickey, 2019). According to Cleaver and De Koning (2015), in the process of institutional bricolage, institutional components from "different origins are continuously reused, reworked, or refashioned to perform new functions" (Cleaver, 2001; De Koning, 2011; Galvan, 1997; Sehring, 2009). For instance, the adivasi practice of *Kalluseve,* despite not being practiced in the recent past, is borrowed by Anna to strengthen their territorial claims under the Forest Rights Act. In processes of borrowing and adaptation the distinction between what is modern and what is traditional becomes blurred, tradition becomes reinvented. Some examples of bricolage practices in BNP are given in Table 1.

Anderson et al. (1998) note that pluralism recognizes the inevitable existence of differing (often conflicting) positions on questions of substance and describes the dynamic interplay between actors where "distinct groups are actively autonomous and independent, but often interdependent, with legitimate claims and different positions on critical substantive issues". Our study provides a window into the lived complexity of this dynamic in a peri-urban landscape experiencing rapid transformation, and the importance of institutional bricolage to local forest management and conservation outcomes. As noted by Cleaver and de Koning (2015), the processes involved in bricolage can generate contextual meaning in institutions and "help to explain their endurance through adaptation over time". In forest governance scholarship, greater attention to the contextually significant factors affecting institutional development remains necessary to connect possible solutions to the lived experiences of communities. More work is needed to "flesh out" the specific forms of institutional bricolage occurring within different contexts with a view to magnifying the voices of marginalized actors (Po et al., 2019) and

Table 1 Examples of bricolage practices being employed by marginalized forest actors in BNP.

Actor	Practice
Anna—*the Iruliga Adivasi*	• Narrates the origin story of his tribe to establish their sacred connection with forest • Invokes traditional practices such as *Kalluseve* for land claims
Ayya—*the watcher*	• Indulges in various rent seeking actions to augment income • Permits villagers to graze in order to buy their goodwill and cooperation • Levies fine to grazers that keep the gates open • To ensure conformity, the fine is made payable at the village temple, where the temple acts as a legitimizing symbol
Ranga—*the elephant*	• Increasing night time raids as adaptation to living in anthropogenic landscapes • Crossing the barriers such as electric fences through active learning
Akka—*a smallholder farmer*	• Cultivating the outlawed Eucalyptus and grazing inside the national park—both as survival strategies

informing pluralistic forest governance strategies. In particular, research focusing on the mundane practices of forest actors (what individuals do in their everyday work) that both construct and resolve institutional complexity (Smets and Jarzabkowski, 2013) is likely to yield important insights for institutional design and adaptation. Fundamental conceptual questions pertain to the function of bricolage in forest management and conservation are also relevant, such as: Is bricolage an end or a means to an end? Is it a short-term strategy? Can bricolage be facilitated? How can bureaucratic arrangements utilize the creative energy that leads to bricolage? Under what conditions do bricolage practices lose their utility? These questions offer a useful starting point for further empirical research aiming to inform pluralistic regulatory strategies, support inclusive and collaborative decision-making processes, and enhance ecosystem governance outcomes in contexts of rapid environmental change.

5. Conclusion

The ethnographic account presented in this chapter highlights some of the practical challenges facing forest ecosystem conservation and management by mapping specific modes through which four marginalized agents navigate complexity and intersect with each other and forest institutions. The results illustrate the potency of institutional bricolage, wherein agents renegotiate their space in the landscape by actively drawing from available resources to ensure workable institutional arrangements. Understanding how diverse institutions are navigated by marginalized agents is key to understanding formal policy implementation and effectiveness. Shifting the focus to actors and their everyday practices can provide a more accurate account of how a wide plurality of institutions function locally to inform governance strategies. As shown in the cases of Anna, Ayya, Akka and Ranga, formal institutions have a hard time influencing the behaviour of local actors. Actors' behaviour is situated in and follows a logic of practice, that is evolved contextually. Much of the impact of the institutions thus depends on these local actors or bricoleurs and the way they reshape institutions. For the agents in our ethnographic account resisting the pressure to be engulfed by the urban, the green space is a grey area, a poorly defined, contested "no man's land" that every man seems to want a piece of. These actors and their actions are the grey area of forest management and conservation in practice, offering a partial glimpse into the lived complexity of multi-level ecosystem governance and challenging governments to meaningfully grapple with plurality to enhance collective outcomes.

Acknowledgements

We are grateful for the organizational support of Murali Sreenivasa of Ramakrishna Mission, Shivanahalli and sincerely thank the research participants, communities and Karnataka Forest Department who supported, and donated their valuable time and inputs to our study. Lingaraj G Jayaprakash would like to acknowledge the funding support he received through the Government of Quebec's Nature and Technology Fund (FRQNT) Merit Scholarship for Foreign Students (2014-18) and the William Dawson Scholar Award, McGill University (G.M. Hickey). The funding source had no involvement in the conduct of this research.

References

Anderson, J., Clément, J., Crowder, L.V., 1998. Accommodating conflicting interests in forestry-concepts emerging from pluralism. UNASYLVA-FAO, pp. 3–10.

Babin, D., Bertrand, A., 1998. Managing pluralism: subsidiarity and patrimonial mediation. UNASYLVA-FAO, pp. 19–25.

Bantalpad, M., Gayathri, A., Krishnan, A., 2017. Seasonal crop raiding of fruit trees by Asian elephants: an insight into foraging preferences from croplands abutting Bannerghatta National Park, Bengaluru, Karnataka, India. Asian Journal of Environment & Ecology 2 (2), 1–12.

Bazeley, P., Jackson, K., 2013. Qualitative Data Analysis with NVivo, second ed. Thousand Oaks.

Benjaminsen, T.A., Lund, C., 2002. Formalisation and informalisation of land and water rights in Africa: an introduction. Eur. J. Dev. Res. 14 (2), 1–10.
Cleaver, F., 2001. Institutional bricolage, conflict and cooperation in Usangu, Tanzania. IDS Bulletin 32 (4), 26–35.
Cleaver, F., 2002. Reinventing institutions: bricolage and the social embeddedness of natural resource management. Eur. J. Dev. Res. 14 (2), 11–30.
Cleaver, F., 2012. Development Through Bricolage. Earthscan, London and New York.
Cleaver, F., De Koning, J., 2015. Furthering critical institutionalism. Int. J. Commons 9 (1).
Convery, I., Corsane, G., Davis, P., 2014. Making Sense of Place: Multidisciplinary Perspectives. vol. 7 Boydell & Brewer Ltd.
Daniel, J.C., 1998. The Asian Elephant: A Natural History. Natraj Publishers.
De Koning, J., 2011. Reshaping Institutions Bricolage Processes in Smallholder Forestry in the Amazon. Wageningen University and Research.
Douglas, M., 1986. How Institutions Think. Syracuse University Press.
Etzold, B., Jülich, S., Keck, M., Sakdapolrak, P., Schmitt, T., Zimmer, A., 2012. Doing institutions. A dialectic reading of institutions and social practices and its relevance for development geography. Erdkunde, 185–195.
Galvan, D., 1997. Institutional syncretism and the articulation of modes of production in rural Senegalese land tenure relations. Afr. Rural Urban Stud. 4 (2–3), 59–98.
Gopalakrishna, S.P., Kaonga, M.L., Somashekar, R.K., Suresh, H.S., Suresh, R., 2015. Tree diversity in the tropical dry forest of Bannerghatta National Park in Eastern Ghats, Southern India. Eur. J. Ecol. 1 (2), 12–27.
Gray, T.N., Phan, C., 2011. Habitat preferences and activity patterns of the larger mammal community in Phnom Prich Wildlife Sanctuary, Cambodia. Raffles Bull. Zool. 59 (2), 311–318.
Gubbi, S., 2011. Watchers and guards — unsung heroes of wildlife conservation. https://www.conservationindia.org/articles/watchers-and-guards-unsung-heroes-of-wildlife-conservation/comment-page-1.
Gupta, A., Ferguson, J., 1997. Culture, Power, Place: Explorations in Critical Anthropology. Duke University Press.
Guruprasad, S.L., Ningaiah, N., Gangadhar, M.R., 2013. Indigenous knowledge on medicinal plants among the Iruliga tribal population of Western Ghats areas, Karnataka, India. J. Anthropol. 9 (1), 195–203.
Ingram, V., Ros-Tonen, M., Dietz, T., 2015. A fine mess: Bricolaged forest governance in Cameroon. Int. J. Commons 9 (1).
Jayaprakash, L.G., Hickey, G.M., 2019. Mistaking the map for the territory: what does the history of Bannerghatta National Park, India, tell us about the study of institutions? Soc. Nat. Resour. 32 (12), 1433–1450.
Johnson, C., 2012. Bricoleur and bricolage: from metaphor to universal concept. Paragraph 35 (3), 355–372.
Kashwan, P., 2017. Inequality, democracy, and the environment: a cross-national analysis. Ecol. Econ. 131, 139–151.
Kirksey, S.E., Helmreich, S., 2010. The emergence of multispecies ethnography. Cult. Anthropol. 25 (4), 545–576.
Kohn, E., 2007. How dogs dream: Amazonian natures and the politics of transspecies engagement. Am. Ethnol. 34 (1), 3–24.
Kottak, C., 2006. Mirror for Humanity. McGraw Hill, New York, NY.
Lenin, J., Sukumar, R., 2008. Action plan for the mitigation of elephant-human conflict in India. Transformation 10, 35.
Locke, P., 2017. Elephants as persons, affective apprenticeship, and fieldwork with nonhuman informants in Nepal. J. Ethnogr. Theory 7 (1), 353–376.

Locke, P., Buckingham, J., 2016. Conflict, Negotiation, and Coexistence: Rethinking Human-Elephant Relations in South Asia. Oxford University Press.
Lowe, C., Münster, U., 2016. The viral creep: elephants and herpes in times of extinction. Environ. Hist. 8 (1), 118–142.
Lynch, O.J., 1998. Law, Pluralism and The Promotion of Sustainable Community-Based Forest Management. UNASYLVA-FAO, pp. 52–56.
Manjunatha, B.R., Annapurna, M., Shylaja, K., 2014. Socio-economic life of Iruliga tribe in Karnataka. Cult. Anthropol. 10 (1), 69–71.
Menon, V., Tiwari, S.K., Ramkumar, K., Kyarong, S., Ganguly, U., Sukumar, R., 2017. Right of Passage: Elephant Corridors of India, second edition. Wildlife Trust of India, New Delhi.
Mosse, D., Lewis, D., 2006. Theoretical approaches to brokerage and translation in development. In: Development Brokers and Translators: The Ethnography of Aid and Agencies. Kumarian Press, USA, pp. 1–26.
Münster, D., Münster, U., 2012. Consuming the forest in an environment of crisis: nature tourism, forest conservation and neoliberal agriculture in South India. Dev. Change. 43 (1), 205–227.
Po, J.Y., Saint Ville, A.S., Rahman, H.T., Hickey, G.M., 2019. On institutional diversity and interplay in natural resource governance. Soc. Nat. Resour. 32 (12), 1333–1343.
Rajeev, B.M.T., 2002. Wildlife Management Plan for Bannerghatta National Park, 2003–04 to 2007–08. Government of Karnataka, India, p. 175.
Real, P.D., 2008. Slums do stink: artists, bricolage, and our need for doses of "real" life. Art J. 67 (1), 82–99.
Sehring, J., 2009. Path dependencies and institutional bricolage in post-soviet water governance. Water Altern. 2 (1), 61–81.
Singh, V.V., 2008. Wildlife Management Plan for Bannerghatta National Park 2008–2013. Forest Department, Government of Karnataka, Bangalore, India.
Smets, M., Jarzabkowski, P., 2013. Reconstructing institutional complexity in practice: A relational model of institutional work and complexity. Human relations 66 (10), 1279–1309.
Sukumar, R., 2006. A brief review of the status, distribution and biology of wild Asian elephants Elephas maximus. Int. Zoo Yearbook 40 (1), 1–8.
Varma, S., Anand, V.D., Gopalakrishna, S.P., Avinash, K.G., Nishant, M.S., 2009. Ecology, conservation and management of the Asian elephant in Bannerghatta National Park, southern India. A Rocha India/ANCF: Asian Elephant Ecology and Conservation Reference Series 1, 13–52.
Vira, B., Dubois, O., Daniels, S.E., Walker, G.B., 1998. Institutional pluralism in forestry: considerations of analytical and operational tools. iUNASYLVA-FAO, pp. 35–42.
Wollenberg, E., Anderson, J., Edmunds, D., 2001. Pluralism and the less powerful: accommodating multiple interests in local forest management. Int. J. Agric. Resour. Gov. Ecol. 1 (3–4), 199–222.
Wollenberg, E., López, C., Anderson, J., 2005. Though All Things Differ: Pluralism As a Basis for Cooperation in Forests. CIFOR.

Further reading

De Koning, J., Cleaver, F., 2012. Institutional bricolage in community forestry: an agenda for future research. In: Forest-People Interfaces. Wageningen Academic Publishers, Wageningen, pp. 277–290.
Freire, P., 2018. Pedagogy of the Oppressed. Bloomsbury Publishing, USA.

CHAPTER FIVE

Posthumanist Pluralities: Advocating for nonhuman species' rights, agency, and welfare in ecosystem governance

Bastian Thomsen[a,b,c,*], Thomas Cousins[a,d], Kellen Copeland[c], Jennifer Thomsen[f], Sarah Coose[g], Abigail Mensah[g], Samuel R. Fennell[b], Anant Deshwal[h], Jose Guzman[b], Shelby Copeland[b], Dane Nickerson[b], Max Duggan[b], Amy Schneider[b], Marley Taylor[b], Asier Hernandez Saez[b], and Andrew Gosler[a,e]

[a]School of Anthropology and Museum Ethnography, University of Oxford, Oxford, United Kingdom
[b]Human Dimensions of Natural Resources Department, Colorado State University, Fort Collins, CO, United States
[c]Environmental Studies Graduate Program, Oregon State University, Corvallis, OR, United States
[d]Department of Sociology and Social Anthropology, Stellenbosch University, Stellenbosch, South Africa
[e]Edward Grey Institute of Field Ornithology, Department of Zoology, University of Oxford, Oxford, United Kingdom
[f]College Veterinarian Medicine and Biomedical Sciences, Colorado State University, Fort Collins, CO, United States
[g]Geography and Tourism Studies Department, Brock University, St. Catharines, ON, Canada
[h]Biology Department, Bradley University, Peoria, IL, United States
*Corresponding author: e-mail address: bastian.thomsen@anthro.ox.ac.uk

Contents

Advances in Ecological Research, Volume 66
ISSN 0065-2504
https://doi.org/10.1016/bs.aecr.2022.04.004

Abstract

Ecosystem governance is a highly contested issue where competing human stakeholder groups wield unequal power over nonhuman species. Pluralistic approaches to ecosystem governance consider a diversity of voices and disciplines to apply translational ecological knowledge to policy. However, the rights, agency, and welfare of nonhuman species – at the abstract (macro) and individual (micro) levels – are rarely considered, distancing nonhumans from human interests. This 'othering', results in nonhumans being treated as subservient to humans in power relations, which can result in poor welfare and even death. This study draws on multispecies ethnography, posthumanism, multispecies livelihoods, decolonial ecology, and translational ecology literature to critically review current paradigms in ecosystem governance and questions how to represent nonhumans equitably in ecosystem governance. This essay also considers qualitative data collected during 15 months of a multi-sited, multispecies ethnographic study that explored wolf-human (*Canis lupus*) relations in the western United States during the Trump Administration. A pluralistic posthumanist wildlife-human coexistence framework for translational ecology is presented that emphasizes nonhumans' rights, welfare, and agency in ecosystem governance.

Never doubt that a small group of thoughtfully committed citizens can change the world. Indeed, it's the only thing that ever has.

Margaret Mead

1. Introduction

Ecosystem governance is too often anthropocentric and centers on conflict between competing human stakeholder groups who speak for and about wildlife, without their consent. Though difficult ontological and ethical questions are at stake in arguments for nonhumans' welfare, rights, and agency (Hamilton and Taylor, 2017; Fennell & Sheppard, 2021; Thomsen, 2021), humans have a moral responsibility to consider these issues in any potential decision that affects nonhumans' interests. By attending to the zoological and ecological uncertainties that inform public debates and policy about wolf-human coexistence in a contemporary U.S. context, an ethic of coexistence can be seen to emerge through highly contentious debates about the cultural and political context of not just wolf, but wildlife in general.

This study takes an anthropological lens to multidisciplinary pluralism in wildlife-human coexistence studies. We also present a posthumanist, multispecies livelihoods approach to translational ecology based in a plurality of animal ethics (detailed below). Sandbrook et al. (2013) state that:

> *There is now widespread agreement that social research is relevant to conservation (Mascia et al., 2003). Extensive efforts have been made to improve communication between natural and social scientists interested in conservation (e.g., Adams, 2008; Büscher and Wolmer, 2007). However, this communication has been described as a 'dialogue of the deaf' (Agrawal and Ostrom, 2006) and is challenging personally and professionally (Brosius, 2006; Campbell, 2005; Fox et al., 2006)...we believe that the role and place of social research in conservation remains a major source of misunderstanding, miscommunication, and contention among conservation researchers (p. 1487).*

Pluralism "emphasizes more ephemeral and transient interdisciplinary connections and underscores the autonomy of the disciplines with respect to one another" and is a "long-standing and complex issu[e] in the philosophy of science" (Persson et al., 2018, p. 1). Though the theories, terms, methods, and analyses used in this paper may be unfamiliar to some readers, we take careful consideration to unpack these concepts from the social sciences, and anthropology more specifically. Each theory used informs the framework (see findings and discussion), and though some readers may not agree with our assessment of certain topics (e.g., nonhuman animal genocide), we do not mean to denigrate any particular human who has suffered from any form of violence. We do challenge natural scientists, however, to openly consider our animal ethics arguments grounded in anthropology and developed from a multispecies ethnographic study, to rethink how we treat each particular individual, no matter the species. Our posthumanist, multispecies livelihoods approach complements compassionate conservation from biology (see Santiago-Avila and Lynn, 2020). The aim of this study is to engage with natural scientists, and proffer insights into a growing, albeit minority interest, across all disciplines to consider the rights, welfare, and agency of particular beings, no matter the species, in a step toward multispecies justice (see Celermajer et al., 2021; Thomsen et al., 2020b, 2021a).

Just as it can be challenging for some social scientists to read conservation biology literature, we appreciate that the language and positionality we present here may seem obscure or inaccessible. To help bridge the natural and social sciences while maintaining true to our anthropological roots in an act of interdisciplinary pluralism, we interrogate common ethical stances and terminology employed to distance humans from other species, implicitly and explicitly, which shape and reinforce socially constructed norms in wildlife-human relations. We confront the effects of speaking for and about wildlife in the abstract to proposition a pluralistic ethical framework based in multispecies livelihoods (Copeland, 2020, 2021; Thomsen et al., 2021a,b;

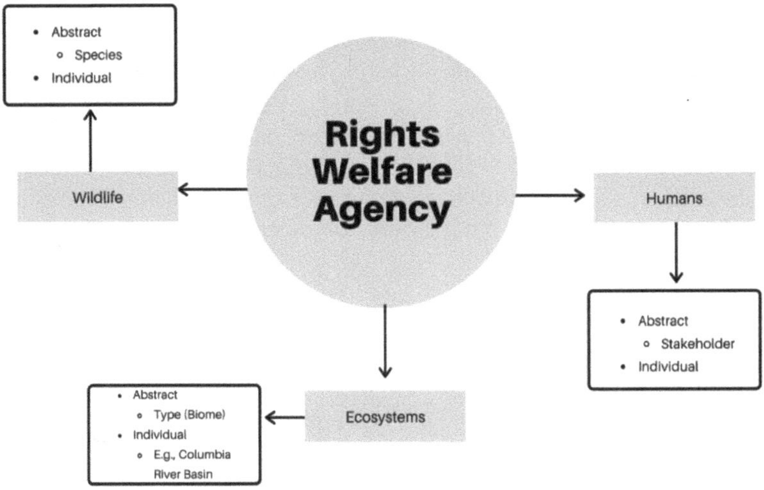

Fig. 1 Posthumanist wildlife-human coexistence framework.

Thomsen and Thomsen, 2020), and translational ecology, which theorizes how to inform science-based decision making in environmental policy, governance, and management (see Enquist et al., 2017; Hallett et al., 2017; Thomsen et al., 2021a). The posthumanist wildlife-human coexistence framework (see Fig. 1 below) is premised on a critique of human-dominated power relations that considers nonhumans' livelihoods and capacity to act in social-ecological systems (SES).

Though most wildlife-human conflict and coexistence studies center on charismatic species such as the gray wolf (*Canis lupus*), wolf is a useful species with which to conceive a wildlife-human coexistence conceptual framework due to its polarizing effects on various stakeholder groups that are steeped in Euro-American history, culture, and politics (Blécourt, 2007; Kunstler, 1991; Lappalainen, 2019; Thomsen, 2021). Wolf politics came to the foreground in Colorado in November 2020, as more than 3,000,000 Coloradans voted narrowly to pass a state ballot initiative, Proposition 114 (Brasch, 2020). The initiative mandates Colorado Parks and Wildlife (CPW) to use science-based evidence and solicit community input to reintroduce wolf to the "Western Slope" of the state by December 31, 2023. Voter demographics were largely split (generally) between pro-wolf, "liberal" (in the U.S. sense of the word) urbanites and anti-wolf, conservative ruralites (Sponarski et al., 2013). The vote was much closer than previous public opinion polls had suggested (Meadow et al., 2005; Niemiec et al., 2020; Pate et al., 1996). For example, researchers based at Colorado

State University conducted a survey in January 2020 of 734 participants, representative in terms of age, gender, and geography throughout Colorado, and found that approximately 84% supported wolf reintroduction (Niemiec et al., 2020). Studies are currently underway to try and determine what may have led to differences between prior public opinion surveys and actual ballots cast.

Starting in December 2020, CPW's Wildlife Commission began the process to develop a plan to implement Proposition 114. In two initial public meetings streamed live on YouTube, one in December 2020 and one in January 2021, commissioners emphasized that diverse stakeholder voices should be included in the process to create an equitable plan that is satisfactory to varied human stakeholders. This study examines the perspectives and concerns of a range of stakeholders in the western U.S. concerning the reintroduction and repatriation of gray wolf in the lower 48 U.S. states since the mid-1990s, and specifically Colorado following Proposition 114. The central question addressed includes how to equitably represent non-humans' rights, welfare, and agency in the governance of 'other' species to foster wildlife-human coexistence?

2. Theoretical background

Each of the following theories informs the posthumanist, multispecies livelihoods framework presented in the findings and discussion section. We commence with social-ecological systems (SES) to emphasize how a unification attempt to bridge the sciences can create greater obstacles, and contribute to scientific imperialism, which postcolonial studies helps to counter. Perrson state that "Unificationism seeks unification and perceives disciplinary boundaries as conventional, representing no long-term obstacle to progress…Both approaches [unificationism and pluralism] have their merits and pitfalls. Unification runs the risk of scientific imperialism, while pluralism can result in insurmountable barriers between disciplines (p. 1). Following the discussion of SES and postcolonial studies, we provide a brief account of posthumanism, which is a relatively new philosophical positioning that champions "more-than-human" rights, agency, and welfare (see Thomsen et al., 2021a), for its influence on multispecies livelihoods. This sections concludes with translational ecology before moving on to the anthropological methods that informed this study and are common in the discipline (see Creswell and Creswell, 2017).

2.1 Social-ecological systems and postcolonial studies

Social-ecological systems emerged in the past three decades as an interdisciplinary attempt to integrate the natural and social sciences and move beyond the commodification and dominance of nonhumans in human-environment systems (Berkes et al., 2000; Fabinyi et al., 2014). Social-ecological systems is considered "a system that includes societal (human) and ecological (biophysical) subsystems in mutual interactions" (Harrington et al., 2010: 2773). Holzer et al. state that recent calls in the literature "suggest that effective conservation policy and management require multiple and integrated forms of knowledge, including knowledge of ecosystems and their function, and the understanding of human societies, which interact with and depend on those systems" (2018, 808). However, SES has been criticized for relying too heavily on the power of institutions and a macro-level focus on the "environmental resilience paradigm" that does not adequately attend to (human) social dynamics (Fabinyi et al., 2014).

Social-ecological systems' growing popularity in theory and practice inspired several efforts to redress social questions of the framework, such as "whose resiliency, whose agency, or whose power?" (Béné et al., 2012; Matin et al., 2018; Olsson et al., 2015). Fabinyi et al. argue that "while understanding the diverse perspectives of different interest groups in SES is important, power relations mean that these different voices can be marginalized or privileged, easy to identify or invisible" Fabinyi et al., 2014: 7). This marginalization of certain groups such as Indigenous, women, ethnic or racial minorities aligns with postcolonial studies (Roy, 2015; Arcari, 2019; Pulido and De Lara, 2018). Postcolonial studies amalgamate social and political ecology, along with ecofeminism to consider the entangled embodiments of historical (post)colonial legacies, systems of oppression and violence, and environmental degradation to unpack new pathways toward socioenvironmental justice (Ferdinand et al., 2020).

Similarly, postcolonial approaches focus on groups who are made subservient to another based on power dynamics (Galastri, 2018; Spivak, 1988; Thomas, 2018; Thomsen, 2021), where "the experiences of subjects [are] marginalized by scientific and political structures of representation" (Watson, 2011: 55). Some postcolonial scholars have embraced nonhumans as postcolonial subjects (see Mitchell, 2002; Arcari, 2019; Thomsen et al., 2021a); Roy argues that nonhumans should be understood as historically oppressed subjects:

> *We have seen that nonhuman animals had an indelible effect on imperial history, and that their subalternity was manifest in at least three distinct ways: they were victims of imperial violence; they were products of imperial regimes of subjectification; and they were usually marginalized in imperial historiography (2015: 70).*

Norton examines the plight of postcolonial subjects in the 'Atlantic World' through a technological framework to argue that "since at least Aristotle technē [technology] was used invidiously by elites to refer to 'the lowly knowledge of how to make things or produce effects, practiced by animals, slaves, and craftspeople' and in contrast with prestigious episteme of 'learned' men invested in abstractions" (2017: 25). This distancing of marginalized subjects cultivates the production of abstract (macro-level) discourse, which stimulates rationalities that propagate violence, oppression, and discordance concerning nonhumans' agency. In the case of U.S. wolves, they experienced similar systematic persecution and murder to other marginalized groups (Fritts et al., 2003; Lynn, 2010), as targets of a eugenic logic employed to justify wolf's extirpation in the U.S. from the 1860s–1940s (Spiro, 2009). Spiro details how Eugenics was used in the case of wolf.

During the study, nearly every interlocutor (post-positivist term for informant or respondent), pro or anti-wolf supporter, could be said to be guilty of some form of "speciesism". Ryder coined the term speciesism in the 1970s to mean discrimination that favors certain species or groups of species over other individuals (Horta and Albersmeier, 2020). "Anti-wolfers" perpetuated European-American cultural stereotypes of wolf as villain to be feared (Lappalainen, 2019; Thomsen et al., 2021a), where some argued that wolves are only good for the '3 S's': shoot, shovel, and shut-up' (Fritts et al., 2003). However, several "pro-wolfers" were quick to launch a "what about" argument concerning the purported crimes of other species', such as "What about bears or cougars? They kill more cattle than wolves ever do." This othering displaces blame without addressing the real fears of livestock ranchers who want to protect their cattle and livelihoods. Posthumanism complements efforts to challenge existing power relations and provides a "more-than-human" theoretical grounding for an equitable, pluralistic (multi-ethics application), wildlife-human coexistence framework.

2.2 Posthumanism

Posthumanism is a post-modern philosophical paradigm that challenges the separation of nature and society to consider all species and individuals equally in wildlife-human relations (Cohen, 2019; Fabinyi et al., 2014; Thomsen, 2021). Cielemęcka and Daigle problematize the concept of human exceptionalism in sustainability to "focus on developing environmentally ethical projects suitable for our times…[and] offer a concept of posthuman sustainability that decenters the human, re-positions it in its ecosystem and, while remaining attentive to difference, fosters the thriving of all instances of life" (Cielemęcka and Daigle, 2019: 67). Posthumanism deconstructs humans' dominance over nature to theorize a 'more-than-human' wildlife-human relations positioning and posits a paradigm of coexistence (Cohen, 2019; Thomsen et al., 2021a,b). However, Braidotti challenges this theoretical standing for its reliance on working within current power structures:

> *The posthuman subject is not postmodern, because it does not rely on any anti-foundationalist premises. Nor is it deconstructivist, because it does not function within the linguistic turn or mediation. Not being framed by the ineluctable powers of signification, the posthuman subject is consequently not condemned to seek adequate representation of its existence within a system that is constitutionally incapable of granting due recognition (2016: 24).*

Whether posthumanism can be displaced or isolated from postmodernism or human exceptionalism is beyond the scope of this study, but we echo Braidotti's efforts to confront systematic oppression, violence, and suffering of nonhumans (Fennell and Thomsen, 2021).

Posthumanism nonetheless challenges anthropocentric notions of superiority, and the power humans exert over nonhumans by way of direct action, such as environmental degradation (Jackson et al., 2020; Min et al., 2020) or the killing of nonhumans for population "control" and "management" (Jiguet, 2020; Castelblanco-Martínez et al., 2021), and indirectly through quotidian norms such as the use of single-use plastics that are discarded and consumed by marine life (Good et al., 2020). We caution, however, that it would be reductive to equate all humans as equally responsible or homogenic in the production of nonhuman (flora and fauna) degradation, which is unmistakably entangled with late-stage global capitalism and dominated by patriarchal, often Western, authorities of the "Global North" who commit "state-corporate crime" (Kramer and Michalowski, 2012). In addition, the uneven distribution of human responsibility is shaped by histories of colonial conquest. Corporate and nation-state hegemonies' anathema of marginalized multispecies groups (Gonzales, 2017) distances subservient subjects to reproduce and perpetuate violence, oppression, and subjugation in favor of profit and power (Alcadipani and de Oliveira Medeiros, 2019; Faber and Schlegel, 2017).

Nor does posthumanism divorce humans from the environment in favor of everything more than human (Wolfe, 2010; Bolter, 2016; Copeland, 2021); rather, it engages multispecies livelihoods and technologies in an assemblage of lifeways (Copeland, 2021; Thomsen, 2021; Thomsen et al., 2021a,b; Thomsen and Thomsen, 2020).

A posthumanist wildlife-human coexistence framework is paramount in theory and practice if wildlife is to persist, as more than 60% of the world's wildlife has perished in the last 50 years (Grooten and Almond, 2018). Manfredo et al. state that humans must "consider wildlife as part of their broader social community, deserving of rights and caring treatment" (2020: 3). But how can humans move beyond caring for nonhumans in the abstract, to create meaningful protections for individuals' interests in practice?

2.3 Multispecies livelihoods and translational ecology

Multispecies livelihoods is a posthumanist approach to wildlife-human coexistence that attends to the rights, welfare, and agency of humans and nonhumans (Copeland, 2020, 2021; Thomsen et al., 2021a; Thomsen and Thomsen, 2020). Thomsen et al. define multispecies livelihoods as:

> *The right for human and nonhuman animal species to not only exist but to secure the necessities of life in a manner that does not infringe on another species' right to live, except for sustenance hunting or legitimate safety concerns to foster optimal conditions for wildlife-human coexistence (2021a: 4).*

A multispecies livelihoods approach has been applied to a range of wildlife-human interactions, particularly when wildlife encroaches on human livelihoods, or vice versa (Copeland, 2020; Thomsen et al., 2021a,b; Thomsen and Thomsen, 2020). Copeland (2021) problematize fisheries management in the North Pacific for anthropocentric practices that value certain "commercially important" species (e.g., salmon [*Salmo salar*]) while diminishing the rights and ecosystem functions of competitor species' (e.g., seals [*Pinnipedia*] and sea lions [*Otariinae*]).

Thomsen et al. (2021b) consider the role of wildlife ecotourism as a conduit for sustainable development and the conservation of the critically endangered Marsican (or Apennine) brown bear (*Ursus arctos marsicanus*). Similarly, Thomsen (2021) visited five wolf ecotourism operators, including four "wolf sanctuaries" and a protected area, Yellowstone National Park (YNP). Thomsen's findings suggested that no WTA—including, and perhaps especially, YNP—was an ideal model of wildlife ecotourism, as wolves did not have the right to exist free of exogenous threats, and could be legally shot, trapped, or harvested if they left the park's artificial boundaries (Paybarah, 2021). Even though YNP is an ethically complex example of "protected area" nature-based tourism, non-captive environments are optimal for the application of multispecies livelihoods frameworks. While wildlife ecotourism is a practical human livelihood to apply a posthumanist multispecies livelihoods approach in practice for its focus on wildlife-human interactions, this study responds to Thomsen et al.'s (2021a) call for a broader posthumanist wildlife-human coexistence framework that could materialize through translational ecology in policy, discourse, scholarship, and governance.

Translational ecology addresses highly complex environmental issues by using cross-disciplinary research concerning "the sociological, ecological, and political contexts of an environmental problem" (Enquist et al., 2017: 541). It champions evidence-based interventions in ecosystem governance that will "inform and improve decision making for environmental management and conservation" (Enquist et al., 2017: 541). Its most promising theoretical contribution is perhaps its ability to account for the "deliberate engagement with end-users and an understanding of the social and cultural contexts in which a research project functions" (Wall et al., 2017: 551). Though collaboration between stakeholders in ecosystem governance can be a mercurial process, Bodin et al. (2020) suggest that developing trust and managing risk may enhance cooperation within stakeholder networks. Hallet et al. state that:

> *Interest in translational ecology (TE) – a research approach that yields useful scientific outcomes through ongoing collaboration between scientists and stakeholders – is growing among both of these groups. Translational ecology brings together participants from different cultures and with different professional incentives (2017: 578).*

While translational ecology may conflict with "mainstream basic scientific research" for its applied approach and engagement with stakeholders, it has nonetheless been embraced across a wide variety of environmental issues (e.g., water rights, forest fire mitigation, owl population management, etc.) to translate theory in practice (Hallett et al., 2017). In this study, posthumanist, multispecies livelihoods, and translational ecology serve as the basis for the pluralistic (interdisciplinary and multi-ethical approach), posthumanist wildlife-coexistence framework.

3. Methodology

3.1 Multispecies ethnography

Multispecies ethnography examines the intersectionality of wildlife-human interactions in shared landscapes (Haraway, 2008; Kohn, 2013; Parathian et al., 2018; Thomsen et al., 2021a,b). It is grounded in anthropology, and its methods rely on interviews, participant observation, and archival data, among others (see Kirksey and Helmreich, 2010). Kirksey and Helmreich (2010) discuss how animals are regarded as part of the landscape, food for humans, and as symbols of culture—*zoe* (bare life)—that have materialized within the realm of *bios* (specific life) alongside humans. Locke (2018) argues that multispecies ethnography is an "anthropology beyond life" that perceives nonhuman life in terms of its political, social, and economic utility (Ingold, 2013). This reframing has permitted various nonhumans to gain legible biographical and political lives, such as Sandra the Orangutan (*Pongo*), who was granted nonhuman personhood in Argentina by Judge Elena Liberatori. Judge Liberatori declared that after living in captivity for more than 20 years at the Buenos Aires Zoo, Sandra had been wrongly imprisoned for most of her life, was a nonhuman person, and entitled to better living conditions and some of the same legal rights as humans (Elassar, 2019). The now 35-year-old Sandra was resettled to the accredited ape sanctuary, Center for Great Apes, in Florida, U.S. in 2019. Multispecies ethnography was useful in this study as it complements posthumanism to deconstruct human exceptionalism while maintaining the core tenets of human agency, and challenges humans to rethink the distribution of consciousness and the capacity to act, with its attendant concepts of rights and responsibilities in shared landscapes (Aisher and Damodaran, 2016). However, this study heed's Kopnina's (2017) warning to multispecies ethnographers that they cannot remain apolitical when writing about nonhumans' plight, which we provide examples for below.

3.2 Data collection

Ethnographic data collection consisted of participant observation, semi-structured interviews, online public lectures and meetings, and archival research. The study was conducted over 15 months in Oregon, Washington, Idaho, Montana, Wyoming, Yellowstone National Park (YNP), Colorado, and California, U.S. Archival research included document analysis from publicly available sources (e.g., Idaho State Wolf Management plan) and internal organizational documents shared with the first author. More than 120 formal and informal interviews were completed with a variety of stakeholders that included U.S. state-level wolf and wildlife agents, cattle ranchers, pro-wildlife environmental nonprofit organizations, conservation scientists (biological and social), pro and anti-wildlife activists, among others. Snowball sampling techniques were used to gain access to key stakeholders actively involved in wolf discourse and governance (Acharya et al., 2013). Collective demographic information of the informants included age ranges from 22 to 78 years old, gender representation of approximately 60% male to 40% female, about 85% had at least a bachelor's degree, but greater than 90% of interlocutors appeared to be racially white. This lack of diversity could possibly be attributed to a potential implicit bias by the first author, but it was not due to a lack of effort to engage minority representation including Indigenous perspectives, where multiple emails and phone calls went unreturned. Perhaps, it may reflect the underrepresentation of ethnic and racial minorities in the conservation field, which was also addressed in CPW's January 2021 public meeting streamed on YouTube.

Nearly half of all interviews were conducted online, e.g., via *Zoom*, to permit a greater distribution of informants across the western U.S., and to account for lockdown restrictions during the COVID-19 pandemic (Hanna, 2012). Hand-written field notes were completed by the first author during and or immediately following interviews, and at the end of each day of participant observation. This allowed for the authors to later reflect and conceptualize the study by considering past perceptions and observations, as "so much of ethnographic fieldwork depends on the researcher's own experience and perspective—i.e., the "I" must be acknowledged—it really does matter where you as that researcher 'stand' relative to the process of your own fieldwork and ultimately to the subject of your study" (Hoey, 2014, p.4).

3.3 Data analysis

Except for public figures who agreed to be publicly identified, all research participants were anonymized to facilitate open discourse and protect individuals' identities. Formal interviews lasted between 26 and 106 minutes and were digitally recorded and transcribed. About 60% of interviews were conducted during participant observation and were rarely digitally recorded. "During their observations, ethnographers routinely use informal or conversational interviews, which allow them to discuss, probe emerging issues, or ask questions about unusual events in a naturalistic manner. Because of the 'casual' nature of this type of interview technique it can be useful in eliciting highly candid accounts from individuals" (Reeves et al., 2008, p.513). Authors four through 12 assisted the first author to ensure that digitally recorded transcriptions were accurate and were then loaded into *NVIVO* software for coding and analysis. Data was coded using an inductive "bottom-up" approach where codes were data-driven, rather than fitting the data to preestablished theoretical codes, which is common in anthropology (Clarke and Braun, 2014; Creswell and Creswell, 2017). Findings were coded thematically, with saturation reached prior to finishing the interviews (Creswell and Creswell, 2017). Due to the extensive amount of data collected, certain themes and findings unrelated to this paper's argument were omitted. We used multiple sources of data to inform the theoretical framework and include some quotes in our analysis to help conceptualize the context in which actors face (Creswell and Creswell, 2017). It is common in ethnographic research dissemination to focus on the accounts and perspectives of the authors (e.g., Yates-Doerr, 2020), and collate the data and present it as a summary with a discussion of its importance; it would be impossible to quantify or present all the data collected in one chapter or article (e.g., Cousins, 2016; Thomsen et al., 2021a), as some anthropologists return to the data collected during their Ph.D. fieldwork decades later to inform their current studies (e.g., Wilbert and Wilbert, 2011).

4. Findings and discussion

4.1 Triggers

Findings reaffirm previous studies on wolf's ability to problematize human interactions and serve as a symbolic vehicle for understanding human cultural or socioeconomic strife (Fritts et al., 2003; Lappalainen, 2019; Thomsen, 2021; Thomsen et al., 2021a). However, this study departs from debates solely about wolf-human conflict to explore how wolf helps to reconsider how humans can coexist with wildlife in a time of increasingly polarized socio-political discourse. Ranchers and state wildlife agents in the Intermountain West of the U.S. (geographical area between the Cascade/Sierra and Rocky Mountain Ranges) speak of "trigger words," including "coexistence." The Montana Livestock Loss Board Executive Director, George Edwards, detailed how terms such as coexistence immediately "trigger" ranchers and put them on the defensive when speaking with people they consider "others," such as environmentalists or urbanites, who do not share their views, beliefs, culture, or lifestyle. Triggers are relatively new in common vernacular and mean a word or topic that evokes strong emotions in a person. Wolf also permitted diverse stakeholders to speak about their views, attitudes, and beliefs about several other species.

This included how ranchers care for domesticated livestock destined for slaughter, or self-proclaimed environmentalists who fight for wolf's reintroduction to Colorado but are quick to berate other nonhumans such as coyotes (*Canis latrans*) or bears (*Ursidae*). However, most utterances of wolf were offered as abstract statements, rather than a reference to a specific individuals' interests—human or nonhuman—demanding a pluralistic conception of wildlife-human ethics in theory, and through translational ecology in practice.

4.2 Animal ethics and posthumanist wildlife-human coexistence framework

The posthumanist wildlife-human coexistence framework is grounded in animal ethics, and embraces virtue ethics at the individual-level, rather than just concentrating on a utilitarian ethical positioning that permeates ecological discourse (Franklin, 2005; Washington and Maloney, 2020). Utilitarianism refers to "a teleological or ends-based theory that focuses on optimum outcomes, ends or consequences of an action" (Fennell, 2012: 240). Simply, it is a focus on prioritizing the greatest good for the greatest number, which is logical at the macro-level when attempting to preserve species from a biodiversity conservation perspective (Fennell and Thomsen, 2021). However, "within this view, it is almost impossible to avoid sacrificing individual beings" (Winter, 2020: 8). Yazevich et al. (2019) criticize utilitarianism for its overt anthropocentricity:

> *In the system of anthropocentric attitudes, nature is perceived as an object of human activity, where the goal of its interaction with nature is the satisfaction of various utilitarian needs. In this logic, the notion of a "pragmatic imperative" is asserted, where it is right and normal to recognize that it benefits and maximizes the benefits to a person (Chernikova, 2007). This position is strengthened with the help of utilitarian ethics, which received its natural spread in the conditions of the modernization and development of the market economy (p. 4).*

Though utilitarianism considers nonhumans as sentient beings with an interest in self-preservation (Winter, 2020), it distances us, as humans, from nonhumans when considering the biopower we exercise over individuals. Several interlocutors from across the U.S., including pro-wolf conservation biologists, expressed a willingness to accept that the hunting and trapping of wolves may be necessary to appease anti-wolfers if a certain (yet to be determined) artificially constructed and anthropocentric threshold of wolves were present on Colorado's landscape. At the ecosystems level, this is a logical, ethically justified utilitarian response to convince anti-wolfers to coexist with these apex predators on shared landscapes. However, the utilitarian ethical lens reinforces an abstract view of wolf rooted in Cartesian dualism, which promulgates sentiments that humans have a moral right to determine the fate of other species.

Ecological studies have also been impacted by social carrying capacities rather than only "objective" ecological data. In 2016, CPW "launched controversial predator control studies that involved trapping and killing" of black bears (*Ursus americanus*) and mountain lions (*Puma concolor*) (Blevins, 2019: 1). On March 30, 2021, U.S. District Court Chief Judge Marcia Krieger, based in Denver, CO, ruled that "the U.S. Fish and Wildlife Service illegally helped pay for a Colorado program to kill dozens of mountain lions and black bear in an experiment to determine if the predators were partly responsible for declining mule deer populations...[and] failed to do a required analysis of the program's environmental effects" (Anderson, 2021: 1). Considering the current ecological crisis (Steffen et al., 2011), such actions engender the question of how it remains ethical to kill nonhumans? If life is valued, how can each particular nonhuman's right to exist be foregrounded from a multispecies livelihoods perspective?

Virtue ethics provide individual-level foci toward nonhumans that examine "what sort of person should I be" (Fennell & Sheppard, 2021: 320). Clearly, the extent to which animals might share in virtue ethics depends on the question of how we understand nonhumans to participate in reason and will. The posthumanist wildlife-human coexistence framework is a coupled multispecies livelihoods and translational ecology conceptual approach that considers the rights, agency, and welfare of three key constituents in wildlife-human relations: wildlife, humans, and ecosystems. Only by equally valuing each constituent at the abstract and individual-levels (utilitarianism *and* virtue ethics), can a pluralistic ethic for wildlife-human coexistence be applied in practice through translational ecology. We apply the conceptual framework to each constituent following Fig. 1 below.

4.2.1 Wildlife

Wolves have been poached (a distancing word used in the U.S. to mean the illegal killing signifying murder from a posthumanist lens) with little recourse, as Montana's newly elected state Governor Greg Gianforte "trapped and killed a wolf near Yellowstone National Park without first taking a mandated trapper education course," on 15 February 2021, where "Morgan Warthin, spokeswoman for Yellowstone National Park, said the Wolf, No. 1155, was born in the park, was estimated to be 6 or 7 years old and had been collared by park biologists in 2018. We use terms the terms 'killing' and 'murder' to indicate a clear ethical perspective with respect to nonhumans' interests. 'Once the wolf left the park, it no longer was considered a Yellowstone wolf'" (Paybarah, 2021: 1). Wolf No. 1155 (another application of human distancing from wildlife by assigning a number to the wolf in lieu of a name) lost its life for leaving its invisible cage of YNP; Governor Gianforte was merely issued a written citation. Paybarah continues: "The one-time certification class, which lasts about three hours, teaches trappers about wolf biology, best practices for trapping and related regulations, Mr. Lemon [spokesman for Montana Fish, Wildlife and Parks (MFWP)] said. 'The class is geared toward the ethical harvest of wolves'" (Paybarah, 2021: 1).

Montana's Wolf Management Plan outlines the minimum criteria required to 'take' or 'harvest' a wolf legally (MFWP, 2020). The words 'taking', and 'harvesting' are used advantageously to mitigate potential reactions by the public regarding the murder and maiming of wolves. They may even promote the killing of wolves depending upon one's perspective. Cambridge Dictionary (2021) defines the word harvest as a verb: "to pick and collect crops, or to collect plants, animals, or fish to eat" Cambridge Dictionary (2021). Though Merriam-Webster (2021) definition includes the application of population control in relation to wildlife harvest, we contest the use of the word by MFWP, or any other actor, to justify killing wildlife without the intent of nutritional consumption. The use of harvest to justify the hunting, trapping, and killing of wolves ethically results in what could be considered systematic state-sanctioned murder. Though murder may be a difficult word for some humans to apply to nonhumans, posthumanism aids us as humans to move beyond reserving words of violence for human-use only, and to apply all terms and concepts of violence by valuing particular individuals, no matter the species.

The United States Department of Agriculture's Wildlife Services (USDAWS) 'killed or euthanized' (their terms) 68,360,467 individual wildlife (all species) in the United States from 1996 to 2020, including 7,123 (Timber) gray wolves, and 19 Mexican gray wolves (*Canus lupus baileyi*) (USDA, 2021a). The gray wolf population is estimated at ~6,000 gray wolves in the lower-48 U.S. states (USFWS, 2017). Table 1 below displays these figures, which only cover federal data, and do not include nonhuman animals killed or euthanized by individual states not using USDAWS, or unreported data on private lands. If we were speaking of humans, this would be called genocide. Van Sittert contends that "by broadening the

Table 1 USDA wildlife service # of all animal species, invasive species and wolves killed/euthanized 1996–2020.

Year	# of All nonhuman species killed/ euthanized	# of invasive nonhuman animal species killed/ euthanized	# of Gray wolves (timber) killed/ euthanized intentional	# of gray wolves (timber) killed/ euthanized unintentional/ nontarget	# of gray wolves (Mexican) killed/ euthanized intentional	# of gray wolves (Mexican) killed/ euthanized unintentional/ nontarget
1996	1,501,901	N/A	0	0	0	0
1997	1,816,539	N/A	0	0	0	0
1998	1,242,284	N/A	183	1	0	0
1999	4,082,575	N/A	172	1	0	0
2000	3,267,839	N/A	174	0	0	0
2001	1,759,469	N/A	118	0	0	0
2002	1,455,759	N/A	192	2	0	0
2003	1,690,227	N/A	184	3	0	0
2004	2,767,152	N/A	190	0	0	1
2005	1,746,248	N/A	251	1	1	0
2006	1,642,823	N/A	277	1	3	0
2007	2,420,270	N/A	338	2	4	0
2008	4,997,172	N/A	396	0	0	0
2009	4,120,291	N/A	478	2	0	0

Continued

Table 1 USDA wildlife service # of all animal species, invasive species and wolves killed/euthanized 1996–2020.—cont'd

Year	# of All nonhuman species killed/ euthanized	# of invasive nonhuman animal species killed/ euthanized	# of Gray wolves (timber) killed/ euthanized intentional	# of gray wolves (timber) killed/ euthanized unintentional/ nontarget	# of gray wolves (Mexican) killed/ euthanized intentional	# of gray wolves (Mexican) killed/ euthanized unintentional/ nontarget
2010	5,008,928	N/A	452	0	0	0
2011	3,752,356	N/A	365	0	0	0
2012	3,352,378	1,765,446	501	2	1	0
2013	4,378,456	2,336,840	319	1	1	0
2014	2,715,467	1,404,838	321	1	0	0
2015	3,206,238	1,524,955	383	1	1	0
2016	2,726,870	1,137,377	415	0	0	0
2017	2,307,122	987,047	355	1	1	0
2018	2,652,405	1,155,630	357	0	0	0
2019	2,240,802	982,064	301	1	1	0
2020	1,508,896	1,075,704	381	0	5	0
Total:	*68,360,467*	*12,369,901 (2012–2020)*	*7,103*	*20*	*18*	*1*

U.S. Department of Agriculture (USDA), 2021, 2020 Program Data Reports viewed 02 April 2021, https://www.aphis.usda.gov/aphis/ourfocus/wildlifedamage/SA_Reports/SA_PDRs.

anthropocentric focus of the emerging genocide literature on colonial settlement frontiers to include other species it becomes possible to track the post-conquest deployment of genocide technologies and practices" (Van Sittert, 2016: 2). It is difficult, at least from a posthumanist lens, to call 68,360,467 wildlife deaths over the past 25 years anything but nonhuman animal genocide.

Further human-nonhuman hierarchies include subjective Western scientific principles that justify murder of certain invasive species, such as Nutria (*Myocastor coypus*) in riparian ecosystems and fail to recognize invasives that serve human interests such as domesticated livestock. Oregon defines invasive species as "non-native organisms that cause economic or environmental harm and are capable of spreading to new areas of the state; not including humans, domestic livestock, or non-harmful exotic organisms" (OregonLaws, 2020). This law demonstrates an anthropocentric value-laden policy privileging humans and domestic livestock, suggesting that individual wildlife could be equally protected under human laws. Wildlife Services began capturing data on the amount of invasive nonhuman animal species they killed or euthanized in 2012. Between 2012 and 2020, 12,369,901 invasive individuals were killed or euthanized (see Table 1), representing 49.3% of the 25,088,634 total nonhuman animal deaths during that time span.

The recruitment of a "scientific" justification for their destruction is conceivably dependent on who serves human interests versus who does not—the latter being subject to state prosecution in Oregon. Human effects on entire ecosystems and individual habitats are treated less seriously than that of nonhuman effects, and blame is passed down through human hierarchies that "speak" for nonhumans. This often occurs directly and indirectly, such as where ocean acidification and resulting climate change effects are perpetuated by human populations at the detriment of "lesser" living ecosystems (Compton et al., 2011; Doney et al., 2009; Radford, 2019). The posthumanist framework already accounts for current practices that favor a species-level approach, and its merits include the recovery of some endangered species who were at the brink of extinction, such as the bald eagle (*Haliaeetus leucocephalus*) (Winder and Watkins, 2020). However, the data presented in Table 1 stress the critical urgency required to protect all nonhuman life, and foreground every species' rights, welfare, and agency at the individual level.

4.2.2 Humans

Few interlocutors could speak about a particular individual wolf outside of wolf WTAs, and if they did, the wolf typically did not have a name.

Occasional stories of individual wolves have captured the imaginations of biologists, ecotourists, ranchers, and the public, such as the charismatic and complex story of YNP wolf "O-six," named for her birth year. In *American Wolf*, Blakeslee (2017) vividly describes the life of O-six and her interactions with hunters, cattle ranchers, and competing packs. He unpacks the intricacies of wolf-human conflict in the western U.S., where wolf is often used as a symbol in human conflict, as Euro-American ruralites fight to maintain their way of life, while environmentalists champion ecological restoration. Indigenous rights and perspectives, including their traditional ecological knowledge (T.E.K.), are rarely considered, if at all, in debates about who has the power to shape contemporary human life and our relations with other species on shared landscapes (Cortes-Vazquez and Zedalis, 2013; Ferguson and Messier, 1997; Shelley et al., 2011).

From the posthumanist conceptual approach to wildlife-human coexistence, humans should not be disparaged, as we are a part of ecosystems, but we must consider two levels of coexistence. The first is to represent all human perspectives equitably, as CPW is attempting to do in its effort to develop a wolf management plan for Colorado. By embracing diverse groups such as Indigenous tribes, cattle ranchers, environmentalists, and conservation scientists, biopower can be dispersed to represent all human groups in wildlife-human relations equitably, at least at the abstract level. Intraspecies battles (i.e., humans) for power will most likely persist, but only by engaging all stakeholders, particularly marginalized groups, and by providing equal platforms for debate can change occur to improve multispecies welfare. The second key criterion is to consider individuals' agency and their collective power. This is evidenced by the passing of Proposition 114, where Coloradoans had the right to vote, and ushered in a new era of democratic conservation after four failed attempts to gain authorization to reintroduce wolves to the state over the past 40 years (Colorado Parks and Wildlife (CPW), 2020). Multispecies livelihoods are not limited to human and nonhuman animals, however, as "we exist in a time of increasing 'plant blindness'... [where] vegetation is experienced as a homogenous 'greenery', rather than an assemblage of particular species (Wandersee and Schussler, 2001)" (Sochacka, 2021: 5). The ecological crisis demands that we consider wildlife-human relations through a pluralistic ethical lens, within and of ecosystems, especially due to humans' unprecedented effects on the biosphere (Brook et al., 2008; Ellis, 2018).

4.2.3 Ecosystems

In 2008, wolf dispersed from YNP, Idaho, and British Columbia, Canada to repatriate parts of eastern Washington and Oregon. This dispersal occurred in the greater Columbia Basin Region (CBR) that extends across parts of seven western U.S. states and Canada to the Pacific Ocean.

"Biodiversity fragmentation risk, or conditions that prevent viable unification of natural areas," (referring to the ecological use of the term signifying habitats for nonhuman species), "include limited spatially protected areas (Chapin et al., 2009), urban sprawl (Radeloff et al., 2005), wildlife-human conflict (Hill, 2015), large-scale industrialization projects (Harte et al., 2010), and human-human conflict (Redpath et al., 2013)" (Thomsen et al., 2021a,b: 9). Keystone species, often apex-predators such as large carnivores including the gray wolf, have been reintroduced to targeted areas such as YNP in 1995-6 to ameliorate biodiversity fragmentation (Halofsky and Ripple, 2008). In Yellowstone, wolves have been credited for altering elk behavior to foster the recruitment of aspen (*Populus tremuloides*), willows (*Salix spp.*), and cottonwoods (*Populus spp.*) to the area (Halofsky and Ripple, 2008; Ripple and Robert, 2012). However, some critics challenge the reintroduction of wolves' levels of impact on ecosystems as being overstated (Barber-Meyer, 2015), and some perceive large carnivores as a threat to human lives (Butler et al., 2011), as well as human livelihoods such as cattle ranching (Sommers et al., 2010). Nonetheless, carnivore reintroduction provides a promising option to consider individual ecosystems welfare and agency in practice.

Human actors' hierarchical interests have problematized similar conservation narratives. These include terrestrial forest biomes, riparian habitat, and their inhabited flora and fauna. The rights of individual riparian habitat within the Columbia River Basin ecosystem can be considered through the federal Clean Water Act (CWA), which requires states to report the quality of its surface waters every 2 years (U.S. EPA, 2020a). The individual welfare of rivers and streams, lakes, reservoirs, estuaries, watersheds, beaches, and oceans were extended by implementing the 2018/2020 integrated report on November 12, 2020 (see DEQ, 2020):

> *Under Section 303(d) of the CWA, states, territories, and authorized tribes are required to develop lists of impaired waters… The law requires that these jurisdictions establish priority rankings for waters on the lists and develop TMDLs [Total Maximum Daily Loads] for these waters - a calculation of the maximum amount of a pollutant that a waterbody can receive and still meet water quality standards (DEQ, 2020; interactive story map).*

While the rights and welfare of water basins can be considered through environmental laws and management, the agency that represents more natural ecological flows, and processes can be overlooked when humans wield unequal power over nonhumans throughout entire ecosystems. Current practices in ecosystem governance often fail to recognize the intrinsic value of nonhumans and are typically assessed for the beneficial use for human populations.

The prioritization of certain nonhumans' rights over others (e.g., cattle over trees) has also created immense tensions between human stakeholders as ecosystems are frequently denied welfare, rights, and agency in the U.S., especially under the Trump Administration. Trees have demonstrated mycelium networks which contribute to learning and memory (Simard, 2018; Piñuela et al., 2021). Nimble policy frameworks must account for flora such as trees to be understood as living individuals with extensive movement and communication that extends throughout nearby habitat and greater ecosystems. This perspective, however, is not simply theoretical. In 2014, after more than 70 years of governmental management, New Zealand ushered in what is believed to be the first instance of granting an ecosystem personhood and the legal rights that accompany it. The former national park, *Te Urewera*, was entrusted to the Indigenous Tūhoe, who "are legally responsible for the rainforest's care. They protect the precious site through an ancient Maori practice known as *kaitakitanga*, which can be roughly translated to mean 'guardianship' and is a way of managing the environment based on a Māori worldview" (Gibson, 2020: 1). From the posthumanist wildlife-human coexistence framework lens, ecosystems must be granted equal rights, welfare, and agency, if individual wildlife and humans are to persist within, and contribute to their welfare.

5. Conclusion

Findings from this study suggest that humans must recognize the grave disparities in power relations between human stakeholders, as well as between species in wildlife-human relations. Power imbalances persist through implicit and explicit biases toward multispecies marginalized groups, are often informed by utilitarian ethics, and reproduced by employing distancing words that "other" submarginalized particular individuals and groups. Tensions between competing human stakeholder groups will endure, especially when considering the current state of U.S.'s extreme-polarization and "cultural wars" that transpired in political discourse during this study. The study was limited by focusing primarily on a charismatic species that may generate different human emotions and conceptions of wildlife than a less-polarizing species such as cottontail rabbits (*Sylvilagus spp.*). It was also limited by the COVID-19 pandemic in some regards, as participant observation was limited, and the study had to be altered to foreground digital ethnographic methods (Hanna, 2012).

Nearly all interlocutors, and corresponding public discourse, shared similarities in that they primarily spoke for and about the wolf, and wildlife in general, in the abstract. This pluralistic, ethical, translational ecology approach demonstrates that the predominant utilitarian ethics paradigm in conservation management is inadequate on its own, evidenced by the horrific posthumanist view of nonhuman genocide that goes unchecked in the U.S. This effort to coalesce utilitarian and virtue ethics complements the compassionate conservation and multispecies justice efforts that champion individual nonhuman rights and welfare in theory and practice (see Santiago-Avila and Lynn, 2020). The posthumanist wildlife-human coexistence conceptual framework is, nonetheless, a simple attempt to challenge the "traditional conservation paradigm" (Santiago-Avila and Lynn, 2020) and foreground marginalized multispecies groups who continue to suffer. However, if this framework can occur in practice through translational ecology, nonhumans could gain previously unimaginable rights, at least in a U.S. perspective. This framework was also pluralistic for its reliance on anthropology and other social science perspectives in translational ecology.

Future studies should test this framework in a myriad of settings and cultures as "numerous factors, including biological, geographic, political, economic, social, institutional, financial, cultural, and historical features, make each conflict or coexistence situation unique" (Madden, 2004: 249). It's more immediate use is perhaps for natural scientists to consider pluralistic, social science approaches to wildlife-human coexistence studies, as this framework may only be capable of materializing in practice if interdisciplinary teams of scholars and practitioners continue to overcome our different approaches, and learn to better communicate with one another. The posthumanist wildlife-human coexistence conceptual framework challenges us as humans to heuristically consider the rights, welfare, and agency of all humans, nonhumans, and ecosystems—especially marginalized multispecies stakeholders—in every decision that affects them. This prompts the question, what would the effects be if nonhumans' interests were equitably considered at the abstract and individual level when creating wildlife policies, and enacting them through management and governance?

References

Acharya, A.S., Prakash, A., Saxena, P., Nigam, A., 2013. Sampling: why and how of it. Ind. J. Med. Special. 4 (2), 330–333.
Adams, W.M., 2008. Thinking like a human: social science and the two cultures problem. Oryx 41, 275–276.
Agrawal, A., Ostrom, E., 2006. Political science and conservation biology: a dialog of the deaf. Conserv. Biol. 20, 681–686.
Aisher, A., Damodaran, V., 2016. Introduction: Human-Nature Interactions Through a Multispecies Lens. vol. 14 Ashoka Trust for Research in Ecology and the Environment/Wolters Kluwer India Pvt. Ltd., p. 293. no. 4.
Alcadipani, R., de Oliveira Medeiros, C.R., 2019. When corporations cause harm: a critical view of corporate social irresponsibility and corporate crimes. J. Business Ethics, 1–13.
Anderson, J., 2021. Judge: US Agency Illegally Paid for Colorado Predator Hunt. viewed 09 April 2021. http://apnews.com/article/fish-wildlife-service-illegally-paid-colorado-predator-hunt-622b30d0c9abf1421e1241fffa38d61e.
Arcari, P., 2019. 'Dynamic' non-human animals in theories of practice: views from the subaltern. In: Social Practices and Dynamic Non-Humans. Palgrave Macmillan, Cham, pp. 63–86.
Barber-Meyer, S.M., 2015. Trophic cascades from wolves to grizzly bears or changing abundance of bears and alternate foods? J. Anim. Ecol. 84 (3), 647–651.
Béné, C., Wood, R.G., Newsham, A., Davies, M., 2012. Resilience: new utopia or new tyranny? Reflection about the potentials and limits of the concept of resilience in relation to vulnerability reduction programmes. IDS Working Papers 405, 1–61.
Berkes, F., Folke, C., Colding, J., 2000. In: Berkes, F., et al. (Eds.), Linking social and ecological systems: management practices and social mechanisms for building resilience. Cambridge University Press, Cambridge, UK.
Blakeslee, N., 2017. American Wolf. Penguin Random House, New York NY.
Blécourt, W., 2007. I would have eaten you too: werewolf legends in the Flemish, Dutch and German Area. Folklore 118 (1), 23–43.
Blevins, J., 2019. Colorado Wildlife Officials are Reluctant to OK Gray Wolf Reintroduction. So Advocates Want Voters To Do It. viewed 01 November 2020. https://coloradosun.com/2019/04/25/colorado-gray-wolf-reintroduction-ballot-proposal/>.
Bodin, Ö., Baird, J., Schultz, L., Plummer, R., Armitage, D., 2020. The impacts of trust, cost and risk on collaboration in environmental governance. People Nat. 2 (3), 734–749.
Bolter, J.D., 2016. Posthumanism. The International Encyclopedia of Communication Theory and Philosophy, pp. 1–8.
Braidotti, R., 2016. Posthuman critical theory. In: Critical Posthumanism and Planetary Futures. Springer, New Delhi, pp. 13–32.
Brasch, S., 2020. Colorado Voters Want Wolves Back in Colorado. Now Comes the Hard Part. viewed 06 November 2020. https://www.cpr.org/2020/11/06/colorado-voters-want-wolves-back-in-colorado-now-comes-the-hard-part/.
Brook, B.W., Navjot, S.S., Bradshaw, C.J.A., 2008. Synergies among extinction drivers under global change. Trends Ecol. Evol. 23 (8), 452–460.
Brosius, J.P., 2006. Common ground between anthropology and conservation biology. Conserv. Biol. 20, 68.
Büscher, B., Wolmer, W., 2007. Introduction. The politics of engagement between biodiversity conservation and the social sciences. Conserv. Soc. 5. http://www.conservationandsociety.org/article.asp?issn=0972-4923;year=2007;volume=5;issue=l;spage= 1;epage=21;aulast=Busch.
Butler, L., Dale, B.W., Beckmen, K., Farley, S.D., 2011. Findings related to the March 2010 fatal wolf 37 attack near Chignik Lake, Alaska. Alaska Department of Fish and Game.
Cambridge Dictionary, 2021. Harvest. viewed 26 March 2021. https://dictionary.cambridge.org/us/dictionary/english/harvest.

Campbell, L.M., 2005. Overcoming obstacles to interdisciplinary research. Conserv. Biol. 19, 574.
Castelblanco-Martínez, D.N., Moreno-Arias, R.A., Velasco, J.A., Moreno-Bernal, J.W., Restrepo, S., Noguera-Urbano, E.A., Jiménez, G., 2021. A hippo in the room: predicting the persistence and dispersion of an invasive mega-vertebrate in Colombia, South America. Biol. Conserv. 253, 108923.
Celermajer, D., Schlosberg, D., Rickards, L., Stewart-Harawira, M., Thaler, M., Tschakert, P., Winter, C., 2021. Multispecies justice: theories, challenges, and a research agenda for environmental politics. Environ. Polit. 30 (1-2), 119–140.
Chapin, F.S., III, Kofinas, G.P., Folke, C., et al., 2009. In: Chapin III, F.S. (Ed.), *Principles of ecosystem stewardship: Resilience-based natural resource management in a changing world*, first ed. 1. Springer Science & Business Media, New York, NY.
Chernikova, I.V., 2007. Epistemology: The Main Problems and the Evolution of Approaches in the Philosophy of Science. Kuzbass University, Kemerovo.
Cielemęcka, O., Daigle, C., 2019. Posthuman sustainability: an ethos for our anthropocenic future. Theor. Cult. Soc. 36 (7–8), 67–87.
Clarke, V., Braun, V., 2014. Thematic analysis. In: Teo, T. (Ed.), Encyclopedia of Critical Psychology. Springer, NY.
Cohen, E., 2019. Posthumanism and tourism. Tour. Rev. 74 (2).
Colorado Parks and Wildlife (CPW), 2020. Wolf Management. viewed 02 November 2020. https://cpw.state.co.us/learn/Pages/CON-Wolf-Management.aspx>.
Compton, J.E., Harrison, J.A., Dennis, R.L., Greaver, T.L., Hill, B.H., Jordan, S.J., Campbell, H.V., 2011. Ecosystem services altered by human changes in the nitrogen cycle: a new perspective for US decision making. Ecol. Lett. 14 (8), 804–815.
Copeland, K., 2020. Commodifying biodiversity: socioeconomic approaches to wildlife human coexistence. In: Leal, F.W., Azul, A.M., Brandli, L., Lange, S.A., Wall, T. (Eds.), Partnerships for the Goals. Encyclopedia of the UN Sustainable Development Goals. Springer, Cham, pp. 1–12.
Copeland, K., 2021. Reimagining innovation for "social" entrepreneurship: nonhuman spaces for the SDGs. J. Int. Council Small Business 2, 1–13.
Cortes-Vazquez, J.A., Zedalis, M., 2013. Identity and native species conservation: similar historical ecologies from Idaho to Spain. Hum. Ecol. 41 (6), 937–945.
Cousins, T., 2016. Antiretroviral therapy and nutrition in Southern Africa: citizenship and the grammar of hunger. Med. Anthropol. 35 (5), 433–446.
Creswell, J.W., Creswell, J.D., 2017. Research Design: Qualitative, Quantitative, and Mixed Methods Approaches. Sage Publications.
Department of Environmental Quality, 2020. Water quality—EPA approved integrated report. State Oregon. viewed 23 March 2021. https://www.oregon.gov/deq/wq/Pages/epaApprovedIR.aspx.
Doney, S.C., Fabry, V.J., Feely, R.A., Kleypas, J.A., 2009. Ocean acidification: the other CO_2 problem. Annu. Rev. Mar. Sci. 1, 169–192.
Elassar, A., 2019. Sandra the Orangutan, Freed From a Zoo After Being Granted 'Personhood,' Settles Into her New Home. viewed 26March 2021. https://www.cnn.com/2019/11/09/world/sandra-orangutan-florida-home-trnd.
Ellis, E., 2018. Anthropocene: A Very Short Introduction. Oxford University Press, https://doi.org/10.1093/actrade/9780198792987.001.0001.
Enquist, C.A., Jackson, S.T., Garfin, G.M., Davis, F.W., Gerber, L.R., Littell, J.A., Shaw, M.R., 2017. Foundations of translational ecology. Front. Ecol. Environ. 15 (10), 541–550.
Faber, D., Schlegel, C., 2017. Give me shelter from the storm: framing the climate refugee crisis in the context of neoliberal capitalism. Capit. Nat. Social. 28 (3), 1–17.
Fabinyi, M., Evans, L., Foale, S.J., 2014. Social-ecological systems, social diversity, and power: insights from anthropology and political ecology. Ecol. Soc. 19, 4.

Fennell, D.A., 2012. Tourism, animals and utilitarianism. Tour. Recreat. Res. 37 (3), 239–249.
Fennell, D.A., Sheppard, V., 2021. Tourism, animals and the scales of justice. J. Sustain. Tour. 29 (2–3), 314–335. https://doi.org/10.1080/09669582.20201768263.
Fennell, D.A., Thomsen, B., 2021. Tourism & animal suffering: mapping the future. Ann. Tour. Res. 91, 103317.
Ferdinand, M., Chaillou, A., Roblin, L., 2020. Why we need a decolonial ecology. Rev. Proj. 2, 52–56.
Ferguson, M.A., Messier, F., 1997. Collection and analysis of traditional ecological knowledge about a population of arctic tundra caribou. Arctic, 17–28.
Fox, H.E., Christian, C., Nordby, J.C., Pergams, O.R.W., Peterson, G.D., Pyke, C.R., 2006. Perceived barriers to integrating social science and conservation. Conserv. Biol. 20, 1817.
Franklin, J.H., 2005. Animal Rights and Moral Philosophy. Columbia University Press.
Fritts, S.H., Mech, L.D., Stephenson, R.O., Hayes, R.D., Boitani, L., 2003. Wolves and Humans. 'Wolves: Behavior, Ecology, and Conservation'. The University of Chicago Press, Chicago, IL, pp. 289–316.
Galastri, L., 2018. Social classes and subaltern groups: theoretical distinction and political application. Capital Class 42 (1), 43–62.
Gibson, J., 2020. The Māori tribe protecting New Zealand's sacred forest. BBC Travel. Retrieved September 29, 2021, from https://www.bbc.com/travel/article/20200608-the-mori-tribe-protecting-new-zealands-sacred-rainforest.
Gonzales, A., 2017. Trumpism, authoritarian neoliberalism, and subaltern Latina/o politics. J. Chicago Stud. 42 (2), 147–164.
Good, T.P., Samhouri, J.F., Feist, B.E., Wilcox, C., Jahncke, J., 2020. Plastics in the Pacific: assessing risk from ocean debris for marine birds in the California current large marine ecosystem. Biol. Conserv. 250, 108743.
Grooten, M., Almond, R.E., 2018. Living Planet Report-2018: Aiming Higher.
Hallett, L.M., Morelli, T.L., Gerber, L.R., Moritz, M.A., Schwartz, M.W., Stephenson, N.L., Woodhouse, C.A., 2017. Navigating translational ecology: creating opportunities for scientist participation. Front. Ecol. Environ. 15 (10), 578–586.
Halofsky, J., Ripple, R., 2008. Linkages between wolf presence and aspen recruitment in the Gallatin elk winter range of southwestern Montana, USA. Forestry 81 (2), 195–207.
Hamilton, L., Taylor, N., 2017. Ethnography After Humanism: Power, Politics and Method in Multi-Species Research. Pelgrave Macmillan, London, UK.
Hanna, P., 2012. Using internet technologies (such as Skype) as a research medium: a research note. Qual. Res. 12 (2), 239–242. https://doi.org/10.1177/1468794111426607.
Haraway, D., 2008. When Species Meet. University of Minnesota Press, Minneapolis, MN.
Harrington, R., Anton, C., Dawson, T.P., De Bello, F., Feld, C.K., Haslett, J.R., Kluvankova-Oravaska, T., 2010. Ecosystem services and biodiversity conservation: concepts and a glossary. Biodivers. Conserv. 19 (10), 2773–2790.
Harte, M., Campbell, H., Webster, J., 2010. Looking for safe harbor in a crowded sea: Coastal space use conflict and marine renewable energy development. In: Shifting Shorelines: Adapting to the Future,The 22nd International Conference of The Coastal Society, June 13-16, 2010 ,Wilmington, North Carolina, USA, 22 (1).
Hill, C.M., 2015. Perspectives of "conflict" at the wildlife–agriculture boundary: 10 years on. Hum. Dimens. Wildl. 20 (4), 296–301. https://doi.org/10.1080/10871209.2015.1004143.
Hoey, B.A., 2014. A Simple Introduction to the Practice of Ethnography and Guide to Ethnographic Fieldnotes. Marshall University Digital Scholar, pp. 1–10.
Holzer, J.M., Carmon, N., Orenstein, D.E., 2018. A methodology for evaluating transdisciplinary research on coupled socio-ecological systems. Ecol. Indic. 85, 808–819.

Horta, O., Albersmeier, F., 2020. Defining speciesism. Philos. Compass 15 (11), 1–9. https://doi.org/10.1111/phc3.12708.

Ingold, T., 2013. Anthropology beyond humanity. J. Finn. Anthropol. Soc. 38 (3), 5–23.

Jackson, M.V., Choi, C.Y., Amano, T., Estrella, S.M., Lei, W., Moores, N., Fuller, R.A., 2020. Navigating coasts of concrete: pervasive use of artificial habitats by shorebirds in the Asia-Pacific. Biol. Conserv. 247, 108591.

Jiguet, F., 2020. The fox and the crow. A need to update pest control strategies. Biol. Conserv. 248, 108693.

Kirksey, S.E., Helmreich, S., 2010. The emergence of multispecies ethnography. Cult. Anthropol. 25 (4), 545–576.

Kohn, E., 2013. How forests think. University of California Press, Berkeley, CA. https://doi.org/10.1525/9780520956865.

Kopnina, H., 2017. Beyond multispecies ethnography: engaging with violence and animal rights in anthropology. Crit. Anthropol. 37 (3), 333–357. https://doi.org/10.1177%2F0308275X17723973.

Kramer, R.C., Michalowski, R.J., 2012. Is global warming a state-corporate crime? In: Climate Change From a Criminological Perspective. Springer, NY, pp. 71–88.

Kunstler, B., 1991. The werewolf figure and its adoption into the greek political vocabulary. Class. World 84 (3), 189–205.

Lappalainen, K., 2019. Recall of the fairy-tale wolf: "Little Red Riding Hood" in the dialogic tension of contemporary wolf politics in the US west. ISLE Interdiscip. Stud. Lit. Environ.

Locke, P., 2018. Multispecies ethnography. In: The International Encyclopedia of Anthropology, pp. 1–3.

Lynn, W.S., 2010. Discourse and wolves: science, society, and ethics. Soc. Anim. 18 (1), 75–92.

Madden, F., 2004. Creating coexistence between humans and wildlife: global perspectives on local efforts to address human–wildlife conflict. Hum. Dimens. Wildl. 9 (4), 247–257. https://doi.org/10.1080/10871200490505675.

Manfredo, M.J., Urquiza-Haas, E.G., Carlos, A.W.D., Bruskotter, J.T., Dietsch, A.M., 2020. How anthropomorphism is changing the social context of modern wildlife 855 conservation. Biol. Conserv. 241, 108297.

Mascia, M.B., Brosius, J.P., Dobson, T.A., Forbes, B.C., Horowitz, L., McKean, M.A., Turner, N.J., 2003. Conservation and the social sciences. Conserv. Biol. 17 (649), 650.

Matin, N., Forrester, J., Ensor, J., 2018. What is equitable resilience? World Dev. 109, 197–205.

Meadow, R., Reading, R.P., Phillips, M., Mehringer, M., Miller, B.J., 2005. The influence of persuasive arguments on public attitudes toward a proposed wolf restoration in the southern Rockies. Wildlife Soc. Bull. 33 (1), 154–163.

Merriam-Webster, 2021. Harvest. viewed 26 March 2021. https://www.merriam. webster.com/dictionary/harvest.

Min, K., Cuiffi, J.D., Mathers, R.T., 2020. Ranking environmental degradation trends of plastic marine debris based on physical properties and molecular structure. Nat. Commun. 11 (1), 1–11.

Mitchell, T., 2002. Can the Mosquito Speak? Rule of experts: Egypt, techno-politics, modernity. University of California Press, Berkeley, CA, pp. 19–53.

Montana Fish, Wildlife, & Parks (MFWP), 2020. Hunting and Trapping Regulations. https://fwp.mt.gov/fishandwildlife/management/wolf/.

Niemiec, R., Berl, R.E., Gonzalez, M., Teel, T., Camara, C., Collins, M., Hoag, D., 2020. Public perspectives and media reporting of wolf reintroduction in Colorado. PeerJ 8, e9074.

Norton, M., 2017. Subaltern technologies and early modernity in the Atlantic world. Colon. Lat. Am. Rev. 26 (1), 18–38.
Olsson, L., Jerneck, A., Thoren, H., Persson, J., O'Byrne, D., 2015. Why resilience is unappealing to social science: theoretical and empirical investigations of the scientific use of resilience. Sci. Adv. 1 (4), e1400217.
OregonLaws, 2020. ORS 570.755—Definition of invasive species—2020. Oregon Rev. Stat. viewed 23 March 2021. https://www.oregonlaws.org/ors/570.755#:~:text=(1)%20As%20used%20in%20this,livestock%20or%20nonharmful%20exotic%20organisms.
Parathian, H.E., McLennan, M.R., Hill, C.M., Frazão-Moreira, A., Hockings, K.J., 2018. Breaking through disciplinary barriers: human–wildlife interactions and multispecies ethnography. Int. J. Primatol. 39 (5), 749–775. https://doi.org/10.1007/s10764-018-0027-9.
Pate, J., Manfredo, M.J., Bright, A.D., Tischbein, G., 1996. Coloradans' attitudes toward reintroducing the gray wolf into Colorado. Wildlife Soc. Bull., 421–428.
Paybarah, A., 2021. Montana governor gets written warning after killing a wolf. viewed 26 March 2021. http://www.nytimes.com/2021/03/23/us/greg-gianforte-wolf-violation.html.
Persson, J., Hornborg, A., Olsson, L., Thorén, H., 2018. Toward an alternative dialogue between the social and natural sciences. Ecol. Soc. 23 (4).
Piñuela, Y., Alday, J.G., Oliach, D., Castaño, C., Bolaño, F., Colinas, C., Bonet, J.A., 2021. White mulch and irrigation increase black truffle soil mycelium when competing with summer truffle in young truffle orchards. Mycorrhiza 31, 1–12.
Pulido, L., De Lara, J., 2018. Reimagining 'justice' in environmental justice: radical ecologies, decolonial thought, and the black radical tradition. Environ. Plan. Nat. Space 1 (1–2), 76–98.
Radeloff, V.C., Hammer, R.B., Stewart, S.I., Fried, J.S., Holcomb, S.S., McKeefry, J.F., 2005. The wildland-urban interface in the United States. Ecol. Appl. 15 (3), 799–805.
Radford, T., 2019. James Lovelock at 100: the Gaia saga continues. Nature 570 (7762), 441–443.
Redpath, S.M., Young, J., Evely, A., Adams, W.M., Sutherland, W.J., Whitehouse, A., Gutierrez, R.J., 2013. Understanding and managing conservation conflicts. Trends Ecol. Evol. 28 (2), 100–109.
Ripple, W.J., Robert, L.B., 2012. Trophic cascades in Yellowstone: the first 15 years after wolf reintroduction. Biol. Conserv. 145 (1), 205–213.
Roy, R.D., 2015. Nonhuman empires. Comp. Stud. South Asia Afr. Middle East 35 (1), 66–75. https://doi.org/10.1215/1089201X-2876104.
Sandbrook, C., Adams, W.M., Büscher, B., Vira, B., 2013. Social research and biodiversity conservation. Conserv. Biol. 27 (6), 1487–1490.
Santiago-Avila, F.J., Lynn, W.S., 2020. Bridging compassion and justice in conservation ethics. Biol. Conserv. 248, 108648.
Shelley, V., Adrian, T., Lisa, N., 2011. Attitudes to wolves and wolf policy among Ojibwe tribal members and non-tribal residents of Wisconsin's wolf range. Hum. Dimens. Wildlife 16 (6), 397–413.
Simard, S.W., 2018. Mycorrhizal networks facilitate tree communication, learning, and Memory. In: Memory and Learning in Plants. Springer, Cham, pp. 191–213.
Sochacka, A., 2021. Lessons from nettles: A case-study on managing an urban green-space for the Planthroposcene, Working Paper. Oxford, UK.
Sommers, A.P., Price, C.C., Urbigkit, C.D., Peterson, E.M., 2010. Quantifying economic impacts of large-carnivore depredation on bovine calves. J. Wildlife Manage. 74 (7), 1425–1434.
Spiro, J., 2009. Defending the Master Race: Conservation, Eugenics, and the Legacy of Madison Grant. UPNE.

Spivak, G.C., 1988. In: Nelson, C., Grossberg, L. (Eds.), Can the subaltern speak? Marxism and the interpretation of culture. University of Illinois Press, Urbana, pp. 271–313.
Sponarski, C.C., Semeniuk, C., Glikman, J.A., Bath, A.J., Musiani, M., 2013. Heterogeneity among rural resident attitudes toward wolves. Hum. Dimens. Wildlife 18 (4), 239–248.
Steffen, W., Perrson, A., Deutsch, L., Zalasiewicz, J., Williams, M., Richardson, K., et al., 2011. The Anthropocene: from global change to planetary stewardship. Ambio 40 (7), 739–761. https://doi.org/10.1007/s13280-011-0185-x.
Thomas, P.D., 2018. Refiguring the subaltern. Polit. Theory 46 (6), 861–884.
Thomsen, B., 2021. Wolf ecotourism: a posthumanist approach to wildlife ecotourism. In: Fennell, D. (Ed.), Routledge's Handbook of Ecotourism, pp. 117–131.
Thomsen, B., Thomsen, J., 2020. Multispecies livelihoods: partnering for sustainable development and biodiversity conservation. In: Leal Filho, W., Azul, A.M., Brandli, L., Lange Salvia, A., Wall, T. (Eds.), Partnerships for the Goals Encyclopedia of the UN Sustainable Development Goals. Springer, Cham. https://doi.org/10.1007/978-3-319-71067-9_99-1.
Thomsen, B., Gosler, A., Cousins, T., 2020b. Democratizing Conservation: A New Paradigm in Conservation management. Working Paper, Oxford, UK.
Thomsen, B., Thomsen, J., Cipollone, M., Coose, S., 2021a. Let's save the bear: a multi-species livelihoods approach to wildlife conservation and achieving the SDGs. J. Int. Coun. Small Business. https://doi.org/10.1080/26437015.2021.1881943.
Thomsen, B., Thomsen, J., Copeland, K., Coose, S., Arnold, E., Bryan, H., Prokop, K., Cullen, K., Vaughn, C., Rodriguez, B., Muha, R., 2021b. Multispecies livelihoods: a posthumanist approach to wildlife ecotourism that promotes animal ethics. J. Sustain. Tour., 1–19. https://doi.org/10.1080/09669582.2021.1942893.
U.S. Department of Agriculture (USDA), 2021a. 2020 Program Data Reports. viewed 02 April 2021. https://www.aphis.usda.gov/aphis/ourfocus/wildlifedamage/SA_Reports/SA_PDRs>.
U.S. EPA, 2020a. About EPA's Work in the Columbia River Basin. viewed 02April 2021. https://www.epa.gov/columbiariver/about-epas-work-columbia-river-basin>.
U.S. Fish and Wildlife Service, 2017. Gray Wolf (Canis lupus) Current Population in the United States. viewed 01 November 2020. https//www.fws.gov/Midwest/wolf/aboutwolves/wolfpopus.html.
Van Sittert, L., 2016. Routinising genocide: the politics and practice of vermin extermination in the Cape Province c. 1889–1994. J. Contemp. Afr. Stud. 34 (1), 111–128.
Wall, T.U., McNie, E., Garfin, G.M., 2017. Use-inspired science: making science usable by and useful to decision makers. Front. Ecol. Environ. 15 (10), 551–559. https://doi.org/10.1002/fee.1735.
Wandersee, J., Schussler, E., 2001. Toward a theory of plant blindness. Plant Sci. Bull. 47 (1), 2–7. https://botany.org/bsa/psb/2001/psb47-1.pdf.
Washington, H., Maloney, M., 2020. The need for ecological ethics in a new ecological economics. Ecol. Econ. 169, 106478.
Watson, M.C., 2011. Cosmopolitics and the subaltern: problematizing Latour's idea of the commons. Theory Cult. Soc. 28 (3), 55–79.
Wilbert, W., Wilbert, C.A.L., 2011. In: Freire, G. (Ed.), Filoterapia warao: fundamentos teóricos. *Prespectivas en salud indıgena: cosmovisión, enfermedad y polıticas públicas*. Abya-yala, Universidad Politécnica Salesiana, Quito, Ecuador, pp. 307–326.
Winder, V.L., Watkins, M.A., 2020. Thirty years of bald eagle population recovery and nesting ecology in Kansas, 1989–2018. J. Rap. Res. 54 (3), 255–264.
Winter, C., 2020. A review of animal ethics in tourism: launching the annals of tourism research curated collection on animal ethics in tourism. Ann. Tour. Res. 84, 102989.
Wolfe, C., 2010. What is Posthumanism? 8 University of Minnesota Press.

Yates-Doerr, E., 2020. Antihero care: on fieldwork and anthropology. Anthropol. Human. 45 (2), 233–244.

Yazevich, M., Kalinina, O., Zhironkina, O., 2019. Ecological consciousness in a shift of scientific paradigms. In: E3S Web of Conferences. vol. 134. EDP Sciences, p. 03004.

Reeves, S., Kuper, A., Hodges, B.D., 2008. Qualitative research methodologies: ethnography. Br. Med. J. 337 (7668), 512–514. http://www.jstor.org/stable/20510696.

CHAPTER SIX

Saving the Sonso Lagoon: contesting entrenched local powers and building practical authority in wetland governance in Valle del Cauca, Colombia

Renata Moreno Quintero[a,*] **and Theresa Selfa**[b]

[a]Department of Social Sciences and Economics, Universidad Autónoma de Occidente, Cali, Colombia
[b]Department of Environmental Studies, State University of New York-College of Environmental Science and Forestry, Syracuse, NY, United States
*Corresponding author: e-mail address: rmorenoq@uao.edu.co

Contents

Abstract

This research focuses on a case study of conflict around the most important wetland of the Valle del Cauca region in Colombia called the Sonso Lagoon. At the end of December 2015, the wetland was being destroyed by neighboring wealthy landowners to grow sugarcane with the acquiescence of local authorities, affecting local rural communities of fishermen and community eco-tourism operators who depend of this wetland for their livelihoods. After 1 year of conflict the wetland was designated a Ramsar site, and restoration of the lagoon was implemented by the environmental authority.

This paper analyzes how this situation turned around; we reconstruct the steps taken to save the wetland, explaining how wetland activists used resources, defined the problem and mobilized networks, and how these strategies connect to particular characteristics of the political context and environmental governance in the region. We demonstrate the importance of taking into account institutional political

Advances in Ecological Research, Volume 66
ISSN 0065-2504
https://doi.org/10.1016/bs.aecr.2022.04.005

arrangements to analyze the ways in which the conflicts are expressed, how the policy agenda is defined, as well as the strategies employed by advocacy groups.

We draw upon perspectives on environmental governance that stress the importance of understanding the institutional context in which agencies are embedded and that constrain their responses to public demands. In the same vein, we use the work of Layzer and Rinfret, (2020) to understand how agency missions, standard operating procedures and past interactions with client groups constrain day-to day decision-making around wetland management in the region. Finally, we use the concept of "practical authority" developed by Abers and Keck (2013) to explain the permissive institutional context that allowed extensive damage to the wetland in the first place. We conclude by arguing that pluralism in environmental governance must address social inequities to ensure long-term ecosystem recovery.

1. Introduction

The Sonso Lagoon is the largest wetland of Valle del Cauca, a region located in South West Colombia (Fig. 1). This lagoon, encompassing

Fig. 1 The Sonso Lagoon and the Cauca River. *Agua de Sonso.*

2,045 Ha, is the last relic of flooded dry tropical forest and in 2017 was designated a Ramsar site (the most important international recognition for wetlands). One year before this designation, the wetland was being destroyed. Two of the neighboring landowners were building dikes, removing vegetation to expand sugarcane crops, and digging ditches to use water from the wetland to irrigate the new plantations, all of which was severely damaging the wetland. It was December 23, 2015, one day before Christmas. By January 11th, 2016 no action had been taken. The regional environmental authority (Corporación Autónoma Regional del Valle del Cauca-CVC) remained silent while more than 20 excavators obliterated 47 hectares of the lagoon, despite public outcry and continued claims by the fishermen of the lagoon. This was a predictable situation given the history of unchecked wetland degradation by landowners in the region, the lack of monitoring and law enforcement by the CVC and the weakness of grassroots organizations of fishermen whose livelihoods depended on wetlands. An overwhelming feeling of impotence had seized all involved in the defense of the wetland at that time. However, on March 7th the CVC publicized that it had started works to restore the wetland with assistance from the municipality of Buga, the National Army, the Police and the Environmental Attorney (Procurador Ambiental) of the region and that the culprits would be punished. How did this situation turn around? What does this case tell us about wetland governance in the Valle del Cauca region in Colombia and the role of environmental governance in highly unequal societies? These are the main questions of this paper.

In the first sections of the paper we reconstruct the steps taken to save the wetland, explaining how wetland activists used resources, defined the problem and mobilized networks, and how these strategies connected to particular characteristics of the political context and environmental management in the region. Then we analyze the problems in the construction of "practical authority" (Abers and Keck, 2013) in wetland governance that allowed for wetland degradation and the Sonso crisis, and finally we draw on the lessons from this case to discuss alternatives to enhance wetland governance in the region.

This research contributes to current debates on environmental governance in Latin American contexts, by providing recommendations to better include underrepresented groups in environmental policy processes. It also contributes to the growing literature on environmental governance by expanding the discussion about democratization processes in Latin America to the environmental field, where struggles over natural resources are also linked to broader challenges to elitism and exclusion that characterize many Latin American democracies. By analyzing the different ways that the state and citizens interact to manage environmental problems in parts of the world where states never had full control of their territory and have limited authority over regional powers (González, 2009), we add important insights to the environmental governance literature.

We understand governance here from a polycentric perspective (Andersson and Ostrom, 2008), as a dynamic process based on interactions among different state and non-state actors at different scales that coordinate actions and make decisions through both formal and informal linkages on particular historical, socio-economic and environmental contexts. Decisions around water, as noted by Zwarteveen (2015), occur in concrete places and are made by specific people with their own identities, bodies, culture and interests. Ecosystem governance is related to how the ecosystems are managed in a specific territory, how different stakeholders are engaged in this process, and how environmental problems are framed, included or excluded from the political agenda. In this paper, we understand pluralism as the diversity of participants and experiences involved in environmental degradation and environmental governance, which usually relies on institutions at multiple levels and scales. We approach this diversity from a particular interest on aspects of power, in addition to the need to understand how institutions and people also have diverse needs, views, interests and behaviors in governing the ecosystems.

In accordance with this approach, Hochstetler and Keck (2007) also stress the importance of studying a particular policy area like the environment within a broader political context that recognizes interactions between different levels of political institutions and among state and society actors, each of which has multiple commitments and connections.

Related to this, Brooks (1993), as well as Keck and Sikkink (1998) showed that external pressures are key factors to consider because they influence what issues become formulated as public policy issues and they limit the possible responses of state elites. Increasing global concerns over wetlands have led to the creation of global conventions such as the Ramsar convention and multi-party joint efforts to protect these ecosystems that put pressures on nation states, as illustrated in the Sonso Lagoon case. In addition, global economic integration processes advance developmental pressures that may be at odds with conservationist efforts. Analyzing the role of conservation and developmental actors in this case is thus important to understanding tensions in wetland governance.

Along with external pressures and global trends, our approach also stresses the importance of the institutional context in which environmental agencies are embedded to explain what constrains agencies' responses to public demands. According to Andersson and Ostrom (2008) the institutional context in the case of ecosystem governance is compounded by the relationships between municipal actors and local resource users as well as national government agencies. The institutional context is also specific to the characteristics of state building in each country. In Colombia, regional environmental agencies have a great deal of autonomy in policy design and implementation. This feature makes the implementation of wetland policies more dependent on negotiations with local powers, which have ample opportunities to influence the final outcome of policies, preventing other actors from exerting equal participation.

Agencies are also influenced by internal factors such as agency missions, standard operating procedures and past interactions with client groups, all of which constrain day-to day decision-making around ecosystem governance (Layzer and Rinfret, 2020). In addition to that, environmental agencies' authority to make binding decisions cannot be taken for granted. Abers and Keck (2013) explain how organizations need to build reputation and recognition based on organizational and technical capacities to position themselves as real authorities, and the degree to which they achieve or fail to achieve this "practical authority" has consequences for the way these agencies relate to other actors in the governance system, with implications for addressing pluralism through collaboration and equal participation.

Finally, Hochstetler and Keck (2007) also call attention to the importance of state-society linkages in ecosystem governance, such as the networks connecting activists in civil society and committed individuals in state agencies. According to them, these networks facilitate issue advocacy and institutional change. Building upon this, our analysis of the Sonso Lagoon case complements the analysis of the institutional context of wetland governance in the region with an analysis of the way actors use resources, define problems, and build networks to understand how the case became salient and was included on the political agenda.

2. Data and methods

This research was designed as an ethnographic case study (Burawoy, 1991). The data on which the analysis is based are drawn primarily from semi-structured interviews, and supplemented by four additional sources. Thirty-five interviews[a] were conducted with a wide range of actors related to wetland governance in Colombia between 2013 and 2015. During this period, we attended many of the meetings of the Cauca River Corridor Project, a state-led project to control additional flooding of the Cauca River and the meetings of the wetland management committee of the Sonso Lagoon in Buga Valle del Cauca. The interview respondents include agency officials, sugarcane growers, grassroots leaders, NGO members, consultants and scientists, representing a broad range of the key actors in wetland governance in the region. We initially assembled a core list of respondents from key participants of the Cauca River Corridor Project; once interviews began I expanded the list through snowball sampling. Interview questions revolved around wetland meanings and uses, relationships to other actors in wetland management, participation in wetland decision-making and conflicts related to wetlands.

The interviews were supplemented with agency reports about the Sonso Lagoon case issued in response to community legal complaints. We also analyzed media coverage of the case during the months of December 2015 and January, February and March 2016, taken from newspapers, television news, and web sources. Finally, we drew on press releases from the sugar sector. We used this information to analyze institutional response and reactions from the sugar sector to the case. We also used social media posts from grassroots websites to analyze the way wetland advocates resorted to social media to advance their denunciations and reach out a broader public.

The information gathered was transcribed and open coded to develop analytic themes that emerged through analysis. These themes included (1) regional political power (2) responses to the destruction of the wetland (3) interaction of state and society actors in wetland governance. I identified theoretical approaches to analyze those themes, from which I selected categories of analysis, and turned to a fine-grained analysis of the data in focused coding (Emerson et al., 1995) to identify subthemes and subtopics according to those categories, trying to follow a process of reflexive or dialectical interplay between theory and data. The categories of analysis were the following: (1) networks (2) problem framing (3) state authority (4) role of economic and political actors (5) decision-making scales in wetland governance. I also coded field notes and secondary data using the same categories. Based on the main themes I constructed the problem and sorted the data sources

[a] Their names, when given, were changed to ensure confidentiality.

according to that; then I re-coded the data according to the categories to analyze each of the sub-themes. The themes and sub-themes developed served as the basis for some of the narrative portions of the paper.

3. Confronting entrenched local powers. De-localizing the conflict and jumping scales

Land resources in Valle del Cauca have been central instruments of prestige and power, and are characterized by long standing dominance by powerful actors such as agricultural and drug mafia entrepreneurs, who have ties to public institutions and the province's politics. One of the enemies that local wetland advocates confronted was a former Congressman who is currently in jail for his links with the Cali Cartel. The fact that he was able to keep "his lands," despite being in jail, is revealing of the de facto power that drug dealers still hold in the territory. In addition, the fact that the crops mentioned were sugarcane also suggests the alignment between the interests of the drug dealers and the sugar elites that represented another powerful economic actor in the region. All these actors have connections to positions of political power in the region. One of the offenders in the Sonso Lagoon case is the cousin of the, by then, Governor of the region. The Governor has been linked to paramilitary groups, and accused of acquiring lands from drug dealers in the region in ongoing judicial processes. At the same time, the Governor was the political godmother of the director of the regional environmental authority (CVC) of that time, which means that he was chosen director as a political favor to the group headed by the Governor. In addition, the Governor was advised by the sugar mill owner Henry Eder who was also President Santos' appointee at the CVC board of directors, closing the circuit of territorial, political and agency control represented in Fig. 2.

This figure is not just representative of the Valle del Cauca region, but is a reflection of state development processes and politics in Colombia, in which the central state has never had full control of the territory and operates in the regions through delegating authority to local and regional power networks with whom it needs to negotiate continuously (González, 2009; Grajales, 2013). The creation of Regional Autonomous Corporations (CARs), which are the regional environmental authorities, falls into what Gellner (1992) calls an "incomplete centralization" characterized by the delivery of power by national authorities to individuals who are supposed to represent them but in practice possess a de facto local base of power. The CARs, created in 1954 by local elites, were given power by the central government as

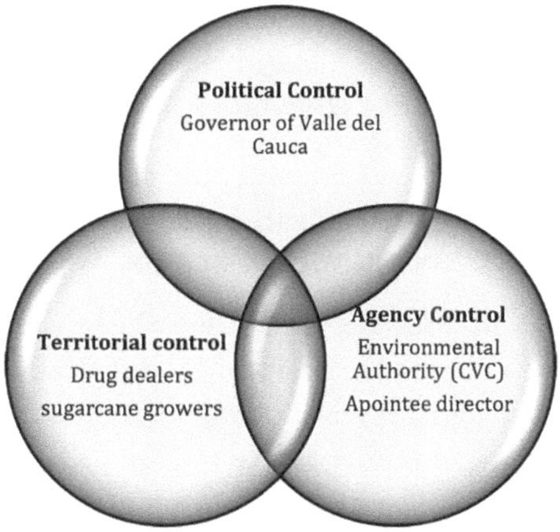

Fig. 2 A close local system of power.

autonomous and independent[b] entities for managing the environment and natural resources within their jurisdictional areas. Since then, CARs have resisted every attempt by the national government to limit their autonomy or to be more extensively controlled by national authorities.

The delegation of power and indirect rule of the State reinforces its fragmentation. This often contributes to making the implementation of modernizing reforms from the center more challenging. The traditional parties managed to control all the state's institutions through party factions, which have hindered any significant modernization of the state.

This configuration of political power helps to explain the type of responses to the Sonso Lagoon crisis, after it became recognized as a problem demanding immediate solution. The fact that the Director of the CVC was a political appointee of the Governor of Valle explains why he was not asked to resign by the Governor, despite the deluge of criticism after the case went public. Instead of stepping away from the agency, a subordinate, the director of the sub-regional CVC office in Buga was fired, not altering the party's grip on the agency.

On the civil society side, this configuration of a close local system of power also helps us to understand the strategies used by wetland advocates. According to Bolívar (2006), social movements in these contexts seek to reinforce direct rule mechanisms of the central state rather than indirect rule

[b] The autonomy is embodied in the independence of the board of directors to appoint its director, to set a preferred type of managerial organization and to hire the staff. Besides, they have their own financial sources.

mechanisms that result from negotiation of the central state with de facto local and regional powers. Likewise, Gibson (2005), who studies struggles between local incumbents and oppositions in authoritarian provinces within nationally democratic countries, argues that incumbents will prevail and prevent change if the scope of conflict remains localized. In contrast, nationalizing the conflict threatens incumbents and provides allies and resources to oppositions.

In the case of the Sonso Lagoon, initial attempts of the groups affected by the wetland damage to reach local officials and local media failed to produce any effective action, demonstrating that any change would be prevented if the conflict remained localized. It was clear that wetland advocates would have to break local boundaries and get national and even international assistance if they wanted to ensure serious attention to their case.

As described by Keck and Sikkink (1998) emerging environmental citizens need to strategically jump scales and engage in transnational networks. This strategy contributes to expanding the scope of conflict and engaging other actors who may step in on the side of the less powerful group (Birkland, 2011: 186). In fact, one of the Sonso Lagoon activists' strategies was to jump scales by resorting to national and international authorities (Fig. 3). This was made through written civil petitions requesting immediate action to stop the damages to the lagoon, the punishment of the wrongdoers and restoration of the wetland, that was directed to the CVC but copied to the Minister of the Environment and the General Attorney for the Environment.

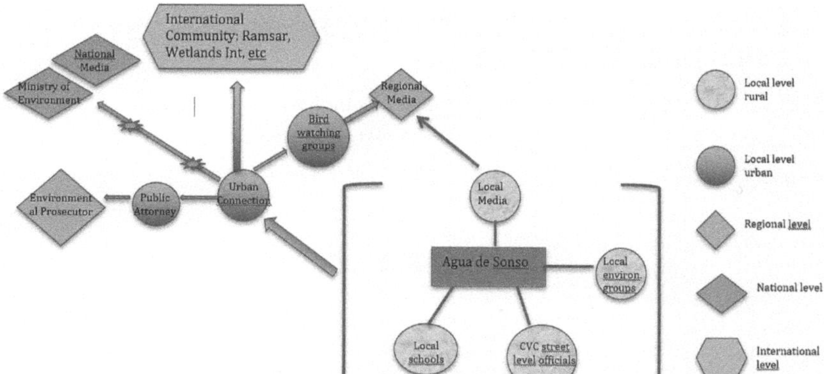

Fig. 3 Breaching boundaries in the Sonso Lagoon case. *The authors.*

With the aid of college professors, emails were sent to several members of the Ramsar secretariat, Wetlands International, and the Inter-American Association for Environmental Defense (AIDA). Each of these organizations wrote in turn urging messages to the Ramsar Secretariat, especially to Maria Rivera, one Colombian member of the secretariat. She replied almost immediately and urged the Ministry of the Environment to step in the case.

Another strategy was to reach out to national media through personal contacts from outside the region. The independent channel Noticias Uno covered the case in the national arena by revealing that the sugar mill that buys from the wrongdoers was Ingenio Castilla (one of the states involved in the case is a supplier of sugarcane for the mill) and urging Asocaña[c] to make a public statement about this case. After this initial coverage, the case got media attention nationwide and local media soon followed the lead. The Sonso Lagoon case was covered 34 times by regional and national media between February and mid-March 2016.[d] The Ministry of the Environment visited Cali on January 29, 2016 to share the commitments made by Colombia at the COP 21. That day the director of the CVC announced the removal of the sub-regional director of the CVC in Buga[e] and the recovery plans for the lagoon.

In order to prevent intervention from central government after the Sonso Lagoon case breached local boundaries of control, local authorities needed to act quickly to demonstrate capacity to solve problems in their jurisdiction. Otherwise, an intervention by central authorities could jeopardize the prevailing power structure shown in Fig. 2. This helps to explain the support obtained by the Director of the CVC from the Governor and their effectiveness in finding resources to start restoring the wetland.

3.1 Local communities transforming weaknesses, identities and mobilizing the state

After all the media attention the Sonso Lagoon case received by mid-February 2016 and the actions being undertaken in reaction to this attention, it seems almost unbelievable that such a small group of people, with a weak

[c] The umbrella organization for the Colombian sugar mills, most of them located in Valle del Cauca.

[d] The Ardila Lulle (the owner of the sugar mill Incauca) channel RCN only covered the case twice and in radio, one to announce that restoration actions were going to be implemented in the wetland (February 11th, 2016) and the other to announce Asocaña rejected the exploitation of the lagoon (February 29th, 2016).

[e] El País, 2016. CVC destituye funcionario por daño ambiental en Laguna de Sonso. January 29th, 2016. http://www.elpais.com.co/elpais/valle/noticias/destituyen-director-regional-dar-por-caso-sonso. Accessed March 19, 2016.

organization and few resources were able to achieve that much. As one of them recalled: "I almost cannot believe what is happening, the dwarf grew up."[f] What allowed such a powerless but courageous group of people to prevail against feared landowners and forced the state to work in restoring a wetland? Strategic allies, social media and a powerful message allowed them to mobilize resources they did not have at their disposal before.

The group comprising the grassroots association Agua de Sonso was formed informally in the 1960s in a makeshift village called Puerto Bertín, nearby the Sonso Lagoon, especially on the shores of the ditch that connected the Lagoon with the Cauca River, comprising 300 lots with less than 1 Ha each. Those communities developed their livelihoods around fishing in the Sonso Lagoon and rivers of the region, complementing this activity with sand extraction from the Cauca and Guadalajara rivers and pig breeding. They were declared by local authorities as a "population at risk" for being settled on a flood prone area and obliged to relocate despite their opposition to the measure.

The connection with urban allies allowed these rural community actors to compensate for their lack of cultural and financial capital. Before the December crisis they had never resorted to formal legal actions, such as civil complaints or *derechos de petición*[g]; they used to bring their claims to CVC officials directly or to the representative of the NGOs at the board of directors of the CVC informally. Low literacy skills prevented them from engaging in this type of activity. Urban connections allowed them to get legal advice. Urban connections allowed the Sonso Lagoon activists to reach out to other better-off and well-organized groups of citizens such as the bird watching groups that in recent years have grown in numbers and were by that time organizing the second Colombian Bird Fair, an impressive event with international guests and a lot of publicity. Finally, connections to urban allies allowed them also to spread the word of what was happening in other languages reaching international audiences. Urban allies also facilitated the use of resources such as drones to make aerial images of the destruction of the lagoon that were then used to make short videos to spur public outrage and to support community complaints before the authorities.

The lack of leverage on the environmental authority by community groups was compensated by the use of social media to create public support

[f] Interview with leader from Agua de Sonso, Buga, March 9th, 2016.

[g] Derecho de petición is a legal figure that allows citizens to request information, action, clarification, etc., to public institutions, which have 15 days to respond.

to their cause. A short video made by a fisherman of the Lagoon showing the destruction of the wetland and agency inattention was shared through Facebook and reached 37,159 views. Pictures and videos of the case were shared on a daily basis and thousands on Facebook watched the news reports. In addition, a smaller version of the *derecho de petición* filed by Agua de Sonso against the CVC and the Ministry was shared on Facebook, asking people to sign it and send it from their personal emails to the mailbox of the CVC, the Ministry of the Environment and the *Procurador Ambiental*.[h] This latter tactic, which is common in the U.S, had never been used before in Valle del Cauca, and the massive public response certainly made an impression on the environmental authorities.

The rapid diffusion of the Sonso Lagoon S.O.S message on social media was facilitated not only by connections to external groups but also by a clear and powerful framing of the problem. According to Layzer (2006: 492):

> *"Environmental advocates are more likely to succeed in proffering the authoritative problem definition if they can tell a story that features loathsome villains, sympathetic victims, a simple causal relationship between them, and an imminent crisis".*

The Sonso Lagoon story had all of these elements. Poor peasants were standing up to defend a valuable ecosystem and the heritage of the region that provided habitat for hundreds of animals, against a powerful insensitive big capitalist landowner motivated only by profit whose land conversion activities were about to cause a rapid destruction of the wetland if no action was taken. It appealed to a generalized but repressed feeling of injustice and past class grievances stemming from historical privileges given to sugar growers at the expenses of small farmers and communities.

The fact that the main villain in the Sonso Lagoon case was a drug dealer made him an identifiable enemy who is unable to contest fishermen framings of the situation due to his illegal and clandestine status. It allowed the wetland activists to define the problem in ways that raised its salience. On the other hand, the fear that illegal actors inspired was so huge that few local advocates dared to show their faces on the cameras at the beginning and many hesitated to keep up the fight because of fear of retaliations. Rapid uptake of the lagoon defense by the media and other civil society groups may have contributed to avoiding putting the first claimants on the spot for a much longer time. Academic actors joined the defense of the Lagoon with a public forum called Rescue Plan for the Sonso Lagoon

[h] It is the office in charge of exercising disciplinary control upon public officials in the environmental management area.

organized by local university professors and other educational institutions. The Assembly of Deputies of Valle del Cauca also joined the wetland defense by calling the Director of the CVC to a public session of accountability about the Sonso case and holding two meetings at the wetland. The last to join were international NGOs such as Colombian Wildlife Conservation Society, Birdlife International and WWF which only issued a press release offering support in the efforts to restore the lagoon.

Different from other cases in which protestors seek out state action, a demonstration was not the first recourse of the activists. The first demonstration only took place in January 30, 2016, organized by fishermen, local environmental groups and schools of the area. The demonstration served to reinforce community links and to reinforce the commitments made by the CVC around wetland restoration efforts.

Recent comments made by external observers of the case praised the emphasis on the scandal through social media for the achievements around the Sonso Lagoon defense. However, it is worth noting that the social media activity was supported by a courageous group of people on the ground, informing the public daily, taking pictures and providing data about the lagoon.

As a marginalized group, Agua de Sonso has limited access to information compared to other better-positioned groups such as sugar growers who have direct communication channels to governments and authorities. However, this was also compensated through low/medium-rank official allies inside the agency. Many years of mandated agency community engagement and participatory policies have created links and friendships among field officials and grassroots organizations that work as networks connecting community actors to agency offices. We see here the importance of networks connecting activists in civil society and committed individuals in state agencies to facilitate issue advocacy and institutional change, as Hochstetler and Keck (2007:19) highlighted for the Brazilian case. These agency allies were very important in providing documentation to support community demands, conveying information about what was happening inside the agency, informing about the reactions of high-rank officials and suggesting tactics to the wetland activists.

Several authors have underscored the role of environmental discourses in framing individual and group identities in the context of local struggles for access to resources (e.g., Pulido, 1996; Rocheleau and Ross, 1995). As mentioned above, in order to justify efforts to resettle them, Puerto Bertin dwellers have been portrayed as "population at risk" and resource destroyers, the latter referring to pig farming and other subsistence/contaminating practices they have resorted to in face of declining fishing stocks in the wetland. In saving the Sonso Lagoon, fishermen contested these identities and cast themselves as wetland stewards, highlighting the importance of the Sonso Lagoon for their livelihoods. By demanding the attention of the authorities to address the misdeeds committed in the wetland, they positioned themselves as the main caretakers and defenders of the wetland.

Saving the Sonso Lagoon also transformed an essentially forgotten wetland into a perceived site of socio-ecological injury, while simultaneously providing the social legitimation to correct that injury. By providing new pressure from below, wetlands activists prompted environmental authorities to do a job they did not have much incentive to do before: protect the wetlands. By doing so community actors achieved what Abers and Keck (2013) call "mobilizing the state," or galvanizing other components of the system to do their job better. CVC officials now are under permanent oversight to comply with the restoration agenda planned for the Sonso Lagoon by community, national and international actors.

In sum, the salience of the case in the media and the intervention of national and international actors shifted the internal balance of power, forcing the response from an agency biased in favor of development, giving more authority to grassroots organizations that were previously isolated and prompting the agency to move in a direction long sought by those community actors.

3.2 The difficult construction of practical authority to manage the environment in the region

We have outlined the main factors that contributed to "save" the Sonso Lagoon from becoming a sugarcane plantation for now, but how did this happen in the first place? Why did two landowners decide to damage the most important wetland of the region without regard to the consequences? Why were they so certain they would get away with that and invest more than US$345,000 on an illegal operation in plain sight of everyone?

The most widely given explanation has been that the drug dealers involved in this case had either bought or threatened (or both) the CVC sub-regional officials, who because of fear or personal interest overlooked the wetland destruction and remained complicit. This explanation would justify the firing of the sub-regional director of the CVC as a way to fix the system and recover institutional legitimacy, but a closer analysis of the case points to a deeper cause rooted in the existence of a permissive institutional context and the absence of "practical authority" by environmental institutions in the region. This section discusses the explanation outlined above and explores the lack of practical authority as an alternative explanation to the Sonso Lagoon case.

Before December 2015 when the damage to the lagoon occurred, the neighboring landowners of the Sonso Lagoon had been enjoying a permissive institutional climate, as construction of levees beyond wetland limits was common as was the practice of moving property fences towards the lagoon. However, an intervention of the magnitude of the December works was unprecedented. This coincided with the election of the Governor of Valle del Cauca Dilian Francisca Toro in 2015. Her assumption of office along with her party's control of the CVC arguably created a beneficial institutional context for these kind of landowners that would ensure immunity from law enforcement, encouraging them to go ahead with their encroachment plans for the lagoon. We contend that if the Sonso Lagoon case had remained a localized conflict, the lagoon would have disappeared.

Scholarly research has underscored the role played by illegal actors in infiltrating and weakening state agencies in the region (Duncan, 2006; Guzman and Moreno, 2007). The region was the birthplace of the powerful drug Cartel de Cali and Cartel of Northern Valle. The first one consolidated its power in the decade of the 1980s and the second after the dismantling of the first in 1995. Drug dealers have acquired lands in the region as a money laundering strategy and because of the prestige attached to landownership (Guzman and Moreno, 2007). The Sonso Lagoon case reveals that even though some drug dealers have been apprehended and imprisoned, they still control the lands they acquired as part of their illegal activities. It is undeniable that their power of intimidation and bribery has affected the functioning of state officials in enforcing the law in the region.[i] However, this pattern of permissiveness to powerful actors by environmental authorities, and an institutional bias toward landownership was not created by drug dealers, but existed well before they became important actors in the region. Instead, drug dealers found a system of unequal environmental law enforcement and weak environmental authorities already in place in the region and only reinforced those existing patterns. State building in the region, before drug trafficking was an important activity, was characterized by the existence of a

[i] During the fieldwork we heard many times CVC officials complaining of not being able to enter a wetland site or to apply sanctions for wetland-related offenses because the landowner was a drug dealer.

powerful elite involved in statecraft that created a vision of government, to a large degree as an extension of their private interests (Guzman and Moreno, 2007).

We argue that this lack of environmental authority is rooted in the incomplete transformation of the CVC from a regional development agency to an environmental agency and the tensions created by this mandated transformation. Similar to the other Autonomous Corporations in the country, the CVC was created to promote regional economic development. From 1954 to 1993, the agency was devoted to infrastructure and agricultural development projects and was attached to the Ministry of Development (1955–1968), the Ministry of Agriculture (1968–1977) and then to the National Planning Department (1977–1993) (Sánchez-Triana et al., 2007). The main priority of the agency was building infrastructure for water resources to be transferred from their sources to support development of other sectors such as agriculture. During all this time the interests of industry, large farmers, cattle ranchers and engineers were well represented at the agency board of directors. An owner of two sugar mills and several farms and companies was the Executive Director of the CVC for a long time. This institutional past created institutional practice and rationale that configured the agency's identity and defined how it related to social actors outside the agency, creating an institutional legacy difficult to change through legal mandate.

Under Law 99 of 1993, the Ministry of the Environment was created and given responsibility for planning national environmental policy. CARs were recast as regional environmental management authorities responsible for implementing national environmental policy, although retaining their autonomy in defining their own priorities and investment programs.

Several authors (Abers and Keck, 2013; Birkland, 2011; Layzer, 2006) have demonstrated that policies do not become translated directly from law to practice, but that implementation agents mediate and interpret them in a process that can seriously alter the final outlook of a law or twist the original goals of policy designers. We argue that the environmental mandate given by law to the CVC conflicted with the developmental mandate of its origins, and hindered the development of practical authority by the agency. Layzer and Rinfret (2020) have shown how attempts to alter policy from outside an agency encounters resistance from its clientele and even its own personnel because of the influence that past policies have in agency practices and culture. In particular, they point out three important elements that are helpful for understanding the CVC's complicated environmental turn, which we discuss in the next sections.

3.2.1 The legacies of past policies shape agencies behavior

According to Layzer and Rinfret (2020) an agency is likely to resist adopting practices that seem contrary to its founding mission, because agencies develop a certain view of their mandates shaped by the legacies of past policies.

The founding mission of the CVC was clearly rooted in a developmental mandate that was especially oriented to assisting large agriculture in the region to get access to water and other resources that under the environmental legislation from 1993 are to be protected and managed differently. The following quote from a CVC official reflects the tensions created by those legacies of past policies.

> *From the sixties to the nineties we devoted ourselves to do all that is contrary to what we do now, dikes, land conversion to agriculture. All the landowners have levee permits issued by us. Many of the things that the law wants to change now were approved by the CVC. In addition, there is a lot of pressure to transform the land to agriculture; the CVC official always said that the landowner has the right to do whatever he wants with his land. (Interview with CVC official, Buga, March 9th, 2015).*

3.2.2 Organizational cultures are shaped by what is considered appropriate behavior inside an agency

Layzer and Rinfret (2020) show how congressional reactions to administrative decisions create a sense of what is appropriate and legitimate behavior among agency employees. Because of the CVC's autonomy, the board of directors of the agency plays a role similar to what we would expect of Congress. The authors contend that over time, this oversight molds the agency's organizational culture and standard operating procedures.

Despite the fact that Law 99 mandated inclusion of new actors, such as representatives of Afro-Colombian, Indigenous and NGOs organizations, in the board of directors of the CVC those groups were and continue to be very weakly represented, while industrialists and sugarcane grower representatives continued to be CVC board members and exert a great deal of influence on the decisions affecting them. Sugar grower representatives in the board of directors have been able to block the approval of wetland management plans as well as certain projects that recommend restrictions to agricultural activities.

Lower ranking officials have learned from these experiences and consequently act to avoid confrontations with members of the sugarcane sector or other large landowners and adopt a permissive behavior towards them. CVC officials expressed several times during fieldwork the futility of going against large landowners' wills because "they will do whatever they want some way or another." As explained by Birkland (2011: 269) street level bureaucrats are not entirely free agents; they are constrained to act in a particular way by the resources available to them, by professional norms, by legal sanctions that can be applied for noncompliance, and, as we see in this case, by informal sanctions and expected behavior for agency directors. This contradicts the hypothesis that stresses bribery as the sole explanation for understanding CVC officials' excessive permissiveness to large landowners. We argue that CVC employees also suffer from a lack of positive incentives to enforce the law against this sector.

3.2.3 Clientele groups constrain agency's choices

According to Layzer and Rinfret (2020), interest groups privileged by past policies in turn create the context for current decisions and constrain an agency's choices among alternatives. That is, despite national environmental mandates that the CVC is required to implement, industrial elites in the Valley are not accustomed to taking action in response to government threats but rather resort to taking advantage of legal loopholes in regulations to stymie coercive efforts (Sánchez-Triana and Ortelano, 2005: 498). In face of clientele opposition to command and control policies in environmental matters, and due to strong influence of economic sectors within the board of directors, strategies such as consultation, partnerships with the productive sector,[j] and voluntary agreements have been privileged for environmental control as shown by Sánchez-Triana and Ortelano (2005). In addition, fines for environmental misconduct have been set too low to motivate firms to comply,[k] enforcement efforts have been of secondary importance, legal procedures to impose the fines are cumbersome and often industries find ways to avoid paying them.

In the absence of counterbalances represented by organized and strong environmental or community interests in the decision-making process within the CVC, those economic interests are more likely to prevail. Economic actors such as sugarcane growers also exert a great deal of influence on the policy processes that directly affect them. They have managed to

[j] The initiative put forward by the CVC in 2002 to protect relicts of dry forest from sugarcane burning practices consists in subscribing partnerships of cleaner production with the sugar sector, the Ministry of the Environment and Asocaña (Salazar et al., 2002: 69).

[k] CVC's fee calculation procedure, which was based on recovery of administrative costs, yielded discharge fees substantially lower than levels required to provide firms with financial incentives to reduce pollution (Sánchez-Triana and Ortelano, 2005: 503).

be the only civil society actors consulted by the agency in designing flood control measures for the region as part of the Cauca River Corridor Project (Moreno and Selfa, 2021).

3.2.4 Practical authority entangled in economic and political webs

In addition to the points highlighted by Layzer and Rinfret, another intervening factor since the introduction of Law 99/93 is the mandated participation of political actors within the CVC through the representation of the municipalities in its board of directors. According to some external observers (Canal and Rodriguez, 2008; Sánchez-Triana et al., 2007), these provisions resulted in the introduction of clientelistic practices within the agency, affecting its technical performance.

> *"Hiring and promotion of staff are based on political criteria rather than technical merit, and this in turn, has increased staff turnover and management disarray... Many members of boards of directors are focused on obtaining contracts from the CAR for various interest groups and or for themselves." (Sánchez-Triana et al., 2007: 56)*

As described by Abers and Keck (2013: 9) implementing laws in ways that endow new arenas with the ability to make binding decisions requires the development of organizational and technical capacities, among other resources. In this case we argue that past policy beneficiaries, especially economic actors, hindered the development of organizational capacities by retaining participation privileges within the agency, while political actors hindered the development of technical capacities by controlling staff hiring and allocation of resources inside the agency. This in turn hindered the agency's ability to resolve problems, which is the basis for building the reputation and recognition an organization needs to position itself as an authority.

Even sugar growers do not recognize the CVC as an authority in the territory, as expressed by this sugar mill engineer who highlights the organizational and technical weaknesses of the environmental management in the region:

> *There are many people that do not waste their time asking for CVC permits to build their levees, they do it and it stays, you go to the CVC and they say yes we have this disciplinary file against him (the transgressor) and he will have to pay a fine, but it does not follow-through, it's just tons of useless papers, so the landowner pounders: the sugar mill went to get the permit and had to cede some land, this other landowner just made his will and he built his levee as he wanted without losing any land, what should I do? The right thing or the wrong thing? This is a*

governmental task (to control levee building and placement), unfortunately it is in the hands of an agency that is not strong either technically or legally because they do not enforce the law equally and they do not have the same technical opinion about all subjects (Interview with a sugar mill engineer, Buga, February 3th, 2015).

This lack of practical authority is translated in landowners' lack of recognition of and refusal to adopt actions that emanate from command and control policies enacted by the environmental agency. Therefore, enforcing environmental norms on powerful local actors, some of whom have a seat at the CVC board of directors, has proven to be a daunting task in the region. The intervention of civil society actors allied with national and international actors in the Sonso Lagoon case shows some hints of how more profound changes could be achieved by altering the balance of power in which the CVC is immersed.

3.3 Contesting elite roles in the region

The role played by sugarcane associations during the Sonso crisis reveals its limited environmental responsibility in the region. This may be due to the described lack of social and state control that can confront the predatory practices exerted by this industry. Even though the lands affected in the Sonso Lagoon were going to be converted into a sugarcane plantation, and this product would be bought by a sugar mill that had signed a contract in advance with the landowner, sugar mills did not take any responsibility for what happened to the wetland. They issued a press release on January 20th in which they rejected responsibility for the damages committed against the Sonso Lagoon and urged the CVC to castigate the wrongdoers and to implement the required restoration activities. Nevertheless, they did not say anything about cancelling the contracts of those sugarcane suppliers who had committed the environmental damages.

As shown by Abers and Keck (2013), it is extremely difficult to get economic actors to act collectively to protect public goods, (in this case wetlands and rivers), particularly if this means monetary or investment burdens for something that they previously received or used without cost. In this case, sugarcane growers have been using water from wetlands for irrigation despite rules prohibiting this use, and lands that are supposed to be set aside as protective forest buffers have been occupied with crops, which has been prohibited by laws dating from the 1970s (Contraloría Departamental del Valle del Cauca, 2009; Escobar, 2005).

The necessary steps to ensure wetland protection include enforcing the laws aimed to protect protective forest buffers, locating levees without interrupting hydrological connections between the river and the wetlands and to ensuring water from wetlands is not used for irrigation. However, the effective application of these rules implies deep changes in the way the sugar industry relates to water resources in the region and a change in their sourcing policies to encourage suppliers to implement better environmental practices in their agricultural activities.

The absence of a discourse and practice of co-responsibility by the sugar industry is arguably one of the factors contributing to environmental degradation in the region, but the adoption of responsible practices towards the environment seems not likely to develop without social pressure and public demand, as has been the case for other industries in the world. The Sonso Lagoon case triggered a public questioning of the role of the sugar sector in the processes of environmental degradation in the region, but it remains to be seen if this challenge is echoed by other social actors, generating incentives for implementing "ethical sourcing" policies by the sugar industry or if this gets stifled as has happened in the past.

4. Conclusions

The Sonso Lagoon case is exemplary in making visible the factors affecting wetland governance in a Colombian region where powerful elites and politicians with links to drug cartels and the sugar industry control the functioning of state institutions at the local level, and in demonstrating how such a sub-system can be challenged to bring about some changes towards the protection of wetlands and a more plural governance system.

In this article we outlined an alternative explanation to corruption in environmental agencies to understand how the most important wetland in Valle del Cauca was being destroyed unchecked. Analyzing the institutional context in which the regional environmental authority of Valle del Cauca is embedded, we pointed to a deeper cause rooted in a pattern of permissiveness to powerful actors by environmental authorities, and an institutional bias toward landownership, that created a system of unequal environmental law enforcement and the absence of "practical authority" by environmental institutions in the region. This pattern of permissiveness was the result of legacies of past policies that privileged developmental values and actors, an organizational culture within the CVC that created a lack of positive incentives to enforce the law against big entrepreneurs, and a decision-making system constrained by clientele groups due to participation privileges retained by economic actors within the agency. These factors along with the interference of political actors in staff hiring and allocation of resources inside the agency hindered the ability of the CVC to develop organizational and technical capacities, necessary to position itself as a legitimate authority in the region.

All the factors mentioned above created a difficult backdrop for including different actors and values in wetland governance in the region. The lack of counterbalances represented by organized and strong environmental or community interests in the decision-making process within the CVC, made economic interests more likely to prevail, especially in the absence of an environmental agency with practical authority to enforce environmental law equally.

However, the Sonso Lagoon case also demonstrated how local marginalized actors in ecosystem governance can alter the balance of power in which environmental agencies are embedded and provide incentives for them to do their job: protecting ecosystems such as wetlands. Grassroots organizations of the Sonso Lagoon achieved this through alliances with low/medium-rank agency officials that compensated their limited access to information, the use of social media to create public support and crafting an appealing message that prompted action in favor of wetlands. The case also highlighted the importance of considering institutional political arrangements to understand state responses to environmental problems and strategies employed by advocacy groups. In the Sonso case we described a close local system of power wetland activists had to break to ensure attention to the wetland problem, which they did through urban and international allies.

A lesson learned from the Sonso Lagoon case is related to the importance of strengthening pluralism for improving wetland governance in the region. The role played by grassroots organizations in defending the Sonso Lagoon showed the contradiction existent in current wetland management in the region, where organizations composed of community actors such as fishermen, small farmers and eco-tourism guides play a very important role in protecting these ecosystems, but at the same time are commonly excluded from decision-making bodies that authorize and regulate such practices.

A proposal to restore wetlands in the region must accompany a proposal to address social inequities, incorporating social groups whose livelihoods depend on wetlands and who have been the most adversely affected by its degradation. We argue that addressing this kind of inequity is important for the creation of legitimate practical authority in the regional environmental management. An agenda in this direction will require (1) an acknowledgement of the social structures that have allowed environmental degradation in the region; (2) the inclusion of marginalized actors such as grassroots organizations in decision-making bodies and procedures; and (3) strengthening grassroots organizational capabilities that allow them to fully participate in wetland management and decision-making.

Another approach to addressing inequities and building practical authority in environmental management is to grant decision-making power to the local wetland management committees in which grassroots organizations participate actively as mandated by the Wetland National Policy issued in 2002. According to observation of those spaces during fieldwork, the committee meetings are indeed attended by stakeholders from different sectors that meet on a regular basis, which in theory should enhance collaboration and participatory outcomes in wetland management. But the committee lacks decision-making capacity. Decisions about the Sonso Lagoon are made outside this committee regardless of the discussions undertaken at this venue. According to several authors (Koontz et al., 2004; Lubell, 2005; Abers and Keck, 2013) when participation takes place without authority for implementation or when deliberation fails to affect outcomes, it can also undermine trust in the system which tends to lose legitimacy.

It is also important to find a way to link those local spaces to other scales of wetland governance as suggested by the importance that national and international actors played in saving the Sonso Lagoon, as a way to introduce a system of check and balances in wetland governance that can work not only during crisis but in a more permanent fashion. This aligns with proposals by scholars of decentralized resource regimes such as Andersson and Ostrom (2008: 80) who argue for the introduction of accountability mechanisms to govern the relationship between the local government and other governance organizations at different levels, such as central government agencies, community-based organizations and non-governmental organizations, as a way to generate positive incentives to invest in natural resource governance since they are rarely generated internally within local governments.

Finally more research is needed to design conditions and incentives that constrain the predatory practices of economic actors, such as sugar growers, from driving unsustainable uses of water resources. Implementing incentives for better agricultural practices instead of governmental subsidies to increase production, the enforcement of restrictions on agricultural encroachment over environmentally fragile areas, and demanding that economic actors adopt sustainable practices either through market mechanisms or other governance mechanisms may help to alter the power imbalances of the institutional context in which environmental agencies in places with weak states are embedded.

References

Abers, R.N., Keck, M.E., 2013. Practical Authority: Agency and Institutional Change in Brazilian Water Politics. Oxford University, New York.

Andersson, K.P., Ostrom, E., 2008. Analyzing decentralized resource regimes from a polycentric perspective. Policy Sci. 41 (1), 71–93.

Birkland, T.A., 2011. An Introduction to the Policy Process: Theories, Concepts, and Models of Public Policy Making, third ed. Routledge, New York.

Bolívar, I, 2006. Transformaciones de la política. Movilización social, atribución causal y configuración del Estado en el Magdalena Medio. In: Archila, M, et al. (Eds.), Conflictos, poderes e identidades en el Magdalena Medio, 1990-2001. Colciencias/ CINEP, Bogotá, pp. 388–427.

Brooks, S., 1993. Public Policy in Canada: An Introduction, second ed. McClelland and Stewart, Toronto.

Burawoy, M., 1991. The extended case method. In: Burawoy, M., Burton, A., Ferguson, A.A., Fox, K.J., Gamson, J., Gartrell, N., Ui, S. (Eds.), Ethnography Unbound Power and Resistance in the Modern Metropolis. University of California Press, Berkeley, pp. 271–287.

Canal, F., Rodriguez, B.M., 2008. Las Corporaciones Autónomas Regionales, quince años después de la creación del SINA. In: Rodríguez-Becerra, M. (Ed.), Gobernabilidad, Instituciones y medio Ambiente en Colombia. Foro Nacional Ambiental, Bogotá DC, Colombia, pp. 303–390.

Contraloría Departamental del Valle del Cauca, 2009. Informe de auditoria gubernamental Modalidad Especial Ambiental a la Gestión Integral del Recurso Hídrico en el Valle del Cauca. Municipios de Guadalajara de Buga, San Pedro, Yotoco y empresas de servicios públicos de acueducto y alcantarillado ACUAVALLE y Aguas de Buga. Vigencias 2007–2008. CDVC-CARNMA—No. 02, September, Cali, Colombia.

Duncan, G., 2006. Los señores de la guerra: de paramilitares, mafiosos y autodefensas en Colombia. Editorial Planeta, Bogotá DC, Colombia.

Emerson, R.M., Fretz, R.I., Shaw, L.L., 1995. Writing Ethnographic Fieldnotes. University of Chicago Press, Chicago.

Escobar, L.A., 2005. Formulación Red de Corredores Biológicos. CVC, Cali.

Gellner, E., 1992. El arado, la espada y el libro. Estructura de la historia humana, Fondo de Cultura Económica, México.

Gibson, E.L., 2005. Boundary control: subnational authoritarianism in democratic countries. World Politics 58 (1), 101–132.

González, F.E., 2009. Espacio, conflicto y poder: las dimensiones territoriales de la violencia y la construcción del Estado en Colombia. Soc. Econ. 17, 185–214.

Grajales, J., 2013. Speaking Law to Land Grabbing' Land Contention and Legal Repertoire in Colombia. The Land Deal Politics Initiative. http://www.iss.nl/fileadmin/ASSETS/iss/Research_and_projects/Research_networks/LDPI/LDPI_WP_17.pdf. Accessed 26 March 2013.

Guzman, A., Moreno, R., 2007. Paramilitary, drug trafficking and state behavior in Valle del Cauca. In: Romero, M., Valencia, L. (Eds.), Parapolitics: The Route for Paramilitary

Expansion and Political Agreements. Corporación Nuevo Arco Iris, CEREC and ASDI, Bogotá DC, Colombia, pp. 165–237.
Hochstetler, K., Keck, M., 2007. Greening Brazil, Environmental Activism in State and Society. Duke University Press, Durham.
Keck, M., Sikkink, K., 1998. Activists Beyond Borders. Cornell University Press, Ithaca and London.
Koontz, T., Steelman, T., Carmin, J., Korfmacher, K., Moseley, C., Thomas, C., 2004. Collaborative Environmental Management: What Roles for Government? Resources for the Future Press, Washington, DC.
Layzer, J., 2006. The Environmental Case, Translating Values Into Policy, second ed. CQ Press, Washington, DC.
Layzer, J., Rinfret, S.R., 2020. The Environmental Case, Translating Values Into Policy, fifth ed. CQ Press, Los Angeles.
Lubell, M., 2005. Do watershed partnerships enhance beliefs conducive to collective action. In: Sabatier, P.A., Focht, W., Lubell, M., Trachtenberg, Z., Vedlitz, A., Matlock, M. (Eds.), Swimming Upstream: Collaborative Approaches to Watershed Management. MIT Press, Cambridge MA, pp. 201–232.
Moreno, R., Selfa, T., 2021. Taming the cauca river: community and sugar landowners' contrasting narratives in addressing flood risk in Valle Del Cauca, Colombia. In: Cortesi, L., Joy, K.J. (Eds.), Split Waters: The Idea of Water Conflicts. Routledge, India, pp. 155–175.
Pulido, L., 1996. Ecological legitimacy and cultural essentialism: hispano grazing in the Southwest. Capital. Nat. Social. 7 (4), 1–22.
Rocheleau, D., Ross, L., 1995. Trees as tools, trees as text: struggles over resources in Zambrana-Chacuey. Dom. Repub. Antipode 27 (4), 407–428.
Salazar, M.I., Gómez, N., Vargas, W.G., Reyes, M., Castillo, L.S., Bolívar, W., 2002. Bosques secos y muy secos del departamento del Valle del Cauca. Corporación Autónoma Regional del Valle del Cauca (CVC), Cali, Colombia.
Sánchez-Triana, E., Ortelano, L., 2005. Influence of organizational learning on water pollution control in Colombia's Valle del Cauca. Water Res. Dev. 21 (3), 493–508.
Sánchez-Triana, E., Ahmed, K., Awe, Y. (Eds.), 2007. Environmental Priorities and Poverty Reduction. A Country Environmental Analysis for Colombia. The World Bank, Washington.
Zwarteveen, M., 2015. Regulating Water, Ordering Society: Practices and Politics of Water Governance (Oratiereeks). Universiteit van Amsterdam, Amsterdam.

Methods and approaches to foster pluralism in translational ecology

CHAPTER SEVEN

Participation as a pathway to pluralism: A critical view over diverse disciplines

Julia Leventon[a,*], Lenka Suchá[a], Barbora Nohlová[a,b], Simeon Vaňo[a,c], and Zuzana Veronika Harmáčková[a]

[a]Global Change Research Institute of the Czech Academy of Sciences, Brno, Czech Republic
[b]Faculty of Humanities, Charles University in Prague, Praha, Czech Republic
[c]Department of Ecology and Environmental Sciences, Constantine the Philosopher University in Nitra, Nitra, Slovakia
*Corresponding author: e-mail address: leventon.j@czechglobe.cz

Contents

Abstract

Wider participation in ecosystem governance has been called for as a means to reach pluralism in biodiversity conservation research and practice. The logical arguments given are that plurality in worldviews, priorities and values (social plurality) should be brought to the table to be reflected in decision-making, via participation. In this chapter, we give critical insight into participation as a pathway to recognizing social plurality in ecosystem governance. We highlight how different disciplines, beyond ecology, have tackled the topic of participation; including development studies and anthropology, governance and policy studies and transformations/transitions research. We look across this diverse literature to understand the ways in which different participatory approaches contribute to ecosystem governance. We outline how governance systems create multiple spaces for participation. We then explore key challenges faced when working within these spaces with the intention of addressing social plurality. We discuss challenges of complexity, inclusion, conflict and power. In doing so, we present an

Advances in Ecological Research, Volume 66
ISSN 0065-2504
https://doi.org/10.1016/bs.aecr.2022.04.006

argument that participation itself must be plural, in order to reflect the positioning within the governance system, and the context of power that it is taking place within. In recognizing that participation as a pathway to plurality cannot be a prescriptive process, we provide seven reflexive questions to steer its design and implementation.

1. Introduction

There is increasing recognition that biodiversity conservation and restoration must incorporate plurality into ecosystem governance. Ecosystem governance is the process and outcomes of decision-making around the management of ecosystems. Plurality is based within an understanding that ecosystems and social systems are not disconnected entities, instead they shape each other in interconnected social-ecological systems. Thus, ecosystem use and change are driven by a variety of contextual socio-economic and institutional factors (Turkelboom et al., 2018). Across a range of social-ecological systems there is a diversity of social systems, in which myriad of values, worldviews, beliefs and positions are recognized and accepted (Connolly, 2005). Such social plurality influences the bundles of ecosystem services that people prioritise and value (Peter et al., 2021), and can extend to our normative visions for future ecosystems, and how these can be achieved (Harmáčková et al., 2021). Relevant, effective ecosystem governance must reflect and incorporate this social plurality in management of different social-ecological systems and their dynamics (Pascual et al., 2021).

To this end, participation is often framed as being a way to incorporate social plurality, and be explicit, transparent and reflexive about values (Aguiar et al. 2020; MacLeod et al., 2021). Following Reed (2008), we define participation as "a process where individuals, groups and organisations choose to take an active role in making decisions that affect them" (p. 2418). Such decisions can relate to concrete project implementation and planning, or more broadly to goal setting (Stringer et al., 2006). Participation has a strong tradition in environmental governance, policy and planning, and international development. Here, the rationale of participation can be: (1) instrumental—to legitimize decisions and improve results; (2) substantive—to incorporate local knowledge and perceptions into decisions; and (3) normative—to increase democracy by addressing incumbent powers (see Wesselink et al., 2011). Particularly this substantive rationale feeds into narratives of the need for social plurality in ecosystem governance. In addition, participation has come to the fore in work from transdisciplinary sustainability science and transformative research.

For example, 'transformative spaces' (e.g. Pereira et al., 2020), have a focus on engaging with people to reflect social plurality while co-creating solutions, all facilitated within social-ecological systems at local scales. Such transformative spaces contribute towards a growing body of transdisciplinary and transformative co-production of knowledge. Such work extends our understandings of participation beyond participation in specific decision making; it bridges between research and practice by integrating across knowledge types, interests and values and therefore directly tackles questions of plurality (e.g. Caniglia et al., 2021; Chambers et al., 2021).

In this chapter, we aim to give critical insight into participation as a pathway to recognizing social plurality in ecosystem governance by looking across diverse traditions of participation. By doing so, we pull together their knowledge to unravel what they collectively contribute to understanding participation, and its strengths and weaknesses, as a pathway to plurality. Our research adds to, and helps to make sense of, existing critiques of participatory processes across disciplines. Indeed, we acknowledge the vast range of literature that studies participation across these different research traditions, highlighting the strengths and limits, and offering critical overview of a range of participatory understandings. Many readers will be well aware of distinctions in Arnsteins' ladder of participation from manipulation to citizens control (Arnstein, 1969) and its numerous revisions. Further, there are well articulated arguments about the tyranny of participation (e.g. Kothari and Cooke, 2001; Williams, 2004) whereby participation disempowers people and contributes to further marginalisation. And there is increasing evidence of the ways in which participation does, or doesn't lead to improved environmental outcomes (e.g. Kochskämper et al., 2018; Daniels and Walker, 2001). Furthermore, there are calls to tailor participatory methods to social contexts, because such contexts determine the entry point of participants engaging with the research process, and forms expectations and biases of all the involved parties (Lapadat et al., 2005). Indeed it is essential to find new participatory approaches that recognize different ways of knowing, being, and interacting, with a view to decolonizing western-centric approaches to participation (Ghosh et al., 2021). We do not wish to repeat these valid arguments. But rather, we intend to add nuance to understanding how these dangers need to be avoided by sensitive navigation of decision-making spaces within ecosystem governance for plurality.

In order to answer our primary aim, we first unpack essential participation literature (Section 2). We do not provide a systematic literature review, but rather an argumentative essay that draws on the three research fields that most directly and consistently address participation: development studies, environmental governance and planning, and transitions and transformations research. We look across them and explore the ways they use and understand participation. We scan broadly through these fields, not limiting ourselves to that which is explicitly labelled as being about ecosystems or biodiversity. Section 2 thus forms a kind of extended introduction to the roots of participation, in order to frame what we mean when we talk about it. It further helps to position the diverse understandings of participation, relative to each other. In Section 3, we bring together the literature to understand participation's role and position within ecosystem governance. Here, we introduce the idea of there being multiple spaces for participation, and we outline the factors that create and shape these spaces. In Section 4, we again integrate across our research fields to discuss how participation can capture social plurality. Again, this is not a systematic review, but an argumentative essay that picks up key pitfalls in participation. In Section 5, we draw on our critiques thus far to discuss the ways in which participation is a pathway to social plurality in ecosystem governance. We (re)iterate the importance of navigating the complexities of multiple spaces for participation within governance systems, and diversity of implementing contexts for participation. We aspire to creating practical recommendations, and with this aspiration in mind, we outline reflexive guiding principles to navigate these complexities. It is our intention that these reflexive principles help in creating participatory processes that are pathways to plurality in ecosystem governance.

2. Participation: Diverse traditions across disciplines

2.1 Development studies and anthropology

The roots of participation can be traced from the work of Paulo Freire on *critical pedagogy*, especially from his book *Pedagogy of the Oppressed* (Freire, 1968). Inspired by the critical exploration of the relationship between the colonizer and the colonized, Freiran tradition argues that people are the carriers of the change, and they are and should be capable to conduct their own analysis of their own reality and contribute to its transformation (Chambers, 1994). The notion of deep societal transformation driven by the people themselves, and not by the external agency, became inspirational for numerous researchers and practitioners active in a wide range of fields and disciplines. As a result, there are countless frameworks and approaches that are using participation for both, research and practice, hence the term participation itself consists of diverse understandings about what it is, how it should be used and practiced, and what are the actual boundaries that draw the line between what is participatory and what is not (Chevalier and Buckles, 2019).

Inspired by Paulo Freire, and strongly linked to the colonial burden of social sciences, historically producing a great asymmetry of roles and power relations (Clair, 2003; Jessee et al., 2015), participation in research and action research emerged as a critical stand to knowledge creation dominated by researchers. Following the postcolonial turn in social sciences, such as writing against culture movement and critical ethnography, participatory approaches are prioritizing the process of knowledge (co)creation over the *knowledge first* (Wittmayer and Schäpke, 2014). Practically, participation translates to the shift in the roles of the participants from passive to active. Before this postcolonial turn, people were usually objects of the research, 'mere contributors', and mostly served as sources of information with only limited or no impact on the research outcomes. The surge of participatory approaches attempted greater emancipation and empowerment of the participants (Clair, 2003; Jessee et al., 2015). In this respect, *participatory action research* (PAR) was conceptualized in order to address the need of direct involvement of researchers with communities by engaged inquiry through sharing of different perceptions of reality. This aims at achieving power among grassroots and that allows for deep societal transformation and empowerment (Rahnema, 1990). In fact, the acronym PAR principally reflects the core elements of participation itself, where *participatory* stands for life in society, *action* for experience and *research* for knowledge production (Chevalier and Buckles, 2019).

Considering the elements and ideological background of PAR, it is not surprising that its principles and ideas eventually proliferated from research to practice, and translated to discourse of *participatory development* that emerged as a criticism of large-scale top-down interventions and projects in the international development sector. Although PAR resonated in developing practice since Freire and his contemporaries, the real surge of participation was during late 1980s and early 1990s when the failure of top-down approach to development became evident. It is especially the work of Robert Chambers on *Participatory Rural Appraisal* (PRA) that gained prominence in the development discourse (Chambers, 1994). Emphasizing the importance of incorporating the voice and knowledge of the poor in the planning and management of development projects, PRA gained popularity among NGOs working in the field of community development.

Such actions were expected to deliver an ownership of local communities over the development projects and their sustainability in a long run (Chambers, 1994). Thanks to its popularity within the NGO community, but also due to lacking emphasis on social aspects of development among international development agencies (Francis, 2001), over the 1990s, the notion of participation became prominent also in governmental development assistance programmes, thus participatory development turned into a dominant discourse promoted by World Bank and alike (Williams, 2004). Further, development agencies have spread participation beyond their own practices, and into those of the governance systems of the countries they are interacting with. This spread has been achieved through an emphasis on fostering Good Governance, which according to many definitions from development agencies (including World Bank, USAID, UNDP) includes focus on processes of participation and transparency (see e.g. Grindle, 2007).

The development studies perspective on participation often spans over ecosystem governance, such as in *community-based natural resource management* (CBNRM), one of the most extended forms of participation at the community level. Here, the rights and responsibilities for managing a resource are given to the communities that are in most contact to, or are most affected by, that resource. This can be seen in managing areas protected for wildlife like, e.g., Zimbabwe's CAMPFIRE project (Shereni and Saarinen, 2020) or joint forest management (Bhattacharya Prodyut et al., 2010). These approaches work very much within existing policy frameworks and targets, but seek to empower communities to manage resources, on the premise that they are best placed to take effective action and decisions over their livelihoods. In some cases, this was also intended as a move to respect local property rights and ways of life. However, in many cases, the initiatives were kept largely as instruments for delivering external policy targets or wishes (e.g. carbon sequestration or nature conservation) in exchange for patchy rewards, without sufficient practical support (see e.g. Nohlová, 2020).

2.2 Environmental governance, policy and planning

The need for participation has been embedded within environmental governance and planning across the world. The emergence of participation in environmental governance and planning cannot be entirely separated from understandings emerging in Section 2.1. as these two often intersect. Following the 1992 Earth Summit in Rio de Janeiro, Agenda 21 pushed countries to strengthen the role of all citizens, and marginalized communities, in creating sustainable futures. While this helped reinforce the participatory turn in international development activities (see above), it also saw the introduction of participation requirements in environmental governance in the minority world.[a] Formal requirements for participating with affected stakeholders, including citizens, civil society and the private sector, emerged in high level legislation, such as the EU Water Framework Directive (e.g. Carter and Howe, 2006). These approaches to participation were often framed around bringing diverse actors into state-led policy-setting or management planning processes. For example, the aforementioned Water Framework Directive requires participation from stakeholders in the setting of river basin management plans.

Linked to these foundations, the push for collaborative governance can be seen as an extension of participation, alongside the push for decentralisation, in environmental policy. On a higher governance level, such decentralisation creates a polycentric system of self-organized authorities that co-operate at multiple scales (Ostrom, 2010) and that participate in and adapt to new developments in policy and practice (Folke et al., 2005). In collaborative governance, stakeholders are brought together to work together on governing specific locations, ecosystems or industries based on consensus (Ansell and Gash, 2008). Within economic sectors, these may occur, for example, as industry-government-civil society partnerships, private governance or self-governance (see e.g. Beyers and Heinrichs, 2020). Alternatively, collaborative governance can be spatially defined, e.g., bringing together farmers and landscape scale coordinating bodies to ensure management across interests at an ecologically-coherent scale (e.g. Westerink et al., 2017). Or, for example, in the case of food policy councils, these collaborations seek to engage across sectors, engaging government, industry and public at local levels to shape place-based strategies to address food system issues (see Bassarab et al., 2019; Sieveking, 2019). Or, in another instance, some local authorities induce collaborative governance in effort to harness a larger social capital and knowledge base for ecosystem stewardship (e.g. Vaňo et al., 2021), which if well promoted by higher authorities, could lead to improved landscape management strategies (Buijs et al., 2019). Key to these collaborative approaches is that they extend

[a] In this chapter, we use the term minority world to denote those countries that are often described using the problematic terms: developed world or global north. Instead, we explicitly recognize that these countries host the minority of world population, despite having a disproportionate influence on resource use.

the time frame, such that there should be longer-term learning and adaptation between the involved actors. However, there may be constraints to the extent to which these approaches can be truly participatory, as the opportunities for learning or shaping agendas is influenced by the broader government policy that they sit within (e.g. Newman et al., 2004).

Most recently, there has been attention on participatory approaches that sit outside of state policy making, and happen independently of it, with the intention of creating impact to policy. A prominent example would be the use of citizens juries to generate place-based recommendations for climate action (see e.g. Ostfeld and Reiner, 2020). While some have occurred at national levels with involvement of the state (e.g. the UK, France, Poland), others have been more local and led by civil society and/or academic actors. The rationale of these citizens juries, in spaces outside of formal policy-making, is to provide additional forums for democracy (Willis, 2020) and explore citizens wishes and policy recommendations when they are provided with in-depth information on a particular topic or location. Currently, these processes are framed as giving public mandates to decision makers to act on climate change, and help to add more depth to public perceptions on climate change and policy responses (Wells et al., 2021). However, there are challenges in ensuring that these are then embedded and picked up within policy and decisions (ibid.).

2.3 Transformations and transitions

Within this section on transformations and transitions, we refer to work that labels itself as transformations research and transitions research as terms that are often used interchangeably (Hölscher et al., 2018), and within transitions, to include social-technical transitions (e.g. Geels, 2018). We note that often, transitions, and social-technical transitions often considers specific sectors (e.g. energy, food), while transformations considers broader social systems. However, both transformations and transitions communities, participation has been more explicitly linked to disrupting or challenging existing power relations in the pursuit of transformation: e.g., transdisciplinary and transformative research (see e.g. Caniglia et al., 2021).

Work here contributes to transforming how ecosystem governance is performed and enacted in place (Pereira et al., 2020). Such approaches could be understood as a form of deep participatory research, as they seek to co-create knowledge by blurring the boundaries between academic and non-academic knowledge and across disciplines. Transformative and transitions research rely on extended participation, and position participants as an inherent part of the whole research process. Active participants and researchers are implementors of the research outcomes, further empowering them, and giving them chance to benefit from the research (Farrell et al., 2019). These approaches therefore remove the distinction between academic research and practice (Fazey et al., 2018). and emphasize the process itself rather than the outcome per se (Wittmayer and Schäpke, 2014).

The manifestations of this form of participation tend to be place based, where long-term collaborations between multiple actors are fostered to find solutions to problems. Examples include real world laboratories and living laboratories (see e.g. Schneidewind et al., 2018; Voytenko et al., 2016). In these, solutions are co-created and tried, using the infrastructure and with actors in the place. They serve multiple functions as sites of innovation, learning, creation and education and interpretation/outreach. Transformative spaces can be an extension of living laboratories, with the focus on generating commitments for long term changes and actions from actors, and specifically to address power imbalances (see Drimie et al., 2018). Indeed, in these spaces, it is the navigation of power relations that creates 'safe-enough spaces' to spark and facilitate meaningful shifts in ideas and behaviours towards sustainability (Pereira et al., 2018).

3. Diverse spaces for participation in ecosystem governance

Looking across these different participation traditions, we can position different approaches to participation within a system framing of ecosystem governance. Ecosystem governance covers the multi-level, multi-sectoral, and multi-interest networks of actors and the range of policies, processes and actions they create. We thus frame ecosystem governance as a multi-level system, whereby policies, plans, programmes and actions for ecosystem governance are implemented over multiple administrative levels (e.g. municipality, region, national, supra-national). Governance systems include state policy, which refers to policies set and implemented by central state across administrative levels ranging from local, regional, national and EU/global levels. But governance extends beyond just state structures, and includes civil society, private sector, academia, and more. Within this framing, people's actions and interactions within a social-ecological system are influenced by policies and processes that transcend local system boundaries and shape resources, practices and goals at the local level, and that include, but are not limited to, state policies and actors (Visseren-Hamakers et al., 2021). Ecosystem governance cuts across multiple topics and sectors, for example food, biodiversity and nature conservation, forestry, energy, water, and health. Thus, what happens at local levels is often a question of interplay, interdependencies and balancing across different topics and interests (Biermann, 2021), adding further complexity. This broader governance system therefore creates multiple spaces for participation (*participation spaces*), along three primary dimensions (see Fig. 1), as described in this section: (1) interest boundary, (2) position relative to the state policies and (3) direction of influence.

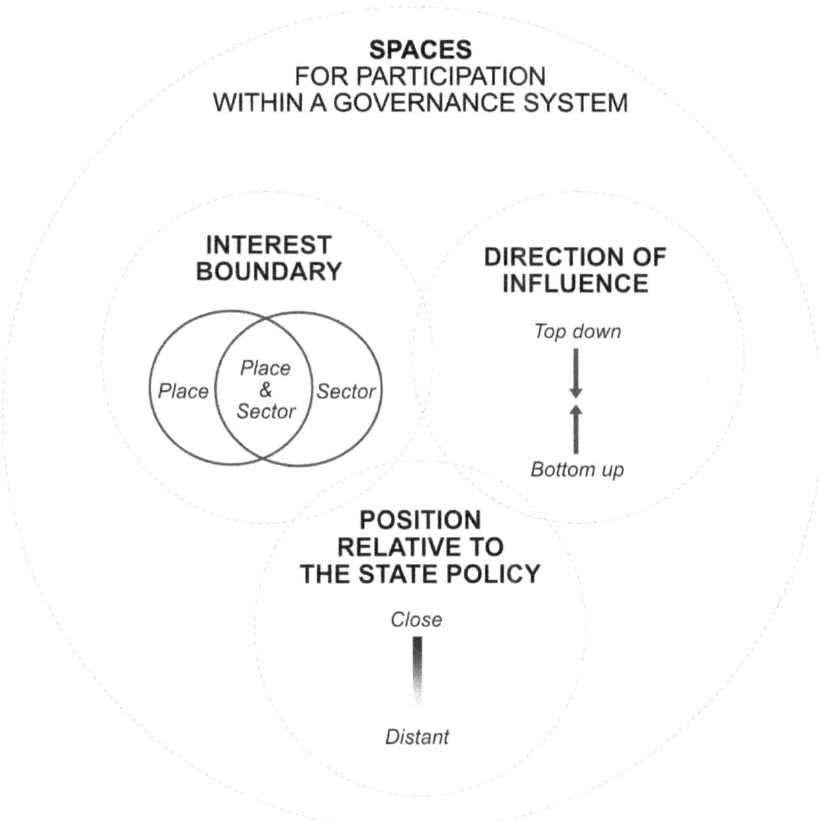

Fig. 1 The three dimensions that shape the space for participation within a governance system: (A) interest boundary; (B) position relative to the state policy; and (C) direction of influence.

The first dimension that shapes participation spaces is the interest boundary. Participation spaces can be shaped by the boundary of the process interest: either by sector (topic) or by a physical place or location. The vast majority of forms of participation are linked to specific places (locations); in some cases, these might be explicitly based on ecosystem boundaries (e.g. river basin management planning). In other cases, they are more likely based around administrative boundaries such as villages, rural areas, and cities (e.g. living labs, food councils). Those that are based on administrative boundaries tend to address multiple topics or sectors. For example, citizens juries tend to engage citizens within a specific place (e.g. a city, or in some cases a country) but are explicitly integrative across sectors and ecosystems by focusing on a specific problem (e.g. climate change) to consider the drivers and solutions from multiple directions. Those that are based on ecosystem boundaries can also integrate across topics and sectors, and/or can be linked to particular topics that shape the ecosystem boundaries (e.g. a river basin, a forest). We can see this in the case of living labs and other transformative approaches, which also often aim to define and solve problems, and therefore integrate across sectors and interests. Conversely, some forms of participation are firmly only within a sector, and are divorced from place entirely; these tend to be those that are focused at industry-civil-society-government collaborations within an industry sector, without direct engagement of citizens. An example would be the collaborative process of the German Textiles Partnership, which engages multiple actors from the textiles sector to participate in industry governance (see e.g. Beyers and Leventon, 2021). It is not tied to a specific place and does not explore place-based implications of decisions, and therefore also has no direct links to individual citizens. Other collaborative approaches are linked to both a specific sector, and a specific place (e.g. food policy councils).

The second dimension defines participatory processes according to their *position relative to state policies* and actors within governance systems. Across the development and governance/policy literature, participation can be defined according to a spectrum of the relationship to state policy. There are participatory processes that are closely tied to state policy and are intended to feed directly into it, such as Participatory Rural Appraisal or input to river basin management plans. At the opposite end of the spectrum are processes with much higher level of participant control over the decisions to be taken and the management of resources, but this control is still sanctioned by, and indeed created to sit within, state policy (e.g. CBNRM, collaborative governance). Governance/policy literature and transformations/transitions research then expands this end of the spectrum further towards creating new spaces for participation in governance beyond those provided by state policy, government and legislative sphere (e.g. independent citizens juries, living/real-world laboratories, transformative spaces). In fact, in some cases, these are intended to be disruptive to these state-provided frameworks.

The third dimension that shapes participation space is its *direction of influence*. This dimension considers where the process seeks to influence decision-making processes; from local level management to international strategies, and levels in between (e.g. municipalities, regions, federal states, countries). Ecosystem governance happens over multiple levels, with agenda-setting decisions often being taken at higher levels, and implementation-focused decisions at the more local levels. The level at which the participatory process interacts with the governance system therefore shapes the kinds of decisions that can be made, and the opportunities to create change. It is tempting to conflate decision-making level with the place or sectoral boundaries of a process. For example, processes based in local places tend to focus on the local level of governance systems. However, this is an oversimplification and in the participation literature, this dimension is not so rigid. Here, some locally focused participation processes also seek to push back up through governance levels to create changes to these broader agendas (citizens juries, living laboratories, transformative spaces). Such forms of participation also do not need to be limited to being local, even if they are defined by place; they could be country-wide or covering large areas. Collaborative governance approaches are also here more flexible in their focus level, seeking local level learning that is also relevant to higher levels of policy.

4. Challenges in participation for achieving plurality

The *complexity* in spaces for participation, across interest, scales and types of decision (Section 3) is an area for potential pitfalls in participation for plurality. People are stakeholders in multiple sectors of interest (e.g. energy, water, biodiversity, food) because they all intersect within people's everyday lives. When participation is arranged by sector, there is a danger of overburdening individuals and organisations that need to participate individually across multiple sectors. From our (the authors') own experiences we increasingly find that civil society organisations, local government and small businesses are feeling overwhelmed with the number of participatory processes they are being asked to join. Transformative spaces, living labs and place-based approaches to participation attempt to overcome such sectoral separation. For example, urban transitions work focusses on cities as places to address sustainability as a holistic concept that considers drivers and barriers from multiple sectors and systems (see Wolfram, 2016; Bulkeley et al., 2016). By focusing on governance of a particular place (often framed as a social-ecological system), they must address challenges holistically, and often systemically (Stringer et al., 2006). However, a focus on place-based participation in this way is also subject to criticism. One source of criticism is in potential for exclusion of people who are not clearly linked to a place, but who are nonetheless affected by decision-making within it. An example would be nomadic communities and refugees (see e.g. Acciaioli et al., 2017). Furthermore, focus on place-based participation can be problematic when it does not address the broader scale political economic factors that shape decisions and actions at local levels (see e.g. Salomaa and Juhola, 2020).

Indeed, this tension in scales is a further complexity challenge for participation. In a multi-level governance system, potential participants are not limited to only one place or scale of governance. We can see that villages (for example) are also embedded within river basins, or protected areas where decisions are made, and again within regional, national and international scales. The decisions and actions at local scale are driven by local decision-makers, politicians, chiefs, with individual citizens included. Although these decisions are stemming from realities on the ground they are, in fact, frequently taken in light of local, national and even global strategies. However, this is not working the other way around. While the sustainability strategies at global level tend to take the form of international agreements, international law, international trade agreements, etc. hence are directly influencing the national/regional/local scales, it is questionable how the values, worldviews, beliefs, and interests emerging from the ground are reflected in these. Indeed, at least in theory, voicing these positions should be streamed through democratic institutions of the respective nations and civil society organisations, yet the reality is disputable. For example, we could consider broad scale processes of designing and modelling global pathways to sustainability. These often imply large-scale shifts in land use and land cover, often entailing a substantial change in the land cover of populated places. They thus necessitate changes in people's places of living and dominant lifestyles. The lack of inclusion of local perspectives in such modelling would therefore suggest a democratic deficit. Yet incorporating localised perspectives in higher level decision making introduces a tension between participation and over-burden, particularly when asking people to participate in multiple processes at multiple scales (Newig and Fritsch, 2009).

However, not allowing space for local voices and realities to be heard at the highest decision-making levels also creates tensions and blocks consideration of local contexts in these decision-making spaces. Tensions around the lack of access to UNFCCC COP26 in Glasgow, 2021 reflected such concerns (see e.g. Steffen and Niranjan, 2021).

Based on the arguments presented thus far, these pitfalls of complexity are clearly linked to questions of *inclusion*: who is present in what spaces of participation? Questions of inclusion relate to who is able to participate, i.e., the questions of who is approached to be included, and whether or not the included participants really represent plurality and diversity of worldviews. Indeed, the presence or absence of viewpoints and the people who represent them has a fundamental influence on both the process of participation and its outcomes (Reed et al., 2009). Therefore, the initial stakeholder analysis and the people and institutions engaged within the process create (or close) the opportunity window for setting agendas (Leventon et al., 2016). However, stakeholder analyses often group people according to imposed categories, hence failing to recognize the diversity within categories, or alternative ways that the participants frame their own identities and interests (Naryanan et al., 2020). Therefore, it is essential to look beyond the narrow definitions of stakeholder interests such as sector or location (Cuppen et al., 2010). Additionally, inclusive participatory processes must consider equal access. For example, caring responsibilities, work duties, burden of over-participation and attitudes such as mistrust towards the process can all influence a participant's ability and willingness to join. Further, White (1996) argues that looking at motivations is critical to understanding representation in a participatory process; some participants will perceive it as obligatory, while others may have variable interests and influences in the process (see also Reed et al., 2009).

Inclusion cannot be separated from questions of *conflict*. While participation is triggered by the premise of navigating plurality, in fact it requires dealing with conflict. Indeed, with plural viewpoints and worldviews, there will be considerable differences in opinions about what problems are (Cuppen, 2012), what solutions are, how to work together and for what purpose (see e.g. Freeth and Caniglia, 2020). Cuppen (2012) outlines that conflict can be constructive when it is manageable, enhances learning, and is authentic (i.e. not stimulated for the sake of the process). Thus, conflicts and challenges should be seen as opportunity for progress, or else they could grow into greater threats over time if avoided, causing mistrust in best intended decisions (Solberg and Miina 1997). As dissensus is unavoidable; the aim is to create a safe space for disagreement and co-operation (Escobar, 2017). This means that we embrace the plurality and transform it into a platform for a process that benefits from differences. Therefore, there is no longer a space for 'competition'. Rather a space for shared ideas and concerns emerges. In this way, constructive conflict embraces the opportunity to address the diversity of participants, and of their ideas and needs. It benefits from the plurality of views in a form of using these different perspectives to create a better solution for everyone (*ibid*).

However, participatory processes can easily close down opportunities for constructive conflict if they are not designed to embrace it. Participatory processes often tend to be consensual, hence leading to solutions, ideas and decisions that are seemingly representative of the people involved in the process (Kothari, 2001). Such a search for consensus often purposefully prevents constructive conflict. This is an established challenge in participation, because it can tend towards suppressing interests, values and viewpoints are for the sake of having a unitary outcome of the process (Mosse, 1994), thus preventing the delivery of plurality. Really exploring plurality within a process requires participants to stretch their comfort zones; being stretched too far can push participants into spaces of deep discomfort (Cuppen, 2012, Freeth and Caniglia, 2020). While trying to avoid such discomfort, conflict is frequently avoided, or managed out of the process, and decisions are made on topics that avoid areas of divergence (Freeth, 2019), and thus decisions aren't made that result in improved sustainability outcomes.

Underlying these challenges of complexity, inclusion and conflict is the cross-cutting issue of *power*. The intention behind many participatory processes is to narrow down the power imbalances and to empower those who are marginalized in decision-making processes. Challenges in achieving this can relate to *inclusion*. For example, who decides who is included, what the purpose is, and whose voices count shape the scope of the participation to consider plurality (see Reed et al., 2017). Indeed, even where the process has inclusive representation of actors, there are not necessarily equal chances to speak or be heard. This relates to the ability to create constructive *conflict* within a process; there are always a people who enjoy more authority (given by their role in society, wealth, race, etc.) and their 'power' can be indirectly imposed on those who are marginalized, thus removing opportunities to voice conflicting viewpoints. This is mostly unintentional, rather linked to the societal role of marginalized people who might simply fear to speak up and oppose those who enjoy a seemingly more powerful position. It is hard to gain trustful participation from actors who are traditionally disempowered, or from those who perceive that a process will be tokenistic. Indeed, instigators of participation must also remain sensitive to the asymmetry of the roles and relationships between themselves and participants as it is shaped by the research process (Collins and Cooper, 2014; Wittmayer and Schäpke, 2014). Such power relationships are also very much contextual, so it also depends on the *position* (socio-cultural background, gender, age, etc.) from which the instigator (and participants) enter the research (Collins and Cooper, 2014).

However, such internal power dynamics are a reflection of the broader political ecology within which the participatory process is positioned (Bixler et al., 2015; Reed et al., 2017). They are always conducted in places with institutional and environmental histories that have shaped power dynamics, governance systems and thus spaces for participation (Bixler et al., 2015). For example, colonial histories can shape the ways in which problems and issues are framed in participatory processes, shifting the actors that fall within the boundaries of the problem frame (see Maat, 2009). Further, imposing participatory governance approaches (such as CBNRM) into institutional contexts that don't match can create disempowerment, elite capture of processes and their benefits, and negative sustainability outcomes (Rihoy and Maguranyanga, 2007). Intervening in such politically framed contexts mean that participation is always political, even if our intentions are not explicitly related to politics (White, 1996). Hence, as researchers and practitioners we need to critically examine the motivations that we have as instigators of participation, and the different, clashing interests emerging from our expectations and those of the participants Furthermore, we need to cautiously consider the aftermath of such processes as the impacts might sometimes do more harm than good.

5. Reflexivity for participation as a pathway to plurality in ecosystem governance

In steering participatory processes that serve as pathways to plurality in ecosystem governance, approaches must be sensitive to the social-political context that they are embedded within (cf. Reed et al., 2017). Participation must address, rather than reinforce, existing power imbalances (cf. Kothari, 2001), and to do so, the context of power must be fully considered. Indeed, if participation seeks to deliver social plurality into the ecosystem governance, it must be understood as a plural process itself. Otherwise, the powerful voices and hegemonic problem framings and worldviews can be reinforced. We risk reducing participation to a tick-box exercise, conducted in a prescribed way, while assuming that it has reflected social plurality. It will not have done so.

However, we can also extend this plurality of participation further by recognizing the multiple spaces of participation within a governance system. When we consider ecosystem governance as a multi-level, multi-sector and multi-interest system, we can see that participation processes fill different spaces therein. These different, or in other words plural, spaces create different opportunities for the outcomes of participation to influence decisions and overall governance. Awareness of this plurality is important because it allows participants to understand their position within the governance system, and the potential role and scope of their involvement in the process in overall ecosystem governance. It ensures that expectations can be managed, thus not risking disempowerment and disengagement from participants. It is also important in highlighting how the outcomes of participatory processes may need to be focused outside of the boundaries of the participants, targeting elsewhere in the governance system. Finally, it is important because it is empowering for participants to understand the pathways of influence from a participatory process (e.g. a citizen's jury) to the decision-makers who should be influenced by their outcomes.

We therefore call for reflexivity in the creation and management of participatory processes in an effort to steer it towards reflecting social plurality (Fig. 2). Reflexivity can refer to the need to consider one's own roles and influence in the process, including expectations, relationships and powers (Whiting et al., 2018). Reflexivity is a process of critical examination of our own, and participant's, worldviews and problem framings, and shifting action in response, rather than remaining fixed to a dominant way of thinking (Augsburg, 2014). Reflexivity is also important for the participants in the process, meaning that the instigators of the process are important in facilitating spaces for societal learning (Wittmayer and Schäpke, 2014). Such reflexivity is a core principle of the more transformative approaches for participation and co-production (e.g. Kläy et al., 2015). Conversely, participatory processes embedded within government policy have been shown to prioritize instrumental learning at the cost of reflexivity (Pallett and Chilvers, 2013). Further, collaborative governance structures that favour efficient decision-making rather than time-intensive co-learning also close down opportunities to practice reflexivity (Beyers and Leventon, 2021).

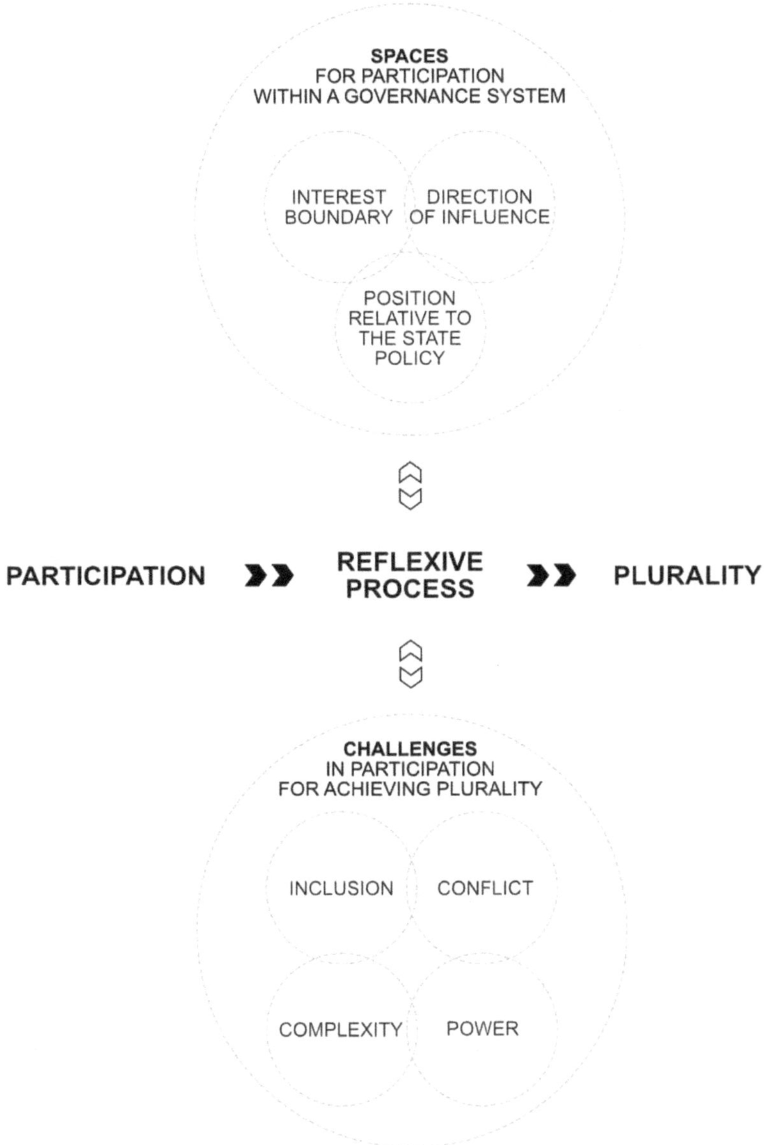

Fig. 2 Reflexivity for participation as a pathway to plurality: Seven reflexive questions should guide participation researchers and practitioners to recognize the plurality of participation, and navigate through the spaces for, and challenges of, participation.

To help foster the necessary processes of reflexivity, we suggest that those involved in participation with the aim of recognizing plurality in ecosystem governance could reflect on the following points:

1. Who is included, and how does this match the participation space?
2. Who holds power within the process, and how is this shaping the process and its outcomes?
3. Who can benefit or be harmed by the process or its outcomes?
4. Who holds power within the broader decision-making context (historical, institutional), and how does that shape answers 1–3?
5. Where in the governance system does the participation target—what is the participation space?
6. How does this participation space fit to others within the ecosystem governance, and what complementarities or complexities does that create?
7. How are we adapting our processes to the answers to these questions?

6. Concluding statements

If participation is to be a pathway to plurality in ecosystem governance, it must grapple with challenges of complexity, inclusion, conflict and power, across a diversity of participatory spaces within a governance system. By looking across different traditions and formats of participation across the literature, we the authors have shown that different framings of participation position themselves in different participation spaces within a broader system of ecosystem governance. There are three dimensions that shape the space for participation within a governance system: (a) interest boundary; (b) position relative to government policy; and (c) direction of influence. The multiplicity of participatory spaces at a range of scales creates complexity in creating inclusive processes that are not creating an over-burden for participants. Furthermore, processes need to promote constructive conflict, and address issues of incumbent power to allow plurality to be fully considered and reflected in decision-making. Power dynamics must be considered both within the process itself, and in terms of the broader political ecology, including the institutional and historical context that the process is embedded within. We provide seven reflexive points to help steer participatory processes towards being pathways to plurality. In noting that social plurality demands plurality in participation, we offer these points as practical suggestions for researchers and practitioners in participation in creating and steering these processes, in an attempt to avoid prescriptive suggestions.

Acknowledgements

We would like to thank Charlie Whitham and Kristýna Falátková for their discussions in early stages of developing this manuscript.

J.L., L.S. and Z.H. recognize funding from Technology Agency of the Czech Republic, grant no: TL05000162: The role of participatory social-ecological research in strengthening international development cooperation.

LS, SV recognize funding from Technology Agency of the Czech Republic, grant no: TL01000238: Urban Adaptation Challenges: Promoting Sustainable Planning Using Integrated Vulnerability Analysis.

BN, ZH recognize funding from JPI Climate through the SOLSTICE Call, project 'Just-Scapes: Environmental justice analysis to advance rural landscape transformations in the face of climate change' (Czech grant number 8F20015).

References

Acciaioli, G., Brunt, H., Clifton, J., 2017. Foreigners everywhere, nationals nowhere: exclusion, irregularity, and invisibility of stateless Bajau Laut in eastern Sabah, Malaysia. J. Immigr. Refug. Stud. 15 (3), 232–249. https://doi.org/10.1080/15562948.2017.1319526.

Aguiar, A.P.D., Collste, D., Harmáčková, Z.V., Pereira, L., Selomane, O., Galafassi, D., Van Vuuren, D., Van Der Leeuw, S., 2020. Co-designing global target-seeking scenarios: a cross-scale participatory process for capturing multiple perspectives on pathways to sustainability. Glob. Environ. Chang. 65, 102198. https://doi.org/10.1016/J.GLOENVCHA.2020.102198.

Ansell, C., Gash, A., 2008. Collaborative governance in theory and practice. J. Public Adm. Res. Theory 18 (4), 543–571. https://doi.org/10.1093/JOPART/MUM032.

Arnstein, S.R., 1969. A ladder of citizen participation. J. Am. Plann. Assoc. 35 (4), 216–224. https://doi.org/10.1080/01944366908977225.

Augsburg, T., 2014. Becoming transdisciplinary: the emergence of the transdisciplinary individual. World Futures 70 (3–4), 233–247. https://doi.org/10.1080/02604027.2014.934639.

Bassarab, K., Clark, J.K., Santo, R., Palmer, A., 2019. Finding our way to food democracy: Lessons from US food policy council governance. Politics Gov. 7 (4), 32–47. https://doi.org/10.17645/PAG.V7I4.2092.

Beyers, F., Heinrichs, H., 2020. Global partnerships for a textile transformation? A systematic literature review on inter- and transnational collaborative governance of the textile and clothing industry. J. Clean. Prod. 261, 121131. https://doi.org/10.1016/J.JCLEPRO.2020.121131.

Beyers, F., Leventon, J., 2021. Learning spaces in multi-stakeholder initiatives: the German Partnership for Sustainable Textiles as a platform for dialogue and learning? Earth Syst. Gov. 9, 100113. https://doi.org/10.1016/J.ESG.2021.100113.

Bhattacharya Prodyut, P., Pradhan, L., Yadav, G., 2010. Joint forest management in India: experiences of two decades. Resour. Conserv. Recycl. 54 (8), 469–480. https://doi.org/10.1016/J.RESCONREC.2009.10.003.

Biermann, F., 2021. The future of 'environmental' policy in the Anthropocene: time for a paradigm shift. Env. Polit. 30 (1–2), 61–80. https://doi.org/10.1080/09644016.2020.1846958.

Bixler, P.R., Dell'Angelo, J., Mfune, O., Roba, H., 2015. The political ecology of participatory conservation: institutions and discourse. J. Political Ecol. 22 (1), 164–182. https://doi.org/10.2458/V22I1.21083.

Buijs, A., Hansen, R., Van der Jagt, S., Ambrose-Oji, B., Elands, B., Lorance Rall, E., Mattijssen, T., Pauleit, S., Runhaar, H., Stahl Olafsson, A., Steen Møller, M., 2019. Mosaic governance for urban green infrastructure: upscaling active citizenship from a local government perspective. Urban For. Urban Green. 40, 53–62. https://doi.org/10.1016/J.UFUG.2018.06.011.

Bulkeley, H., Coenen, L., Frantzeskaki, N., Hartmann, C., Kronsell, A., Mai, L., Marvin, S., McCormick, K., van Steenbergen, F., Voytenko Palgan, Y., 2016. Urban living labs: governing urban sustainability transitions. Curr. Opin. Environ. Sustain. 22, 13–17. https://doi.org/10.1016/J.COSUST.2017.02.003.

Caniglia, G., Luederitz, C., Wirth, T., Fazey, I., Martín-López, B., Hondrila, K., König, A., Wehrden, H., Schäpke, N.A., Laubichler, M.D., Lang, D.J., 2021. A pluralistic and integrated approach to action-oriented knowledge for sustainability. Nat. Sustain. https://doi.org/10.1038/s41893-020-00616-z.

Carter, J., Howe, J., 2006. Stakeholder participation and the water framework directive: the case of the Ribble pilot. Int. J. Just. Sustain. 11 (2), 217–231. https://doi.org/10.1080/13549830600558564.

Chambers, R., 1994. Participatory rural appraisal (PRA): challenges, potentials and paradigm. World Dev. 22 (10), 1437–1454. https://doi.org/10.1016/0305-750X(94)90030-2.

Chambers, J.M., Wyborn, C., Ryan, M.E., Reid, R.S., Riechers, M., Serban, A., Bennett, N.J., Cvitanovic, C., Fernández-Giménez, M.E., Galvin, K.A., Goldstein, B.E., Klenk, N.L., Tengö, M., Brennan, R., Cockburn, J.J., Hill, R., Munera, C., Nel, J.L., Österblom, H., Pickering, T., 2021. Six modes of co-production for sustainability. Nat. Sustain. 4 (11), 983–996. https://doi.org/10.1038/s41893-021-00755-x.

Chevalier, J.M., Buckles, D.J., 2019. Participatory Action Research: Theory and Methods for Engaged Inquiry, second ed. Routledge. https://doi.org/10.4324/9781351033268.

Clair, R.P., 2003. In: Clair, R.P. (Ed.), Expressions of Ethnography: Novel Approaches to Qualitative Methods. State University of New York Press. https://www.sunypress.edu/p-3811-expressions-of-ethnography.aspx.

Collins, C.S., Cooper, J.E., 2014. Emotional intelligence and the qualitative researcher: international journal of qualitative. Methods 13 (1), 88–103. https://doi.org/10.1177/160940691401300134.

Connolly, W.E., 2005. Pluralism. Duke University Press. https://doi.org/10.1515/9780822387084.

Cuppen, E., 2012. Diversity and constructive conflict in stakeholder dialogue: considerations for design and methods. Pol. Sci. 45 (1), 23–46. https://doi.org/10.1007/S11077-011-9141-7/TABLES/1.

Cuppen, E., Breukers, S., Hisschemöller, M., Bergsma, E., 2010. Q methodology to select participants for a stakeholder dialogue on energy options from biomass in the Netherlands. Ecol. Econ. 69 (3), 579–591. https://doi.org/10.1016/J.ECOLECON.2009.09.005.

Daniels, S., Walker, G., 2001. Working through Environmental Conflict: The Collaborative Learning Approach. https://digitalcommons.usu.edu/unf_research/45.

Drimie, S., Hamann, R., Manderson, A.P., Mlondobozi, N., 2018. Creating transformative spaces for dialogue and action: reflecting on the experience of the southern Africa food lab. Ecol. Soc. 23 (3). https://doi.org/10.5751/ES-10177-230302. Published Online: Jul 16, 2018. https://doi.org/10.5751/ES-10177-230302.

Escobar, O., 2017. Pluralism and democratic participation: what kind of citizen are citizens invited to be? Contemp. Pragmatism. 14 (4), 416–438. https://doi.org/10.1163/18758185-01404002.

Farrell, C.C., Harrison, C., Coburn, C.E., 2019. "What the hell is this, and who the hell are you?" role and identity negotiation in research-practice partnerships. AERA Open 5 (2), 233285841984959. https://doi.org/10.1177/2332858419849595.

Fazey, I., Schäpke, N., Caniglia, G., Patterson, J., Hultman, J., van Mierlo, B., Säwe, F., Wiek, A., Wittmayer, J., Aldunce, P., Al Waer, H., Battacharya, N., Bradbury, H., Carmen, E., Colvin, J., Cvitanovic, C., D'Souza, M., Gopel, M., Goldstein, B., Wyborn, C., 2018. Ten essentials for action-oriented and second order energy transitions, transformations and climate change research. Energy Res. Soc. Sci. 40, 54–70. https://doi.org/10.1016/J.ERSS.2017.11.026.

Folke, C., Hahn, T., Olsson, P., Norberg, J., 2005. Adaptive governance of social-ecological systems. Annu. Rev. Env. Resour. 30, 441–473. https://doi.org/10.1146/annurev.energy.30.050504.144511.

Francis, P., 2001. Participatory development at the World Bank: the primacy of process. In: Cooke, B., Kothari, U. (Eds.), Participation: The New Tyranny? Zed books.

Freeth, R., 2019. Dark Mountain: issue 15. In: In Dark Mountain Issue 15: In an Age of Fire. Dark Mountain project (p. 201). https://dark-mountain.net/product/dark-mountain-issue-15-pdf/.

Freeth, R., Caniglia, G., 2020. Learning to collaborate while collaborating: advancing interdisciplinary sustainability research. Sustain. Sci. 15 (1), 247–261. https://doi.org/10.1007/S11625-019-00701-Z/TABLES/3.

Freire, P., 1968. Pedagogy of the Oppressed.

Geels, F.W., 2018. Disruption and low-carbon system transformation: Progress and new challenges in socio-technical transitions research and the multi-level perspective. Energy Res. Soc. Sci. 37, 224–231. https://doi.org/10.1016/j.erss.2017.10.010.

Ghosh, B., Ramos-Mejía, M., Machado, R.C., Yuana, S.L., Schiller, K., 2021. Decolonising transitions in the global south: towards more epistemic diversity in transitions research. Environ. Innov. Soc. Trans. 41, 106–109. https://doi.org/10.1016/J.EIST.2021.10.029.

Grindle, M.S., 2007. Good enough governance revisited. Dev. Policy Rev. 25 (5), 533–574. https://doi.org/10.1111/j.1467-7679.2007.00385.x.

Harmáčková, Z.V., Blättler, L., Aguiar, A.P.D., Daněk, J., Krpec, P., Vačkářová, D., 2021. Linking multiple values of nature with future impacts: value-based participatory scenario development for sustainable landscape governance. Sustain. Sci. 1, 1–16. https://doi.org/10.1007/S11625-021-00953-8/FIGURES/5.

Hölscher, K., Wittmayer, J.M., Loorbach, D., 2018. Transition versus transformation: What's the difference? Environ. Innov. Soc. Trans. 27, 1–3. https://doi.org/10.1016/j.eist.2017.10.007.

Jessee, N., Collum, K.K., Schulterbrandt Gragg, R.D., 2015. Community-based participatory research: challenging "lone ethnographer" anthropology in the community and the classroom. Pract. Anthropol. 37 (4), 9–13. https://doi.org/10.17730/0888-4552-37.4.9.

Kläy, A., Zimmermann, A.B., Schneider, F., 2015. Rethinking science for sustainable development: reflexive interaction for a paradigm transformation. Futures 65, 72–85. https://doi.org/10.1016/J.FUTURES.2014.10.012.

Kochskämper, E., Challies, E., Jager, N. W., & Newig, J., 2018. Participation for Effective Environmental Governance: Evidence from European Water Framework Directive Implementation. Routledge.

Kothari, U., 2001. Power, knowledge and social control in participatory development. In: Cooke, B, Kothari, U (Eds.), Participation: The New Tyranny? ZED Books, pp. 139–152.

Kothari, U., Cooke, B., 2001. The Case for Participation: The New Tyranny? Zed Books.

Lapadat, J.C., Mothus, T.G., Fisher, H., 2005. Role relationships in research: noticing an elephant: international journal of qualitative. Methods 4 (2), 1–19. https://doi.org/10.1177/160940690500400201.

Leventon, J., Fleskens, L., Claringbould, H., Schwilch, G., Hessel, R., 2016. An applied methodology for stakeholder identification in transdisciplinary research. Sustain. Sci. 11 (5), 763–775. https://doi.org/10.1007/S11625-016-0385-1/TABLES/3.

Maat, H., 2009. Is participation rooted in colonialism? Agricultural innovation systems and participation in the Netherlands indies. IDS Bulletin 38 (5), 50–60. https://doi.org/10.1111/J.1759-5436.2005.TB00408.X.

MacLeod, C.J., Brandt, A.J., Collins, K., Dicks, L.V., Catriona MacLeod, C.J., 2021. Giving stakeholders a voice in governance: biodiversity priorities for New Zealand's agriculture. People Nat. 00, 1–21. https://doi.org/10.1002/PAN3.10285.

Mosse, D., 1994. Authority, gender and knowledge: theoretical reflections on the practice of participatory rural appraisal. Dev. Change 25 (3), 497–526. https://doi.org/10.1111/J.1467-7660.1994.TB00524.X.

Naryanan, P., Dheeraj, S., M., & Bharadwaj, S., 2020. Participation, social accountability and intersecting inequalities: challenges for interventions to build collective identity with De-notified, nomadic and semi-nomadic tribal communities in India. Community Dev. J. 55 (1), 64–82. https://doi.org/10.1093/CDJ/BSZ034.

Newig, J., Fritsch, O., 2009. Environmental governance: participatory, multi-level – and effective? Environ. Policy Gov. 19 (3), 197–214. https://doi.org/10.1002/EET.509.

Newman, J., Barnes, M., Sullivan, H., Knops, A., 2004. Public participation and collaborative governance. J. Soc. Policy 33 (2), 203–223. https://doi.org/10.1017/S0047279403007499.

Nohlová, B., 2020. Raising the eff ectiveness of conservation programs: a case study of TKCP environmental and developmental activities in the YUS area, PNG. Czech Slovak J. Humanit. 10 (1), 91 (http://csjh.upol.cz/archive/anthropologia-culturalis-1-2020/).

Ostfeld, R., Reiner, D.M., 2020. Public views of Scotland's path to decarbonization: evidence from citizens' juries and focus groups. Energy Policy 140, 111332. https://doi.org/10.1016/J.ENPOL.2020.111332.

Ostrom, E., 2010. Beyond markets and states: polycentric governance of complex economic systems. Am. Econ. Rev. 100 (3), 641–672. https://doi.org/10.1257/aer.100.3.641.

Pallett, H., Chilvers, J., 2013. A decade of learning about publics, participation, and climate change: Institutionalising reflexivity? Environ. Plan A: Eco Space 45 (5), 1162–1183. https://doi.org/10.1068/A45252.

Pascual, U., Adams, W.M., Díaz, S., Lele, S., Mace, G.M., Turnhout, E., 2021. Biodiversity and the challenge of pluralism. Nat. Sustain. 4 (7), 567–572. https://doi.org/10.1038/s41893-021-00694-7.

Pereira, L.M., Karpouzoglou, T., Frantzeskaki, N., Olsson, P., 2018. Designing transformative spaces for sustainability in social-ecological systems. Ecol. Soc. 23 (4). Published Online: Nov 19, 2018 | Doi:10.5751/ES-10607-230432, https://doi.org/10.5751/ES-10607-230432.

Pereira, L., Frantzeskaki, N., Hebinck, A., Charli-Joseph, L., Drimie, S., Dyer, M., Eakin, H., Galafassi, D., Karpouzoglou, T., Marshall, F., Moore, M.L., Olsson, P., Siqueiros-García, J.M., van Zwanenberg, P., Vervoort, J.M., 2020. Transformative spaces in the making: key lessons from nine cases in the global south. Sustain. Sci. 15 (1), 161–178. https://doi.org/10.1007/S11625-019-00749-X/FIGURES/1.

Peter, S., Le Provost, G., Mehring, | Marion, Müller, T., & Manning, | Peter., 2021. Cultural worldviews consistently explain bundles of ecosystem service prioritisation across rural Germany. People Nat. 00, 1–13. https://doi.org/10.1002/PAN3.10277.

Rahnema, M., 1990. Participatory action research: the "last temptation of saint" development. Alternatives: Global, Local, Political 15 (2), 199–226. https://doi.org/10.1177/030437549001500204.

Reed, M.S., 2008. Stakeholder participation for environmental management: a literature review. Biol. Conserv. 141 (10), 2417–2431. https://doi.org/10.1016/j.biocon.2008.07.014.

Reed, M.S., Graves, A., Dandy, N., Posthumus, H., Hubacek, K., Morris, J., Prell, C., Quinn, C.H., Stringer, L.C., 2009. Who's in and why? A typology of stakeholder analysis methods for natural resource management. J. Environ. Manage. 90 (5), 1933–1949. https://doi.org/10.1016/J.JENVMAN.2009.01.001.

Reed, M.S., Vella, S., Challies, E., de Vente, J., Frewer, L., Hohenwallner-Ries, D., Huber, T., Neumann, R.K., Oughton, E.A., Sidoli del Ceno, J., van Delden, H., 2017. A theory of participation: what makes stakeholder and public engagement in environmental management work? Restor. Ecol. 26, S7–S17. https://doi.org/10.1111/REC.12541.

Rihoy, E., Maguranyanga, B., 2007. Devolution and democratisation of natural resource management in southern Africa: a comparative analysis of CBNRM policy processes in Botswana and Zimbabwe. Occasional Papers Series (PLAAS) 18. http://hdl.handle.net/10566/4424.

Salomaa, A., Juhola, S., 2020. How to assess sustainability transformations: a review. Glob. Sustain. 3. https://doi.org/10.1017/SUS.2020.17.

Schneidewind, U., Augenstein, K., Stelzer, F., Wanner, M., 2018. Structure matters: real-world laboratories as a new type of large-scale research infrastructure: a framework inspired by Giddens' structuration theory. GAIA – Ecol. Perspect. Sci. Soc. 27 (6), 12–17. https://doi.org/10.14512/GAIA.27.S1.5.

Shereni, N.C., Saarinen, J., 2020. Community perceptions on the benefits and challenges of community-based natural resources management in Zimbabwe. Dev. South. Afr. 38 (6), 879–895. https://doi.org/10.1080/0376835X.2020.1796599.

Sieveking, A., 2019. Food policy councils as loci for Practising food democracy? Insights from the case of Oldenburg. Polit. Gov. 7 (4), 48–58. https://doi.org/10.17645/PAG.V7I4.2081.

Solberg, B., Miina, S., 1997. Conflict management and public participation in land management. In: Proceedings of the International conference, Joensuu, Finland, 17–19 june 1996. European Forest Institute.

Steffen, S., & Niranjan, A. (2021). COP26: 'One of the whitest' climate conferences in years, say environmentalists | Environment | All topics from climate change to conservation | DW | 03.11.2021. https://www.dw.com/en/cop26-climate-conference-glasgow-delegates-representation-global-south/a-59708405.

Stringer, L.C., Dougill, A.J., Fraser, E., Hubacek, K., Prell, C., Reed, M.S., 2006. Unpacking "participation" in the adaptive Management of Social–ecological Systems: a critical review. Ecol. Soc. 11 (2). https://doi.org/10.5751/es-01896-110239.

Turkelboom, F., Leone, M., Jacobs, S., Kelemen, E., García-Llorente, M., Baró, F., Termansen, M., Barton, D.N., Berry, P., Stange, E., Thoonen, M., Kalóczkai, Á., Vadineanu, A., Castro, A.J., Czúcz, B., Röckmann, C., Wurbs, D., Odee, D., Preda, E., Rusch, V., 2018. When we cannot have it all: ecosystem services trade-offs in the context of spatial planning. Ecosyst. Serv. 29, 566–578. https://doi.org/10.1016/J.ECOSER.2017.10.011.

Vaňo, S., Stahl Olafsson, A., Mederly, P., 2021. Advancing urban green infrastructure through participatory integrated planning: a case from Slovakia. Urban For. Urban Green. 58, 126957. https://doi.org/10.1016/J.UFUG.2020.126957.

Visseren-Hamakers, I.J., Razzaque, J., McElwee, P., Turnhout, E., Kelemen, E., Rusch, G.M., Fernández-Llamazares, Á., Chan, I., Lim, M., Islar, M., Gautam, A.P., Williams, M., Mungatana, E., Karim, M.S., Muradian, R., Gerber, L.R., Lui, G., Liu, J., Spangenberg, J.H., Zaleski, D., 2021. Transformative governance of biodiversity: insights for sustainable development. Curr. Opin. Environ. Sustain. 53, 20–28. https://doi.org/10.1016/J.COSUST.2021.06.002.

Voytenko, Y., McCormick, K., Evans, J., Schliwa, G., 2016. Urban living labs for sustainability and low carbon cities in Europe: towards a research agenda. J. Clean. Prod. 123, 45–54. https://doi.org/10.1016/J.JCLEPRO.2015.08.053.

Wells, R., Howarth, C., Brand-Correa, L.I., 2021. Are citizen juries and assemblies on climate change driving democratic climate policymaking? An exploration of two case studies in the UK. Clim. Change 168 (1–2), 1–22. https://doi.org/10.1007/S10584-021-03218-6/TABLES/1.

Wesselink, A., Paavola, J., Fritsch, O., Renn, O., 2011. Rationales for public participation in environmental policy and governance: Practitioners' perspectives: environment and planning a. Economy and Space 43 (11), 2688–2704. https://doi.org/10.1068/A44161.

Westerink, J., Jongeneel, R., Polman, N., Prager, K., Franks, J., Dupraz, P., Mettepenningen, E., 2017. Collaborative governance arrangements to deliver spatially coordinated Agri-environmental management. Land Use Policy 69, 176–192. https://doi.org/10.1016/J.LANDUSEPOL.2017.09.002.

White, S.C., 1996. Depoliticising development: the uses and abuses of participation. Dev. Pract. 6 (1), 6–15. https://doi.org/10.1080/0961452961000157564.

Whiting, R., Symon, G., Roby, H., Chamakiotis, P., 2018. Who's behind the lens?: a reflexive analysis of roles in participatory video research. Organ. Res. Methods 21 (2), 316–340. https://doi.org/10.1177/1094428116669818.

Williams, G., 2004. Evaluating participatory development: tyranny, power and (re)politicisation. Third World Quarterly 25 (3), 557–578. https://doi.org/10.1080/0143659042000191438.

Willis, R., 2020. Too Hot to Handle?: The Democratic Challenge of Climate Change. Bristol University Press. https://doi.org/doi.org/10.46692/9781529206036.

Wittmayer, J.M., Schäpke, N., 2014. Action, research and participation: roles of researchers in sustainability transitions. Sustain. Sci. 9 (4), 483–496. https://doi.org/10.1007/S11625-014-0258-4/FIGURES/1.

Wolfram, M., 2016. Conceptualizing urban transformative capacity: a framework for research and policy. Cities 51, 121–130. https://doi.org/10.1016/J.CITIES.2015.11.011.

CHAPTER EIGHT

Pluralistic approaches in research aim to advance farming and freshwater restoration in the Great Lakes basin

Catherine Febria*, Candy Donaldson, Jessica Ives, and Katrina Keeshig

Healthy Headwaters Lab, Great Lakes Institute for Environmental Research, Department of Integrative Biology, University of Windsor, Traditional Territory of the Three Fires Confederacy, Windsor, ON, Canada

*Corresponding author: e-mail address: catherine.febria@uwindsor.ca

Contents

Advances in Ecological Research, Volume 66
ISSN 0065-2504
https://doi.org/10.1016/bs.aecr.2022.04.007

Abstract

The launch of the UN Decade of Restoration calls for the imperative acceleration of efforts to curb ecosystem degradation through local and global pathways. Freshwater restoration poses many challenges due to the complexities of human-induced stressors which require coordinated efforts, across land and water and have for the most part been unsuccessful in achieving restoration goals. Pluralism is one pathway that has potential to overcome challenges to offer transformative research-based solutions. We present two case studies from the Laurentian Great Lakes basin where an intentional pluralistic approach was taken to develop science-based research that was community-responsive and locally relevant while also offering insight to other contexts regionally and globally. These two examples—the Farm and Freshwater Ecology Research Network and the Indigenous Knowledge Circle—engaged local farming and Indigenous communities within a single collaborative program. Here we describe our framework for decentering science to instead centre Indigenous and local knowledge systems and relationships; in doing so, we discuss how this approach improved our research questions and goals while also offering more impactful insights for local practitioners, stewards, and communities. We share short-term versus long-term benefits, co-benefits, and trade-offs in pursuing this effort. Thus far, a plural approach that included sustained investment in relationship-building and a commitment to reciprocal knowledge exchange and co-production are providing critical threads in support of coordinated restoration efforts in the short and long term.

1. Introduction

Efforts to achieve environmental sustainability are increasingly pointing towards holistic approaches. As global challenges become more complex and "wicked", diverse pathways leading to transformative change are needed (Díaz et al., 2019). Much of the evidence on restoration points towards a gap in the translation of science into practice (Palmer, 2008), the lengthy recovery times of ecosystems (Moreno-Mateos et al., 2017), and the need to engage and maximize social dimensions in restoration (Jones et al., 2018; Suding et al., 2015). The recent launch of the UN Decade of Ecosystem Restoration (2021–2030; UN Environment Program, 2019) hopes to accelerate efforts at a range of scales making now an opportune time to identify, engage, and align local science-based efforts and restoration practice with global ones. Herein lies a key opportunity to reframe and re-imagine research efforts to improve chances of ecosystem restoration success.

In the context of the Laurentian Great Lakes of North America, environmental challenges are amplified by layers of complexity. When it comes to management of freshwater resources in the Great Lakes region, restoration solutions for the Lakes themselves are often land-based and require individuals to take collective action. Great Lakes ecosystem health is primarily governed through top-down approaches, with governmental agencies at multiple levels (municipal, provincial/state, tribal, and federal) in charge of various components of environmental management (Alsip et al., 2021). Lakes and their watersheds are heavily influenced by human activities and while the Lakes themselves are hydrologically connected, they are spatially diverse, both within and among lakes. This results in numerous overlapping social, cultural, and economic layers that together build a heterogeneous landscape of local contexts that influence environmental restoration actions on the ground.

The Great Lakes are often regarded as being bi-national waters but are in fact multi-national when we recognize First Nations, Métis, and Tribal Bands. Now, more than ever, there is growing awareness and responsiveness to re-affirming the rights of Indigenous Peoples in the Great Lakes, across Turtle Island/North America and beyond. Within this tapestry of governance, cultural, and ecological dimensions, restoration solutions cannot necessarily be transplanted from one local context to another. In addition, the history of settler colonialism, and the roles of science and governmental institutions must be acknowledged in both the ways in which environmental management is conducted and in terms of the impact of these structures on communities and their ability to engage in research, science, and policy. Sound science-based research on ecosystem restoration are dependent on social factors from funding and new technologies to the translation of knowledge into policy and practice to deliver the desired ecological, social and cultural outcomes desired. Thus, good science must be responsive to and situated within the appropriate social contexts and ecological contexts.

Ultimately, successful restoration efforts must engage socio-ecological complexity and consider a multitude of possible pathways and outcomes. In the region of the Great Lakes basin where this work was situated (i.e., southwestern Ontario, Canada), the dominant landscapes are settled (i.e., urbanized) and agriculturally intensified, with remnant natural habitat left in unceded Indigenous Territories (e.g., Bkejwanong Territory or Walpole Island First Nation) where levels of native biodiversity, including species at risk, are likely to be the highest (McElwee et al., 2020; Schuster et al., 2018, 2019). Most commonly, the region is known to be an active, highly productive, and economically significant agricultural region for Canada. With this in mind, we describe a process for intentionally, ethically, and meaningfully engaging a multitude of views and priorities in grounding and informing freshwater restoration research. We reject common Western scientific approaches as the only ones needed in addressing widespread biodiversity decline, habitat loss, and pollution. In contrast, engagement and social dimensions may be key in achieving planetary sustainability and impactful management decisions (Ban et al., 2018; Jones et al., 2018; Suding et al., 2015).

Pluralism in its broadest sense refers to a multiplicity in ways of knowing and doing that cannot be reduced to a single principle or theory. Pluralism has been applied to multiple environmental applications (e.g., management ethics and economics; Turner, 1988) and offers a more inclusive approach to addressing socio-ecological complexity in research. Specifically, we engaged and operationalized pluralism alongside complementary terms that include actionable science (Palmer et al., 2016) and translation ecology (Chapin, 2017; Enquist et al., 2017), which fall under the broader umbrella term of knowledge co-production (Djenontin and Meadow, 2018; Latulippe and Klenk, 2020; Norström et al., 2020). This also complements growing efforts to honour multiple knowledge systems, such as two-eyed seeing (Rayne et al., 2020; Reid et al., 2021), whereby the strengths of both Indigenous and western ways of knowing are considered and brought together to re-imagine ways of delivering better freshwater science outcomes. Through this lens, our working definition of pluralism is the intentional consideration of diverse ways of knowing, being and doing. Rather than attempting to simplify social and cultural dimensions, or prioritizing science needs, we sought to embrace complexity as part of a more holistic, strategic, and partnership-driven approach to research and actions on the ground. Rather than focusing on a single project outcome, we invested in re-imagining the traditional science-based research process to help support longer-term relationships that could support and influence multiple projects over time. Our long-term hypothesis is that a plural approach that de-centres science to make room for Indigenous and local knowledge systems will result in meaningful, sustained, and more impactful community-responsive research outcomes.

Through two different but complementary case studies we explore how pluralism was operationalized to set our team on an ethical path that invests time and resources into community-based connections that will improve the chances of beneficial societal outcomes and research excellence. While the process is still in its infancy (<3 years at the time of publication), we have received numerous requests from science organizations, communities and practitioners to share knowledge about our process and transformative approach to research. Moreover, reflecting on the process enables the establishment and strengthening of foundational knowledge within our research, stemming from a community perspective. This can only be accomplished through relationship and capacity-building, strengthening of networks through early and ongoing engagement with the communities involved in restoration. Much like ecological food webs, there is great strength in diversity which comes from repeated interactions over time (Paine, 1966). In parallel, the incorporation of storytelling and efforts to understand local practices and research needs ensures that greater understanding can be achieved. As a team, we saw the need to share early insights now as part of our role as treaty partners and committed allies to share how actions in support of restoration efforts may only be a ripple, but are essential in supporting eventual and long-term restoration success. Moreover, according to the literature, restoration has largely been perceived as unsuccessful (Bernhardt et al., 2005; Jones et al., 2018), due to a lack of monitoring and mismatches in restoration approaches, however perhaps also due to biases towards publishing only significant results (i.e., file drawer effect) (Fraser et al., 2018) and a general trend towards less risky research in the sciences (Rzhetsky et al., 2015). In reality, restoration is a social undertaking, and when engaging communities in this work, our intention in sharing stories of successes, works in progress and lessons learned are part of the work required to build trust between and among partners, and ensure equitable, ethically-engaged and inclusive research practices. Thus, here we describe qualitative and quantitative examples of progress towards achieving the elusive goal of restoration success in the hopes that more research groups will see the value of incorporating pluralistic aspects in their research programs.

1.1 Positionality of the authors

Positionality allows us to further contextualize pluralism. The authors share an understanding that diverse ways of knowing nature contribute to a deeper understanding of the complex issues facing ecosystem degradation and the operationalization of pluralism will lead to transformation in research-based outcomes. We also share an understanding that a pluralistic approach offers the opportunity to reframe and co-innovative research, which will improve chances of successful ecosystem restoration. We share similar values as treaty-partners on the land and waters of Turtle Island, otherwise known as North America and the colonized countries of Canada the United States of America (USA). Through a diverse set of lived experiences as immigrant (Febria), settler (Ives), and Indigenous (Donaldson, Keeshig) individuals, each trained in western science and academic institutions, we align through a shared kinship to the Great Lakes basin ecosystem as part of a global ecosystem in need of restoration at local to global scales. We also recognize our treaty responsibilities in taking a pluralistic, decolonial approach to science-based research efforts in the Great Lakes. We are individually and collectively putting our values into action, and seek to respond to the Truth and Reconciliation Commission of Canada's 94 Calls to Action (Truth and Reconciliation Commission of Canada, 2015). While we respect and honour the traditional name for the Great Lakes as the five freshwater seas, or Nayaano-nibiimaang Gichigamiin in Anishinaabemowin/Ojibwe, we will refer to this basin as the Great Lakes throughout the paper.

2. Reframing research as a community-engaged pursuit: The Healthy Headwaters Lab

Launched in 2019, the Healthy Headwaters Lab (HHL) initiated a multi-faceted research program to pursue science that connects land, water, and people to ensure freshwater sustainability for the benefit of future generations. As a result, much of the research taken on by the team focuses on ecosystem restoration at a range of scales, in partnership with local communities and a range of collaborators. Consistent with research environments nationally and beyond, we are committed to amplifying dimensions of diversity and inclusion in research excellence and beneficial societal outcomes (i.e., Dimensions Charter; Government of Canada, 2019). Thus, critical to achieving HHL's mission was a clearly articulated process outlining how we would ethically engage Indigenous communities and local knowledge holders (e.g., farmers, conservation practitioners) to guide, inform, and ground our research design and outcomes (Fig. 1). We deeply understood that engagement is not only about speaking to Indigenous and local community partners, but also listening and engaging in reciprocal knowledge exchange. In situating freshwater research in agricultural context, this meant understanding farm operations and farming challenges; in dealing with our Indigenous community partners, it involved offering paid opportunities, scientific equipment, and training as the community required. Broadly, we entered into relationships not with a pre-designed question, but a commitment to learning and listening as part of a knowledge co-production journey, recognizing the diversity, depth, and breadth of Indigenous and local knowledge systems.

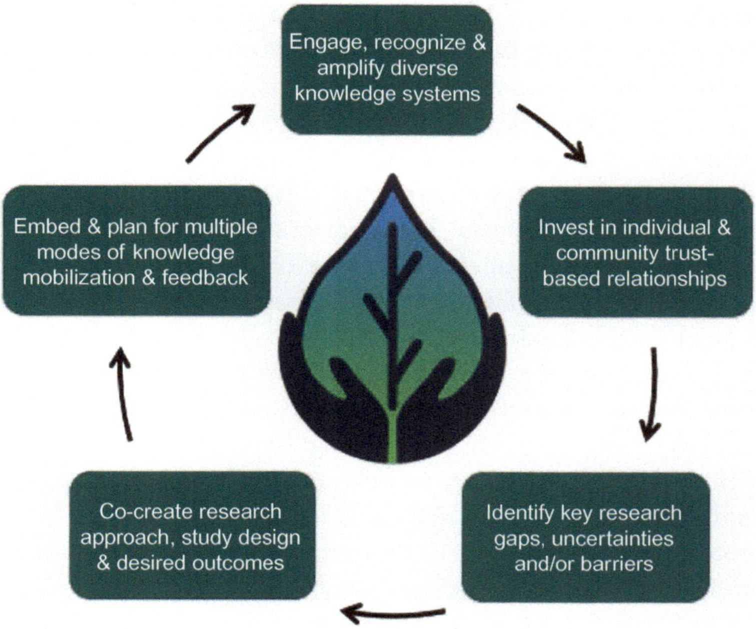

Fig. 1 Key principles of the Healthy Headwaters Lab research process to foster pluralism in freshwater restoration science.

Contrary to most science-based research efforts, our first year's efforts focused on relationship-building, and sought to build trust over time through formal and informal modes of communication and community participation. Through numerous initial small-group and individual meetings with Indigenous and local farming community members we shared HHL's research mission and intent to value local partnerships. In turn, we met with individuals and groups, and held informal conversations about broad themes such as farming and freshwater sustainability in groups of two or three team members. In sharing the kind of research expertise we could offer, we also walked farm fields, looked at machinery, presented at local association meetings, and walked the land with Knowledge Keepers. In repeating these events over the year, we uncovered research priorities, science-based uncertainties, and barriers and/or enablers to science uptake. The broad range of decision-making contexts from the farm-field to whole watersheds, legal frameworks, and other factors were explored and taken into consideration when formulating research questions. For example, traditional approaches to ecosystem studies in response to land use have considered broad land use land cover (LULC) categories (e.g., forest cover, agriculture, urban), and within agricultural landscapes the predominant approach for restoration has been to characterize farming in coarse terms (e.g., till versus no-till). In reality, agricultural landscapes are a nuanced mosaic of management actions involving production versus non-productive vegetation (Case et al., 2020) as part of an agroecosystem (Isaac et al., 2018). In our farm-based research efforts, we initiated a field study on 12 different farm plots situated on Brookston clay-plain, low-gradient soils to assess the impact of tillage regime and forest restoration on properties of soil organic carbon at a range of scales (Weller, 2022). This helped generate regional data, the first dataset of its kind to show the decadal timescales required to restore soil organic carbon, and had implications for nutrient management in the region. In another example, our team invested in creating numerous Indigenous-led roles that spanned project management, grant writing, storytelling and featured environmental artist to build greater capacity on a number of fronts in support of community-led, driven and responsive research efforts.

Grounding these efforts was a HHL Vision and Mission, operationalized by key elements that transcend all projects, including: (1) Strategic alignment of science with decision-making contexts and needs, (2) Holistic and decolonial approaches to pursuing science that is community-driven, and (3) Partnerships across disciplines, communities, and knowledge systems to overcome barriers for mobilization of knowledge into policy and practice. We operationalized these dimensions into a process (Fig. 1) that would engage knowledge systems through sustained investment in trust-based relationships. These relationships led to the identification of key research needs which were then incorporated into research design and outcomes, and these steps are intended to iteratively repeat as the research program grows. As part of the implementation of the research, plans for knowledge mobilization and feedback were accounted for and developed from the start.

2.1 Element 1: Strategic

We instituted programs dedicated to engaging with and learning from individuals who are deeply connected to the landscapes in which our research takes place (i.e., understanding farmer knowledge in agricultural landscapes, and Indigenous connection to, and stewardship of, territorial lands and waters). Hence, we invested in relationships among and across land, water, and people. Initial phases of program building involved individual and group meetings, knowledge syntheses (Eveleens and Febria, 2021; Van Goethem et al., n.d.), the hiring of community members in critical connector roles, and workshops to better understand the existing local knowledge base, gaps, and research priorities (see case studies below; Fig. 2). As alluded to earlier, before collecting a single sample, we spent much of the first year (2019) reviewing existing datasets, forming partnerships, and meeting with individuals and groups to ensure that the research being pursued would be consistent with local decision-making needs. Over time, we strategically pursued institutional, provincial, federal, and international opportunities to implement multiple, interconnected projects. This approach required a blended team, each with individual networks and expertise to build upon. The result is an intentionally diverse team with expertise in farming, teaching, communications, complementary technical skills from ecosystem ecology, wildlife, and biodiversity identification, and varied analytical skills. Relationships, networks, and skillsets were developed over time, and for the majority of our initial development, under the cloud of the global COVID-19 pandemic. Without these essential starting networks and skillsets that came with our initial hires, development and implementation of the projects would have been greatly hampered or reduced. In fact, key activities were affected, and the full potential of our strategy was hampered by the pandemic itself and an inability to meet in person for a majority of our first 3 years. Overall, however, building on already strong networks helped to navigate challenging and uncertain times.

2.2 Element 2: Holistic

Despite the limitations of the pandemic, investing in a process of reciprocal knowledge exchange ensured that social dimensions were explored alongside ecological dimensions in support of a more holistic approach to research. Evidence on restoration has emphasized the importance of engaging Indigenous and local knowledge systems in restoration (e.g., Jones et al., 2018; Suding et al., 2015). The incorporation of local agricultural knowledge and expertise into HHL was made possible through the creation of a Farm Advisory Board (FAB) and research network (see Case Study 1) in parallel with the establishment of an Indigenous Knowledge Circle (IKC; see Case Study 2). We found that striving for a research program that is holistic and pluralistic created a synergy that is more representative of the issues and environments on which our work is focused. For example, the FAB included representation of Indigenous community members engaged in farming. Moreover, we sought opportunities to mobilize knowledge through storytelling and a communications position in HHL, which

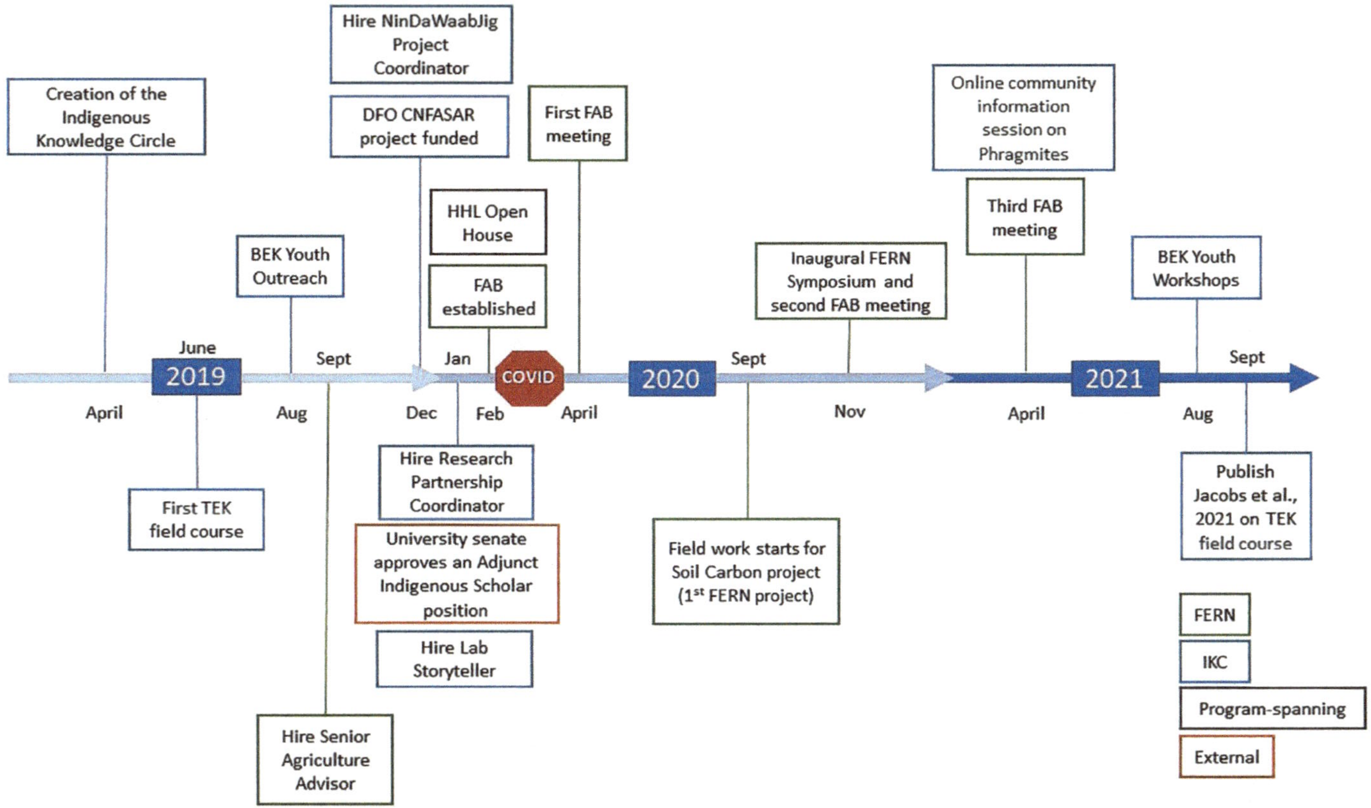

Fig. 2 Timeline of key events illustrating the diverse ways in which local communities, Knowledge holders, and partners were engaged.

acknowledges the importance of long-term data and knowledge keeping consistent with, and in some cases beyond, western science timeframes. Within HHL's agricultural projects (Case Study 1), we consider a whole of farm context, including management actions at the field scale, soil health, health of the surrounding ecosystems, and the farmer's relationship with the land as well as economic considerations when making farming decisions. In parallel, Indigenous knowledge systems point to the importance of considering all aspects of our human experience, our mental, physical, and spiritual well-being. Furthermore, individual well-being cannot be separated from the well-being of the land and water, as all are inextricably connected. Indigenous knowledge holders have pointed to the importance of healing relationships, to the land, water, each other, and ourselves. As such, our work with Indigenous partners and farmers places as much emphasis on community well-being as it does on environmental (soil, water) health. Considering all issues, a shared history, and future aspirations, we aim to contextualize issues alongside social, political, historical, and environmental drivers to explore possible research pathways and prioritize which research-based efforts to pursue.

In parallel, by pursuing a more holistic approach to how and why we pursued research partnerships, these efforts aligned with greater calls for research excellence and support for justice, equity, diversity, inclusion, and reconciliation with Indigenous peoples as key attributes leading to beneficial societal outcomes and impact (Powell, 2018a, 2018b; Wong et al., 2020). As we navigated through academic settings including conferences, teaching, and publications, we consistently prioritized inclusion of our partners in the co-development, co-presentation, and co-authorship of work (Fig. 2; e.g., Jacobs et al., 2021) and the production of plain language summaries of our research in the form of newsletters and posters (Browne et al., 2021). In meetings and workshops, we helped launch, initiate, or organize panels that included Indigenous knowledge keepers and community members that are not normally included in such settings. In the case of the International Association for Great Lakes Research Conference in 2020, we co-created the first Traditional Ecological Knowledge panel. We also helped plan and participate in the International Joint Commission's first Indigenous Knowledge workshop (2021), and contributed efforts in support of the University of Windsor's Faculty of Science Adjunct Indigenous Scholar role. Thus, holism applied not only to the science we pursued but was the intention we carried into all efforts.

2.3 Element 3: Partnership

Partnership is a method of further embedding pluralism within a research program in that it provides a way to meaningfully, and formally, incorporate external views, rather than merely considering input from diverse viewpoints. Ongoing partnerships are a key tool to embed external perspectives into our work and include mutually agreed upon terms, goals, and objectives. Recognizing that our community and research partners came from diverse lived experiences, offering a wealth of knowledges on various aspects to inform the research, we invested in a process that sought to identify shared values, goals, and vision for working together. In the formation of the HHL, several formal and informal partnerships were a priority. Common among these partnerships was the desire to find mutually beneficial ways of working together and incorporate multiple viewpoints into research and research outputs. This involved discussing points of disagreement and the articulation of boundaries between our team and each partner (see Table 2, "conflict among value systems"). For example, not all farmers were intent on changing their farm practices but were open to having research conducted on their farm fields, riparian buffers, and drains. To establish and maintain trust, we co-developed personalized formal legal agreements (i.e., memorandums of understanding, or MOUs). A key aspect of these MOUs was acknowledging that data ownership lays with the owner of the site sampled. A challenge we have encountered is that true and equal partnership is difficult to create and maintain, not only through a pandemic, but also within the current paradigm of western science approaches to research (see "power dynamics" within Table 2), True partnership requires ongoing unlearning, awareness training, and ongoing maintenance. Inevitably, one cannot prepare for all challenges a project may face until work is underway. Working together to understand and overcome challenges requires continuous effort and communication and, in an ideal situation, builds stronger relationships (see Table 2, "strengthened relationships among academics and communities"). Even with mutually agreed upon goals and objectives, differing perspectives will introduce motivations or challenges experienced by one party and not the other. Transparency and communication are essential in working through these aspects of partnership (Table 1). Another aspect of partnership relates to our ever-evolving understanding of our roles as allies to Indigenous partners. There seems to be a growing awareness of, and commitment to, communicating and collaborating with Indigenous communities, yet there remains a clear gap between words and meaningful action across all levels of

Table 1 Critical components of a community-based, pluralistic research program.

Component	Details
Formal connector roles	Formal, paid roles for connectors such as the Senior Agricultural Advisor (FERN) and multiple community-based members of the lab (NinDaWaabJig Project Coordinator and Research Partnership Coordinator; IKC). Such roles are essential for establishing initial trust in the research and for providing insight into the community's needs.
Dedicated time for relationship building	Dedicated time to form relationships without pressure to "achieve goals" or follow agendas.
Equity, diversity, inclusion, Indigeneity, reconciliation with Indigenous Peoples, decolonization	These terms are increasingly applied in academic settings (grants, teaching, service) as elements of research excellence and innovation. While they all mean different things, our approaches to ethically and equitably pursue partnership-based research aligns with systemic changes underway. Our research—and the pursuit of restoration ecology, science and practice—simply cannot wait for institutional and academic culture changes to take place. We also acknowledge that currently much of this work is taken on by women and marginalized communities in academia, notably Black, Indigenous and Peoples of Colour. In our work, we invest time and resources into ensuring that members of the team are taking the time to learn/unlearn their relationships with each of these terms and their own positionality, privileges and biases when conducting this work, all centralized through a formalized lab manual and many other threads to support all those involved in our research program.
Transparency and ongoing communication	True partnership between researchers and community partners cannot exist without transparent and ongoing communication around goals, timelines, methods, and roles. This is especially critical as the challenges faced by pluralistic research (see Table 2, especially "differing concepts of time", "power dynamics", and "conflict among value systems").
Funding mechanisms	Research agencies, especially in the sciences, do not typically fund Connector roles, Knowledge brokers, or traditional Knowledge Keepers as part of science teams or the scientific process. These are often treated as extension roles. We argue that co-investment in knowledge mobilization, community engagement, and capacity building is critical to research excellence and beneficial societal outcomes.

academia and government. It is imperative that we continue to use our position and influence to create awareness and understanding of the need and potential for reconciliation between Indigenous and non-Indigenous people in now-called Canada. Through ongoing partnership and networking, we continue to bring the voices and inspiration of women and underrepresented minorities, with Indigenous voices in particular, into traditionally white-male dominated spaces, challenging the status-quo and shining a necessary light on issues of equity, diversity and inclusion.

The HHL has obligated dedicated funds to support ongoing partnership maintenance, primarily in the form of salary for connector roles (e.g., Community Research Partnership Coordinator and Senior Agricultural Advisor) and funding for community events (e.g., an annual community symposium, see Case Study 1). In the long-term, we continue to gather quantitative and qualitative data to test the hypothesis that investment into partnerships can help bridge differences in ways of knowing and thus enhance efforts to collaboratively restore human-impacted ecosystems. By engaging in processes to understand diverse views, learn from others, and find respectful ways of working together, we hope to demonstrate that freshwater restoration science efforts can be accelerated.

3. Case study 1: Farm and Freshwater Ecology Research Network (FERN)

3.1 Goal of FERN

Despite decades of efforts to restore the Great Lakes, water quality and overall ecosystem health remain degraded (Sterner et al., 2017). What is increasingly clear is that key solutions to Great Lakes water quality issues lie on the landscape, in managing phosphorus loads from entering unmitigated into river networks (Joosse and Baker, 2011). That being said, there is no one-size-fits-all solution for farmers and practitioners (Macrae et al., 2021). Southwestern Ontario's dominant land use is agriculture, which negatively affects water quality both in the headwaters and ultimately in the Great Lakes (Eimers et al., 2020; Wilson et al., 2019). Balancing the need to protect freshwater, which supplies drinking water to communities around Lake Erie, and the livelihoods of a farming community that produces food for people both within and outside of southern Ontario is nuanced and delicate. We observed a status quo position from the broader public and science providers that vilifying the agricultural community as a source of critical non-point pollution instituted an antagonistic relationship between farmers and environmental restoration, whereas relationship building and engaging the farming community in research co-design can create opportunities for mutual understanding and long-term positive change.

Sustainable, long-term solutions require buy-in from the farming community and thus we sought to examine ways to farm sustainably while also respecting an individual farmer's livelihood. Considerable work has been done on best management practices and sustainable farming (Angers et al., 1997; Cusser et al., 2020; Lam et al., 2016), but uptake within the agricultural community is lacking (Baumgart-Getz et al., 2012; Macrae et al., 2021; Prokopy et al., 2008). By involving the farming community in the research process from the beginning, through FERN, we hope to identify and address the key barriers to a successful balance of environmental health and farmer prosperity. This process of trust building and locally focused research (e.g., Weller, 2022) is slowly opening up conversations about the broader farming context and the extent to which conservation measures are being adopted and by whom. Notably, our farming partners consist mainly of multi-generational settler farmers running family operations, small-scale Indigenous growers, representation from the greenhouse industry, and mostly cash crop farming. Missing from these views are multinational corporations, fruit growers, and wineries, to name a few. To address these gaps we will continue to engage new members and perspectives to ensure our research is locally rooted and relevant.

3.2 Network development

The FERN was developed slowly and intentionally, beginning in 2019. Throughout the spring and summer of 2019, the focus was placed on relationship building, with no projects started until September, when the first graduate student was enrolled. In the initial relationship-building stage a series of one-on-one and small group meetings were held with local farmers (Fig. 3A). Such meetings had no formal agenda but consisted of visiting the farmer on their land, walking their fields, and informally learning about their farm, history, and interests, particularly with respect to sustainability. Often refreshments were provided, and the farmers were encouraged to bring up topics of interest to them.

Critical to our initial relationship-building stage was the early hire of a Senior Agricultural Advisor (Table 1, Fig. 2). The Senior Agricultural Advisor was a member of the local agricultural community with a deep social network to draw upon. They acted as a trusted conduit between our newly

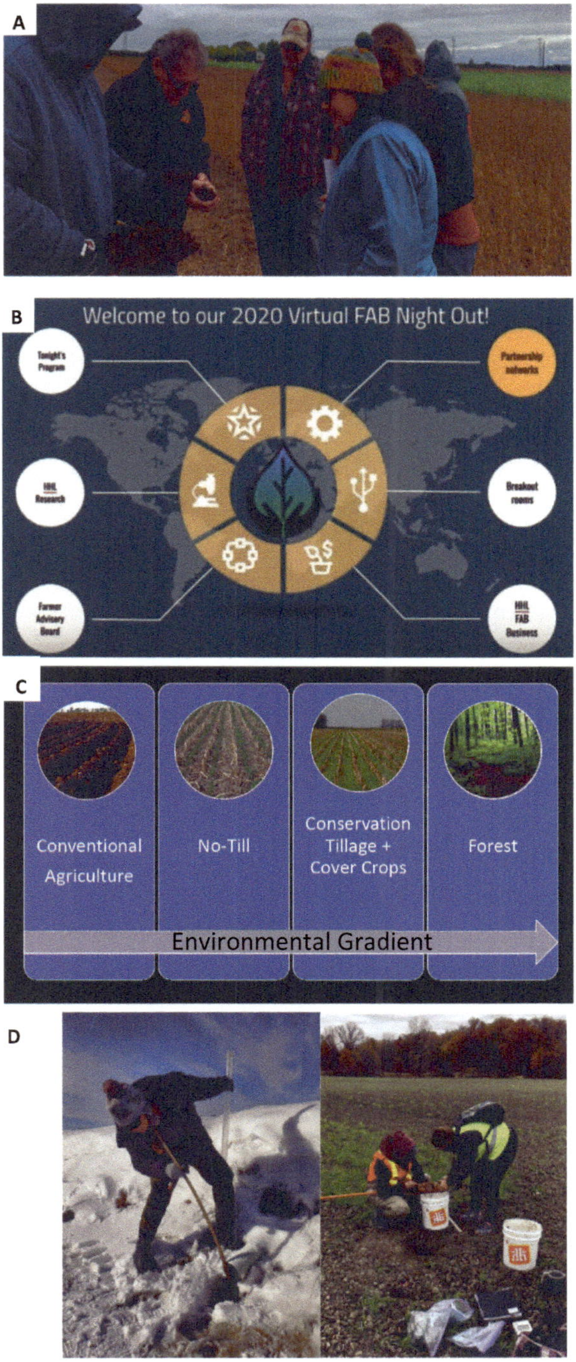

Fig. 3 See figure legend on opposite page.

established academic research group and the existing local network of sustainably minded farmers. Creating an embedded role within our team for a member of the community was fundamental to building trust with farmers, researchers, and conservation practitioners all working in the same region. Not only did our Advisor help to legitimize the role of a freshwater restoration science lab in the local community, they also legitimized the need for local community experts to be integrated into research teams and institutions. Budgeting to make this role an official, paid role was essential for demonstrating the role's value, especially to other academics, the farming community, and the Advisor themself. It continues to be a challenge to find ways to fund such a role in traditional research grant schemes, with greater emphasis going towards student training, research equipment, and science outputs (e.g., publications). One compromise has been to pivot this role to be held by early career researchers who have farming experience and their own networks in the local farming community.

3.3 Advisory board

To provide a formal avenue for community direction, HHL created a Farmer Advisory Board (FAB) in late 2019 and early 2020. The FAB was designed as a non-binding advisory group to serve as a dedicated point of contact between the local agricultural community and the HHL, with membership spanning multiple community perspectives. The FAB purpose was explicitly defined to "assess current and advise on potential work being done by the Healthy Headwaters Lab (Great Lakes Institute of Environmental Research, University of Windsor) and its partners regarding agriculture's impact on water quality and ecosystem health in Southern Ontario". An initial Terms of Reference and board membership were created with input from the Senior Agricultural Advisor through late fall 2019. Initial membership included local farmers, Indigenous community members, and representation from the greenhouse sector.

Fig. 3 Examples of how local farming communities and context were engaged in research including local visits to farms undertaking diverse management practices (A), virtual public event during COVID to connect researchers with local farming communities and FAB members (B), research contextualized to reflect the diversity of management practices across Windsor-Essex farms situated on Brookston Clay soils (Weller, 2022), which together facilitate year-round access to farm and drain networks for research (D). *Photo credits: Healthy Headwaters Lab.*

The FAB was designed to allow regular communication between FAB members and the core HHL team. Two recurring board meetings per year (Aug/Sept and Feb) allow standing opportunities to provide research program updates to the FAB, receive feedback on pre-determined strategic topics, and provide an opportunity for open discussion from the FAB on any topic of interest. Once per year, associated with the late summer FAB meeting, a public FERN Symposium is planned to allow wider community engagement with the HHL research program. Most FAB meetings in 2020 and 2021 (see Fig. 2) were held virtually to accommodate health concerns around the COVID-19 pandemic. The fall FAB meeting in 2021 was held in-person, and feedback from the board members was that this was the most engaging meeting thus far. The inaugural FERN Symposium was also held virtually and delayed to November 2020 due to COVID-19 scheduling issues (Fig. 3B).

To complement meetings and research undertaken across the region, and as a key mechanism for disseminating information with interested individuals, we publish an annual newsletter summarizing key research questions being addressed, individuals involved, and broad level trends and results from our research (Browne et al., 2021). As defined by our formal partnership MOUs, in years when sampling was conducted on their land, each individual farmer also receives an individualized summary of findings specific to their soils and water followed with opportunities to meet and engage with the research team to address questions and ideas for future research and follow-up monitoring. If requested, and when possible, additional measurements are taken for the farmer, such as soil organic carbon, soil conductivity and infiltration rate.

3.4 Looking forward

Across farming landscapes, despite the range of actions individual farmers and communities can take with the intention of improving soil, freshwater and ecosystem health, changes take considerable advance planning and resources (e.g., crop selection, machinery required). Moreover, there is great scientific uncertainty about whether a given best management practice would work in a given local context. In the case of Essex County and southwestern Ontario, low-gradient, clay plain soils are a fundamentally different farming context than elsewhere across Ontario (Van Esbroeck et al., 2016). Thus, working in tandem with the local agricultural community to understand social, economic, and cultural considerations alongside the identification of key science questions requires investment in pluralistic processes in a dedicated, ongoing way into the future.

Logistically, during 2020 and 2021 this was a challenge as most FAB meetings and the first FERN symposium were held virtually due to the global COVID-19 pandemic. This has hindered meaningful in-person engagement and deepening of connections with our local farming community. While we were lucky to have our first (2019) season unaffected, allowing us to build the critical first relationships, future success will require additional investment in relationship building and maintenance.

4. Case study 2: Indigenous Knowledge Circle (IKC)

4.1 Goal of IKC

The motivation for the Indigenous Knowledge Circle (IKC) stemmed from years of informal relationship building between IKC co-founder/inaugural member of HHL, Candy Donaldson, and members of the Aboriginal Education Council (AEC) at the University of Windsor. The AEC provides guidance on improving educational support for Indigenous students, increasing Indigenous representation and completion rates, and raising sensitivity and awareness of Indigenous cultures, issues, and realities. The broad vision of the IKC subsequently came from the union of Febria (HHL founder) and Donaldson, and their shared values and vision for decolonial teaching and research practices in science. Some of the earliest conversations with local Indigenous representatives centred around the issue of representation, in research partnerships with local communities, inclusion in research labs and post-secondary studies. Thus, one of the main goals of the IKC was to help address the historic exclusion of, and ongoing underrepresentation of Indigenous voices in post-secondary education, within the department of Integrative Biology and the Great Lakes Institute for Environmental Research in particular.

In keeping with the findings of the Truth and Reconciliation Commission (Truth and Reconciliation Commission of Canada, 2015) and Universities Canada Principles of Indigenous Education (Universities Canada, 2015), the University of Windsor has committed to building stronger, more meaningful partnerships with Indigenous students, scholars, and communities. The HHL has taken concrete action towards this goal by formalizing the IKC as a means to create, promote, and foster increased representation and training of Indigenous researchers and practitioners in community-based freshwater research locally and abroad.

Fig. 4 Examples of diverse ways in which members of the Indigenous Knowledge Circle worked with and listened to community members to pursue teaching, research, and stewardship activities. This included an Open House Retreat with Indigenous and non-Indigenous researchers and members of the community (A), working with local Bkejwanong Eco-Keepers Indigenous youth group to conduct habitat restoration at Bkejwanong Territory/Walpole Island First Nation (B), being invited to and participating in ceremony at Bkejwanong Territory (C), Examples of Newsletters and Surveys to invite participation in the prioritization of research activities and stewardship actions (D), Recruitment of and support of an Indigenous artist to engage in research (E; Mariah Alexander, 2021). *Photo credits: Healthy Headwaters Lab; Images and text are illustrative examples only; for legible examples, please visit the lab website, www.HealthyHeadwatersLab.ca.*

Broadly, the IKC aims to weave together diverse experiences, skill sets, networks, and resources to address common sustainability challenges through community-based solutions and seeks to build capacity in the next generation of Indigenous scientists (Fig. 4). Operationally, this has included:

- Recruitment and/or support of Indigenous scientists and practitioners,
- Creation of boundary-spanning roles, including flexible paid positions for community members,

- Co-creation of a Traditional Ecological Knowledge Field Course weaving together Indigenous and western science perspectives,
- Co-development of a community-based research grant that prioritizes community priorities and capacity-building,
- Promotion and amplification of Indigenous-led resources, creative outputs, artwork, and story-telling initiatives,
- Participation in and development of panels, seminar series, and workshops that honour and/or centre Indigenous knowledge, and experiences working within western science and Indigenous-led initiatives,
- Network growth of like-minded allies, both Indigenous and non-Indigenous, to strengthen, and expand engagement, learning and partnership opportunities.

Similar to FERN, the IKC benefitted from pre-existing personal relationships developed by various members of the team. A compelling addition to IKC was the creation of a non-traditional role in HHL: the Lab Storyteller. Rooted in Indigenous knowledge systems that have existed since time immemorial, we deeply understood the importance of storytelling in preserving and sharing knowledge about natural ecosystems and people since before western science arrived. A small grant to fund a summer undergraduate student was launched as the pilot for exploring what role storytelling could play in braiding Indigenous knowledges and perspectives with western science in the lab's teaching and research efforts. This enabled us to recruit an incredibly talented Indigenous student enrolled in the Integrative Biology department who also had gifts in commercial photography, marketing, and branding. The summer lab storyteller position launched alongside the first offering of a land-based course on Traditional Ecological Knowledge (Jacobs et al., 2021) and the creation of the lab's website and social media strategy. Truly merging the lab storyteller role with western science, we continue to gather data on the storytelling strategy, pivoting our teaching and storytelling into an evidence-based body of work in progress.

Through a federally funded aquatic species at risk grant (2019–2023) HHL has worked alongside research partners at Walpole Island First Nation/Bkejwanong's Natural Heritage Centre (NinDaWaabJig) to explore how to navigate restoration deliverables through a process of community-based decision making to support and amplify community priorities and long-term goals related to species at risk. Before project initiation, HHL created a flexible paid Project Coordinator position for a Bkejwanong community member. This position, funded through the grant, would act as a connector role between NinDaWaabJig and HHL that would oversee project deliverables with guidance from members of both teams. However, due to personal circumstances of the original hire, the role was readvertised and filled by an Indigenous hire outside of the Bkejwanong community. Given that the new staff member was not from Bkejwanong, the slow, intentional transfer of knowledge and relationship building at the outset was crucial in building trust and understanding. Together we aimed to co-develop an approach to manage invasive *Phragmites australis* from aquatic habitats across Bkejwanong Territory (Walpole Island First Nation), while also engaging the community on the locations they would like to see prioritized, the methods with which removal should be achieved, and what species should be monitored. Unexpectedly, this led to a broader conversation about western science concepts such as "invasive species" and concepts held by the community that plant beings were here for a reason—to teach us—and we have an obligation to first understand its presence, explore its potential utility, and find a solution that presented the least harm to all living beings, including the water. Having launched in late 2019, our efforts were placed on hold or reduced greatly due to the restrictions of the global pandemic, which have persisted, in varying degrees, until the time of writing (>2 years). During the pandemic, in-person gatherings were temporarily replaced with a mail out newsletter, online and mail-in surveys, and virtual meetings to engage community members, introduce the project, and listen to questions, concerns, and input. Although we received excellent feedback, participation rates were much lower than would be expected through in-person community engagement sessions and represented a small demographic of the larger community. Having the Project Coordinator position filled by a member outside of the community posed a challenge to effectively moving ahead with planning, organizing, and executing field activities. Looking forward, we have hired an additional staff member from the community to assist in community interviews that will engage between 20 and 30 community members identified as having close ties to the lands and waters. Halfway through the grant, we continue to work with the community to understand their needs, consider long-term goals and barriers to implementation, and opportunities to work alongside each other. Our teams meet as often as once a week to continue to co-develop a strategy that is designed to engage community members in a way that supports restoration and sustainability goals beyond the limited grant timeline.

The IKC and our Phragmites collaboration with NinDaWaabJig (Walpole Island First Nation's Natural Heritage Centre), have enriched and helped solidify the lab's strategic approach to partnership in action.

These working examples represent shared values in working and learning together as an *ethosystem*, a series of symbiotic relationships and processes at play to support and develop research partnerships and decolonial networks, all based around guiding principles of openness, equity, authenticity, respect, and diversity. Our shared commitment to decolonizing research methodologies and learning reciprocally from the land and one another have led to a richer understanding of connection to place, the importance of relationships, obligations, responsibilities, and interconnectedness beyond that of western science interpretations. Many of these principles are intrinsically embedded in Indigenous pedagogy.

Research-wise, the greatest example of our research process is in the relationship developed between HHL, NinDaWaabJig, and Bkejwanong's youth environmental mentorship program, the Bkejwanong Ecokeepers (BEK). Through ongoing listening, learning, sharing of meals and partnering on numerous grants that ranged in value and shared leadership roles (Box 1), authentic relationships based on trust and reciprocity have created opportunities for both teams to further develop their knowledge and capacity, critical steps needed to build towards coordinated restoration efforts across the region. For example, bringing students to a nearby river system to identify freshwater mussel species at risk introduces them to ecological monitoring skills and restoration opportunities. Moreover, one of the main mechanisms is working directly with youth to inspire and empower them to identify threats to freshwater resources and to support the development of community-led responses (Fig. 4B). Some of the proposed initiatives include restoration of recreational areas, educational signage, community information workshop(s), culturally based science communication initiatives, and protection of turtle nesting habitat. Another mechanism was to provide the community and youth with an offering of services and research expertise available to provide education, training, and capacity building opportunities (e.g., water quality testing, benthic invertebrate sampling, turtle identification and habitat characteristics, freshwater mussel sampling and identification). In instances when expertise beyond HHL was required, we reached across our networks to bring in experts from the University community and community of farmers and practitioners (e.g., FAB/FERN). Likewise, members from the Bkejwanong community joined our FAB to offer their views, insight, and expertise to the HHL team but also to engage with settler farmers. The collective efforts represent a significant investment of time and resources not typically accounted for in traditional research grants but has deepened our relationship with the community and led to

BOX 1 Visualizing HHL networks—An analysis of funding.
Rethinking and reframing the ways in which freshwater restoration science could be pursued, we initiated the lab with an awareness of the importance in gathering data to help assess the importance in social infrastructure and community relationships as a key mechanism for research excellence and impact. An analysis of the allocation of funding was one way through which we could assess and evaluate our effectiveness in achieving and enhancing strategic, holistic, and partnership-based science. Since 2019, we tracked the number and nature of research grants including the funding amount, source (e.g., institutional, provincial, federal sources), as well as in-kind support provided to our community partners at Walpole Island First Nation/Bkejwanong Territory and the Walpole Island Heritage Centre/NinDaWaabJig. We generated a list of nodes for each grant source and project name, as well as an association matrix linking each grant source to either the FERN, IKC or core HHL lab operations. We visualized the network of funded grants for the period of 2019–2021 using igraph in R software (version 4.1.0). In that time period grant awards totalled more than $3million CAD with $1.6million CAD going directly to HHL or NinDaWaabJig. Each HHL project was connected to either IKC or FERN or both. The majority of the funded projects were directed to IKC and/or NinDaWaabJig (85% of all grants) as opposed to FERN-based projects (46%). NinDaWaabJig secured numerous funds of their own, with direct linkages to IKC or HHL to provide research services or training. Grant sources were diverse and balanced between institutional (dark blue nodes), provincial (orange nodes) and federal sources (red nodes), with the majority led by HHL and others led by NinDaWaabJig. Larger, multi-year grants were from federal sources, which provided the most stable source of support to foster, grow and maintain critical partnerships and research networks. COVID-19 hampered efforts to strategically grow FERN networks given our limited ability to meet on site and hold events whereas many of our IKC efforts were able to continue virtually despite the pandemic. This likely was due to HHL team members who lived and worked from Bkejwanong in addition to several grants that were secured during and due to the need to pivot Indigenous communities and programs virtually. Network visualizations also enable us to strategically assess opportunities and prioritize efforts for the year(s) ahead including current efforts to support and strengthen FERN in the short-term. We have also gathered data on in-person meetings including number and nature event, number of people engaged or attending, as part of our ongoing analysis of research impact and excellence. Together, this quantitative and qualitative dataset will contribute to conversations about how funding can be used strategically to support pluralistic research methodologies that underlie community-engaged research and the mobilization of knowledge to ensure truly impactful and beneficial outcomes that are needed to achieve transformational outcomes in science and society (Fig. B1).

BOX 1 Visualizing HHL networks—An analysis of funding.—cont'd

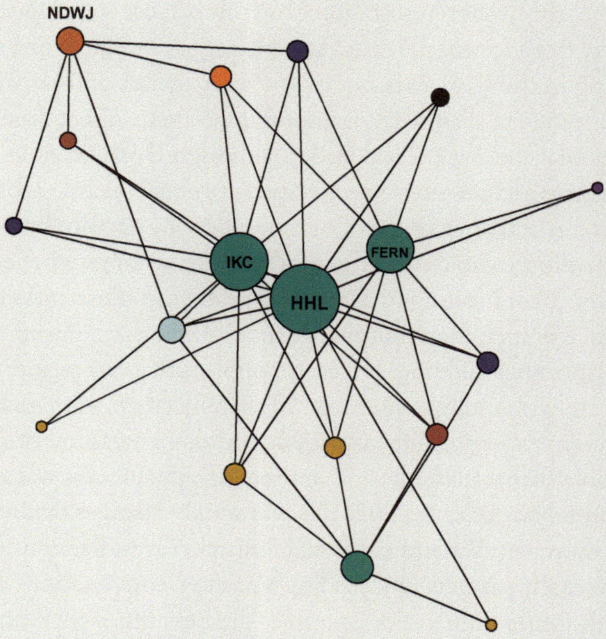

Fig. B1 Network association of key programs (HHL, FERN, IKC; teal nodes) all interconnected via funding supporting specific projects (multicolour nodes). Also included is NinDaWaabJig—Walpole Island Heritage Centre (NDWJ; orange node), a dedicated community partner and co-PI in multiple grants. The relative size of the nodes corresponds to the grant amount (from the $1000s to $100,000s). Linkages connect grants with key HHL programs and associations with either the FERN, IKC, core HHL lab operations or some combination.

additional grant submissions that include the lab as partners in potential future collaborations. By embedding core relationship-building elements within the design of research projects we have been able to explore opportunities in a way that honours our commitments to community and our rights and responsibilities as treaty partners.

The IKC as a central part of the HHL and in relationship with Indigenous communities has been a continuous process of listening, reflecting, learning, unlearning, and exploring ways of incorporating knowledge gained into actionable research and beneficial outcomes across multiple levels, the inclusion of numerous Indigenous roles engaged in science-based research efforts being one. The IKC remains committed to creating decolonial spaces and approaches to freshwater restoration. Intentional use of language such as pivoting away from formal meetings to gathering as a circle is one such example. Gathering in the IKC does not follow a set agenda and remains an open forum for everyone to share what is important to them at that time. The circle also indicates that there is no real leader (although Donaldson does facilitate) and that all voices and members perspectives are equal regardless of status, age, education, etc. Another example of our intentionally decolonized approach is the conscious effort to hold meetings outside of Institutions whenever possible with a preference for meeting in nature. Another example mentioned earlier is the importance of including multiple generations in our community engagement and relationship building. It is commonplace that our relationships in the community are with multiple members of a family circle, with each individual offering different expertise, gifts, and ideas. Although measures of success seem less quantifiable than with traditional approaches, the process was as important as the outcome (Jacobs et al., 2021). It was simple changes that fostered trust and created more familiar and comfortable spaces for Indigenous members to engage in research partnerships, leading to better conversations and helping build towards future efforts. For example, the benefits of investing resources and developing relationships with youth engaged in BEK were reciprocal. We, too, observed a growing sense of responsibility to train, support, engage and empower the youth as restoration stewards. There are no immediate measurable outcomes to link our approach to improved restoration outcomes directly, however our team has grown to include those involved in BEK, and in 2021 our team helped to deliver multiple workshops in the BEK summer program, which has led to planned collaborations and publications. Thus, we have collected narratives through surveys, storytelling, and art, a pluralistic outcome. Another way we have reflected on notions of impact, success and excellence has been through the development and expansion of our network and the number of Indigenous scientists, artists, storytellers, and team leaders we engage through various grants led by or in partnership with our Indigenous community partners (Box 1). The creation of scholarships for students and paid staff positions were other ways way of supporting long-term, community-led, and Indigenous-led research and conservation goals. These roles were community-based or a mixture between being community-based and university-based.

4.2 Looking forward

There is much that western science and settler communities have yet to learn and understand about relationships with the natural world, and the leading role that Indigenous communities have played in conservation (IPBES, 2018). In looking at the evidence (Ban et al., 2018; IPBES, 2018; Suding et al., 2015), Indigenous communities' priorities should be recognized if not elevated to ensure accelerated successful restoration. The initial years of the IKC (2019–2021) were primarily focused on relationship-building, partnership grant writing, and training in ecological monitoring (e.g., benthic sampling, mussel sampling, removal of invasive *Phragmites*, identification of turtles and their habitats). With solid foundational relationships in place, the aim is to continue to engage in habitat assessment, restoration activities, and aquatic sampling with BEK, HHL and a broad suite of partners (e.g., practitioners, government, other Nations and their youth groups). Moreover, Indigenous scholar R.W. Kimmerer has repeatedly shared wisdom on the reciprocity with nature, and how acts of restoration are acts of healing and reconciliation (Kimmerer, 2013). Thus, we will continue to focus our efforts on restoration for future generations. Our time spent with youth will continue to be guided by their interests, pursuing activities highlighting specific interests and skillsets among the team leaders and participants. Working together, side-by-side, primarily on the land to support these interests and identify funding opportunities for ongoing projects is a top priority.

The IKC would not have been possible without seed funding and support from the University (Box 1), an investment that immediately resulted in grants to HHL and the University within the first year. Additionally, the University has supported decolonial approaches across several departments, with Indigenous scholars and allies leading institutional change on a number of fronts. HHL has been part of some efforts in partnership with Bkejwanong (Jacobs et al., 2021) and will continue to advance efforts to bring Indigenous ways of knowing Nature into science courses and research, while also holding space in the Equity-Diversity-Inclusion Task Force and Resistant Pedagogies among other efforts including grants to support trainees (e.g., FishCAST). To date, efforts by the IKC have resulted in network expansion of like-minded allies and increased support from the University and greater representation of Indigenous scholars and Knowledge Keepers in science and teaching spaces. However, systemic changes are still needed, and we continue to leverage our efforts within IKC and beyond to promote—through our own actions and commitments—examples of positive change at all scales from individuals, departments, labs, organizations, and communities. At minimum, we are seeing greater representation of Indigenous ways of knowing and amplification of efforts in support of reconciliation. In the long-term we aspire to seeing Indigenous-led efforts, driven by and for Indigenous communities.

4.3 Responsibilities and obligations

In Canada and across Turtle Island/North America, we collectively share a troubled history built upon systemic injustices towards those with less privilege, which include but also extend beyond Black, Indigenous and People of Colour. At the time of writing, the global COVID-19 pandemic has made clear the inequities in our communities, especially in Canada where there has been growing awareness of a collective responsibility as treaty partners to respond to the 94 Calls to Action by the Truth and Reconciliation Commission of Canada (Government of Canada, 2015). Research environments do not operate in a vacuum and the mechanisms to invest in diversity and inclusion aim to invigorate innovation in sciences are increasing (Government of Canada, 2019). Research teams within public institutions primarily receive public funds to operate. It therefore seems incumbent that research processes are revisited to ensure that reconciliation efforts, research excellence and beneficial impacts are all linked, amplified, and well supported. What has been described here are concrete examples and extensions of the multiple ways in which natural scientists in Canada are encouraged to move forward towards reconciliation (Latulippe and Klenk, 2020; Wong et al., 2020). Thus, one clear outcome from our efforts was an articulated process that embedded an ethics of care, collective responsibilities, and obligations of the HHL team in engaging and amplifying local and Indigenous knowledge with our research.

5. Navigating challenges

Through this process and the case studies described above, we have experienced initial success in uniting often disparate communities and views—academics, settler farmers, and Indigenous knowledge holders—because of a common theme to address freshwater and farm restoration and sustainability challenges for the benefit of future generations. More often than not, however, we have come across challenges that are not always easy to address through the expertise of the team, or because of the historical power dynamics associated with academic institutions

(Tuck and Yang, 2012). Nevertheless, a commitment to more ethical and equitable research processes in support of restoration science and practice provided opportunities to keep moving forward. In a practical sense, however, progress has also been limited both by the short timeframe of our efforts and by restrictions on in-person gathering resulting from the global COVID-19 pandemic. Moreover, the longevity of our research efforts relies on our ability to continue to fund and support diverse and dynamic networks within and across communities and our ability to maintain and grow positive relationships. We have approached building a sustainable research program through strategic connecting and complementarity of new grants, layering in expertise and topics so that students, staff, and projects are connected (Box 1).

Given that research projects and funding cycles are often short (i.e., 1–5 years), our decision to invest in partnerships with long-term stewards of the land are in fact better for restoration efforts, particularly since restoration trajectories are decadal (Jones et al., 2018) and the mobilization of science into practice is often a key barrier to restoration success (Palmer, 2008). Short project windows can lead to selection of activities based on feasibility rather than likelihood of long-term impact. The establishment of partnerships can allow for capacity building and longer-term partnerships that can be responsive to grant opportunities projects as they arise. Ecologically, long-term relationships are better suited for monitoring and assessing restoration at the appropriate scales (i.e., multi-year and decadal scales of ecosystem recovery; Jones et al., 2018; Moreno-Mateos et al., 2017) and also accelerate those trajectories (Bennett et al., 2016).

For many research teams, organizations, and institutions, the investment in such social infrastructure may be daunting, unfeasible, and unappealing. However, we have observed positive outcomes in the short timeframe during which the framework was developed and initially tested. In reviewing the two programmatic case studies here (FERN and IKC), we have identified numerous benefits and challenges associated with pluralistic and community-based restoration research. From these have emerged several overarching themes, all of which contain both challenges and benefits (Table 2). Two themes ("Investment–time" and "Investment–financial") revolve around finite resources needed to conduct effective research, while three more ("Research excellence", "Social outcomes", and "Knowledge mobilization") relate to how community-based research is conducted, and how the outcomes are communicated, perceived, and valued.

Table 2 Challenges and benefits of pluralism identified while operationalizing the Farm and Freshwater Research Network (FERN), Farmer Advisory Board (FAB), and Indigenous Knowledge Circle (IKC).

Theme	Aspect of pluralism	Details
Investment–time	*Challenges*	
	Longer timeframes	Time must be budgeted for relationship and trust building with communities. This is often not compatible with time-constrained deliverables of grants and projects
	Differing concepts of time	Community timelines are often out-of-sync with research timelines. For example, within the FERN, members have expressed frustration with delays in developing and funding proposed work. On the other hand, within the IKC, community partners often do not have the resources to meet short granting timelines.
	Benefits	
	Establishment of trust	Trust grows with time, allowing relationships to weather disappointing research outcomes/failed results. Authentic relationships within the team, within and across communities, over time can lead to new research avenues and grant opportunities, expansive networks, mobilization of knowledge. Supports co-development of research (below).
	True co-development of research	Time investment and trust can offer deeper understanding in ways of knowing, bridge different viewpoints and values, western science and Indigenous knowledge systems. Researchers can more deeply account for and balance science with needs of the community/ies.
Investment—financial	*Challenges*	
	Rare funding mechanisms	Mechanisms to bridge research with communities are rarely funded. Community-based research often takes longer than the traditional granting cycle and thus it can be difficult to produce tangible deliverables to demonstrate traditional research success. Funding long-term connector roles (e.g., Senior Agriculture Advisor) is not standard for research grants, but is essential for building strong relationships with communities.

Table 2 Challenges and benefits of pluralism identified while operationalizing the Farm and Freshwater Research Network (FERN), Farmer Advisory Board (FAB), and Indigenous Knowledge Circle (IKC).—cont'd

Theme	Aspect of pluralism	Details
	Benefits	
	Opportunity to rethink funding mechanisms	Burgeoning interest in pluralistic research and community-based work provides the opportunity to rethink traditional funding mechanisms (e.g., Biological Heritage National Science Challenge).
Research "excellence"	*Challenges*	
	Inherent biases within science	Community-based work may not produce the same type or number of concrete deliverables traditionally valued by western science, resulting in lower perceptions of impact or success.
	Reproducibility of results	Work may be more difficult to replicate or compare given the intrinsic connection with local context
	Lack of training/ expertise	Most teams lack expertise in pluralistic work or knowledge co-production methodologies and training in such methods is scarce. Community partners may not have experience or resources required to navigate institutional processes, methodologies, writing styles/requirements etc.
	Benefits	
	The process as progress	Gathering evidence and examples of transformative change will ultimately show different paths and outcomes are impactful and sustainable when they are community driven and led; A reflexive research praxis allows for insights at local and global scales
	Greater capacity built in early career researchers	Exposure to a greater diversity of perspectives broadens the skillset of early career team members while also introducing them to a greater suite of professional options.
	Effectiveness of results	Results of work are more likely to be successful or effective given that they have already been filtered through the local context, mobilized into local decision-making.

Continued

Table 2 Challenges and benefits of pluralism identified while operationalizing the Farm and Freshwater Research Network (FERN), Farmer Advisory Board (FAB), and Indigenous Knowledge Circle (IKC).—cont'd

Theme	Aspect of pluralism	Details
Societal outcomes	*Challenges*	
	Power dynamics	Western scientists often perceive or present themselves as "experts" and the communities to be "beneficiaries" of their work, rather than understanding that communities are also experts with a different background and perspective. Funding agencies generally provide funds to the academic partner, leaving the community partners beholden to the academic partners for funding.
	Conflict among value systems	Addressing complex issues is in itself a challenge, and when partners have different value systems it can become contentious, especially if valued outcomes are in opposition
	Benefits	
	Increased academic visibility of overlooked groups and perspectives	Inviting community partners to traditionally academic and/or colonial arenas (conferences, academic journals, etc.) can expose the academic community to different ways of thinking and doing, facilitate greater network development, and provide access to previously inaccessible funding and capacity building for often overlooked groups. In this way, increased academic visibility can be mutually beneficial for academia and communities.
	Strengthened relationships among academics and communities	Addressing complex issues with multiple perspectives represented can help identify shared values, questions, and interests, which can help to build networks, leading to greater implementation of replicated studies
	Increased effectiveness of results	Results of work are more likely to be successful or effective given that they have already been filtered through the local context.
Knowledge mobilization	*Benefits*	
	Immediate integration into practice and policy	co-development of goals from the start and diverse research team ensures that no/little time is lost in translating research into action

With finite resources (both time and funding) to support restoration research, researchers often must make difficult decisions about what to prioritize. Given the inherent biases within science about what constitutes "research excellence", these limited funds are generally allocated to concrete deliverables (e.g., student stipends, science research costs, publication costs, conference travel). Unfortunately, some critical components of community-based research (Table 1) are resource-heavy while not immediately resulting in traditional notions of impact, outcomes, and deliverables. Research grants, especially in the sciences, rarely fund networks and community engagement as a core component of research programs. For example, relationship building may take months or years before developing enough trust to address complex and potentially contentious issues in a multi-perspective way. Such a time investment before seeing outcomes is difficult to fit within the framework of a traditional grant. To weather the short-term nature of research funding, we leveraged seed funding to permit flexibility and delayed fieldwork until critical relationships were initiated. To ensure that partnerships were prioritized beyond the initial relationship-building phase, HHL worked with the university to redirect and re-allocate funding from multi-year grants to formalize investments in social infrastructure to support both FERN and IKC. To solidify these efforts, research staff hired in formal connector roles within these programs (Table 1) were engaged in grant writing and partnership development to leverage, extend, and amplify research and community building efforts (Box 1).

6. Assessing early impacts and outcomes

Common questions regarding our slower research path and journey towards ethical and equitable partnerships in a western science research lab often focus on the nature of our results: if they are significant and what are the outcomes. Given the plurality embedded in our approach, we value and expect a plurality of outcomes in addition to societal outcomes that are not determined by us but by the communities our research looks to serve. Underlying this approach is a recognition that social and ecological dimensions of restoration must be addressed together (Suding et al., 2015). Some of these dimensions include scale and socio-ecological mismatches in freshwater restoration (Hand et al., 2018; Palmer, 2009; Sayles and Baggio, 2017), the adoption (or lack thereof) of agricultural management practices (Baird et al., 2016; Baumgart-Getz et al., 2012), and conducting research in a way that is ethical and just (Wong et al., 2020).

In farming contexts elsewhere, such as in Aotearoa New Zealand, Febria et al. (2020) documented how partnerships generate co-benefits for agricultural stream restoration. Over 5 years of place-based research in one focal watershed, more than a dozen actors leveraged individual actions to coordinate and amplify restoration efforts from a single tributary to across a catchment. Also in Aotearoa New Zealand but at a national scale, re-thinking partnerships is becoming actively championed where biodiversity and restoration are concerned (e.g., Living Water Partnership with Dept of Conservation & Fonterra, and, Biological Heritage National Science Challenge) (Norton et al., 2016). In the Great Lakes basin, particularly where this work is situated, the growing recognition that there no one-size-fit-all solution to agricultural management (Macrae et al., 2021) is also coupled with the need to mobilize knowledge through multiple modes of knowledge transfer, engagement, and facilitation (Chapin, 2017; Wood et al., 2014). We have been keeping track of our external efforts through tangible outcomes—publications, grants, events—and continue to document the resources required to grow and connect our research at a range of scales (e.g., Box 1). Peer-reviewed publications now extend beyond those of traditional western science (Barrett et al., 2021; White et al., 2021) to now include case studies on the co-benefits of partnerships in agricultural stream restoration (Febria et al., 2020), restoration syntheses (Eveleens and Febria, 2021), place-based environmental science and ecological research (Eveleens, 2022; Weller, 2022), and also land-based teaching outputs (Jacobs et al., 2021). In addition, we have produced a suite of public resources and plain language summaries for the public on our lab website (amplified on social media channels), on Figshare.com website(https://figshare.com/projects/Healthy_Headwaters_Lab_University_of_Windsor/57503) and a soon to be launched open access journal-type page hosted by the University library, both of which gather quantitative data on impact and number of downloads, while also laying out clear copyright and citation information for users.

Moreover, in writing this article, we intentionally scrutinized the nature of our work and the questions we pursued to identify key aspects, as well as challenges and benefits of pluralistic work (Table 2). These aspects included time investments (Table 1), which helps address the challenge of the mismatch between the time needed to initiate and foster trust-based relationships versus the short timeframes of projects (Table 2). We see examples of strategic partnership investments elsewhere, such as Living Water Partnership (www.livingwater.net.nz), a multi-million, decade-long initiative to implement freshwater restoration in agricultural catchments, and federal science funding agencies such as Biological Heritage National Science Challenge (www.biologicalheritage.nz). Co-investment in research, communities, and practice offers opportunities for mobilizing research into practice, while also creating spaces and opportunities for new expertise and leaders to emerge at this new, interdisciplinary, collaborative nexus. However, more commonly in Turtle Island/North America, grant agencies continue to be fixed to disciplinary focus, yet often require community partners. Most grants signal a preference for Indigenous partners, however considerable time and authentic partnership is required to ensure that the community's priorities are heard and valued rather than tailored to meet a funding agency's call. In our approach, we invest time in understanding both the ecological science needs and the social context within which it fits, requiring considerable human resources to allow for the flexibility and reciprocal knowledge exchange. Reconsidering funding mechanisms to support knowledge brokers and connector-type roles exist elsewhere (e.g., Aotearoa New Zealand's Biological Heritage National Science Challenge) and would make such an approach more feasible for others to pursue and normalize in research communities.

The push towards multiple ways of doing was due to the limitations of network-building during the global pandemic. Our in-person abilities to connect greatly hampered all aspects of our operations and the team. Impacts of the pandemic have been shown to disproportionately affect people based on gender, career-stage, care-giving status, and other overlapping dimensions (Fulweiler et al., 2021; Kramer, 2020; Maas et al., 2020). The adjustments made to projects and our modes of engagement also generated meaningful insights about more inclusive forms of engagement and the perceived barriers of virtual meetings versus learning on the land and meeting in person.

As a research team situated within a university, frequent and ongoing debates over research "impact", "success", and "excellence" are commonplace, as is growing documentation of the inherent biases in traditional metrics of impact both in pan-science publications and outputs (Arismendi and Penaluna, 2016; Larivière et al., 2013). Often, community-based work is viewed as less rigorous given the local context of the work which may limit reproducibility. However, our empirical research on the ground is more likely to influence conservation strategies in the low-gradient, clay-plain agricultural setting of the region, which poses unique challenges for conservation (Lam et al., 2016; Van Esbroeck et al., 2016; Weller, 2022). Our Indigenous community partners share that the process of decentering learning onto the land, away from academic institutions, is progress (Jacobs et al., 2021), and all of these threads contribute to the growing need to pursue less extractive forms of research to ensure beneficial societal outcomes (Powell, 2018a,b). The benefits from a plural approach have included the alleviation of power dynamics between researchers and the community, and through relationship-based engagement, the ability to discuss and navigate differences in views and values. Additional co-benefits include an incredibly diverse team with a wide range of skills and networks. As with many of the efforts we have initiated, the impact of having an intentionally diverse team will continue to reveal itself in time.

7. Conclusions: What can be achieved through pluralism in freshwater restoration science

There is great urgency to accelerate social and ecological outcomes on local and global scales. This paper outlines how a plural approach to freshwater restoration in the Great Lakes Basin has required a multi-prong approach centred on communities to advance scientific research. Under this same umbrella, we have invested in land-based stewards and social networks (e.g., farmers, the public, Indigenous communities) to pursue a more ethical and equitable approach to research collaborations. While it has taken a few years to initiate and grow the FERN and IKC networks, these efforts have been promising at legitimizing our efforts and in mobilizing knowledge. By implementing multiple modes of engaging, through an ethical and equitable approach to partnership, we have documented multiple lines of evidence: a growing network of individual MOUs and research agreements, grant types with varied scope, and a range of publications and public outputs. Ultimately, our experience with FERN and IKC affirms that relationships and connections come before research priorities. Given a long history of colonial-mindset and institutions, and extractive approaches to utilizing natural ecosystems for the benefits of human prosperity, restoring social and ecological relationships are essential to restoring freshwater ecosystems. Investing in research teams and processes to support place-based, community-engaged research will lead to transformative change in how we ensure a shared, prosperous, and sustainable freshwater future.

Acknowledgements

We thank the members of the FERN/FAB and IKC for their trust in this ongoing and evolving relationship and research process. Funding for this work was provided through a Start-up grant from the University of Windsor, a Canada Research Chair Grant in

Freshwater Restoration Ecology from NSERC Canada and support from the Faculty of Science and Office of Research and Innovation Services, a Canada Nature Fund grant for Aquatic Species at Risk, and a University of Windsor Grant for Women. We gratefully acknowledge that this work was conducted on the Traditional Territory of the Three Fires Confederacy of First Nations—the Ojibway, Odawa and Potawatamie—and we will continue to honour our shared obligations towards Truth and Reconciliation with Indigenous Peoples.

References

Alsip, P.J., Hartig, J.H., Krantzberg, G., Williams, K.C., Wondolleck, J., 2021. Evolving institutional arrangements for use of an ecosystem approach in restoring Great Lakes areas of concern. Sustainability 13, 1532. https://doi.org/10.3390/su13031532.

Angers, D.A., Bolinder, M.A., Carter, M.R., Gregorich, E.G., Drury, C.F., Liang, B.C., Voroney, R.P., Simard, R.R., Donald, R.G., Beyaert, R.P., Martel, J., 1997. Impact of tillage practices on organic carbon and nitrogen storage in cool, humid soils of eastern Canada. Soil Tillage Res 41, 191–201. https://doi.org/10.1016/S0167-1987(96)01100-2.

Arismendi, I., Penaluna, B.E., 2016. Examining diversity inequities in fisheries science: a call to action. BioScience 66, 584–591. https://doi.org/10.1093/biosci/biw041.

Baird, J., Jollineau, M., Plummer, R., Valenti, J., 2016. Exploring agricultural advice networks, beneficial management practices and water quality on the landscape: a geospatial social-ecological systems analysis. Land Use Policy 51, 236–243. https://doi.org/10.1016/j.landusepol.2015.11.017.

Ban, N.C., Frid, A., Reid, M., Edgar, B., Shaw, D., Siwallace, P., 2018. Incorporate indigenous perspectives for impactful research and effective management. Nat. Ecol. Evol. 2, 1680. https://doi.org/10.1038/s41559-018-0706-0.

Barrett, I.C., McIntosh, A.R., Febria, C.M., Warburton, H.J., 2021. Negative resistance and resilience: biotic mechanisms underpin delayed biological recovery in stream restoration. Proc. R. Soc. B Biol. Sci. 288. https://doi.org/10.1098/rspb.2021.0354. 20210354.

Baumgart-Getz, A., Prokopy, L.S., Floress, K., 2012. Why farmers adopt best management practice in the United States: a meta-analysis of the adoption literature. J. Environ. Manage. 96, 17–25. https://doi.org/10.1016/j.jenvman.2011.10.006.

Bennett, E.M., Solan, M., Biggs, R., McPhearson, T., Norström, A.V., Olsson, P., Pereira, L., Peterson, G.D., Raudsepp-Hearne, C., Biermann, F., Carpenter, S.R., Ellis, E.C., Hichert, T., Galaz, V., Lahsen, M., Milkoreit, M., López, B.M., Nicholas, K.A., Preiser, R., Vince, G., Vervoort, J.M., Xu, J., 2016. Bright spots: seeds of a good Anthropocene. Front. Ecol. Environ. 14, 441–448. https://doi.org/10.1002/fee.1309.

Bernhardt, E.S., Palmer, M.A., Allan, J.D., Alexander, G., Barnas, K., Brooks, S., Carr, J., Clayton, S., Dahm, C., Follstad-Shah, J., Galat, D., Gloss, S., Goodwin, P., Hart, D., Hassett, B., Jenkinson, R., Katz, S., Kondolf, G.M., Lake, P.S., Lave, R., Meyer, J.L., O'Donnell, T.K., Pagano, L., Powell, B., Sudduth, E., 2005. Synthesizing U.S. river restoration efforts. Science 308, 636–637. https://doi.org/10.1126/science.1109769.

Browne, E., Weller, L., Nolan, S., Ives, J., Febria, C.M., 2021. FERN Newsletter November 2021. figshare. https://doi.org/10.6084/m9.figshare.16920223.v1.

Case, B.S., Pannell, J.L., Stanley, M.C., Norton, D.A., Brugman, A., Funaki, M., Mathieu, C., Songling, C., Suryaningrum, F., Buckley, H.L., 2020. The roles of non-production vegetation in agroecosystems: a research framework for filling process knowledge gaps in a social-ecological context. People Nat. 2, 292–304. https://doi.org/10.1002/pan3.10093.

Chapin, F.S., 2017. Now is the time for translational ecology. Front. Ecol. Environ. 15, 539. https://doi.org/10.1002/fee.1737.

Cusser, S., Bahlai, C., Swinton, S.M., Robertson, G.P., Haddad, N.M., 2020. Long-term research avoids spurious and misleading trends in sustainability attributes of no-till. Glob. Change Biol. 26, 3715–3725. https://doi.org/10.1111/gcb.15080.

Díaz, S., Settele, J., Brondízio, E.S., Ngo, H.T., Agard, J., Arneth, A., Balvanera, P., Brauman, K.A., Butchart, S.H.M., Chan, K.M.A., Garibaldi, L.A., Ichii, K., Liu, J., Subramanian, S.M., Midgley, G.F., Miloslavich, P., Molnár, Z., Obura, D., Pfaff, A., Polasky, S., Purvis, A., Razzaque, J., Reyers, B., Chowdhury, R.R., Shin, Y.-J., Visseren-Hamakers, I., Willis, K.J., Zayas, C.N., 2019. Pervasive human-driven decline of life on Earth points to the need for transformative change. Science 366. https://doi.org/10.1126/science.aax3100.

Djenontin, I.N.S., Meadow, A.M., 2018. The art of co-production of knowledge in environmental sciences and management: lessons from international practice. Environ. Manag. 61, 885–903. https://doi.org/10.1007/s00267-018-1028-3.

Eimers, M.C., Liu, F., Bontje, J., 2020. Land use, land cover, and climate change in southern Ontario: implications for nutrient delivery to the lower Great Lakes. In: Crossman, J., Weisener, C. (Eds.), Contaminants of the Great Lakes, the Handbook of Environmental Chemistry. Springer International Publishing, Cham, pp. 235–249, https://doi.org/10.1007/698_2020_519.

Enquist, C.A., Jackson, S.T., Garfin, G.M., Davis, F.W., Gerber, L.R., Littell, J.A., Tank, J.L., Terando, A.J., Wall, T.U., Halpern, B., Hiers, J.K., Morelli, T.L., McNie, E., Stephenson, N.L., Williamson, M.A., Woodhouse, C.A., Yung, L., Brunson, M.W., Hall, K.R., Hallett, L.M., Lawson, D.M., Moritz, M.A., Nydick, K., Pairis, A., Ray, A.J., Regan, C., Safford, H.D., Schwartz, M.W., Shaw, M.R., 2017. Foundations of translational ecology. Front. Ecol. Environ. 15, 541–550. https://doi.org/10.1002/fee.1733.

Eveleens, R.A., 2022. Does it Take a Community to Save a Species? Examining the Use of Community Interactions to Restore Unionid Species at Risk. University of Windsor, Windsor, ON.

Eveleens, R.A., Febria, C.M., 2021. A systematic review of the global freshwater mussel restoration toolbox. Aquat. Conserv. Mar. Freshw. Ecosyst. https://doi.org/10.1002/aqc.3750.

Febria, C.M., Bayfield, M., Collins, K.E., Devlin, H.S., Goeller, B.C., Hogsden, K.L., Warburton, H.J., Harding, J.S., McIntosh, A.R., 2020. Partnerships generate co-benefits in agricultural stream restoration (Canterbury, New Zealand). Case Stud. Environ. 4. https://doi.org/10.1525/cse.2020.1229632.

Fraser, H., Parker, T., Nakagawa, S., Barnett, A., Fidler, F., 2018. Questionable research practices in ecology and evolution. PLoS One 13, e0200303. https://doi.org/10.1371/journal.pone.0200303.

Fulweiler, R.W., Davies, S.W., Biddle, J.F., Burgin, A.J., Cooperdock, E.H.G., Hanley, T.C., Kenkel, C.D., Marcarelli, A.M., Matassa, C.M., Mayo, T.L., Santiago-Vàzquez, L.Z., Traylor-Knowles, N., Ziegler, M., 2021. Rebuild the academy: supporting academic mothers during COVID-19 and beyond. PLoS Biol. 19, e3001100. https://doi.org/10.1371/journal.pbio.3001100.

Government of Canada, 2019. NSERC—Dimensions—Charter [WWW Document]. URL. https://www.nserc-crsng.gc.ca/NSERC-CRSNG/EDI-EDI/Dimensions-Charter_Dimensions-Charte_eng.asp. accessed 10.14.21).

Hand, B.K., Flint, C.G., Frissell, C.A., Muhlfeld, C.C., Devlin, S.P., Kennedy, B.P., Crabtree, R.L., McKee, W.A., Luikart, G., Stanford, J.A., 2018. A social–ecological perspective for riverscape management in the Columbia River Basin. Front. Ecol. Environ. 16, S23–S33. https://doi.org/10.1002/fee.1752.

IPBES, 2018. Assessment Report on Land Degradation and Restoration.

Isaac, M.E., Isakson, S.R., Dale, B., Levkoe, C.Z., Hargreaves, S.K., Méndez, V.E., Wittman, H., Hammelman, C., Langill, J.C., Martin, A.R., Nelson, E., Ekers, M., Borden, K.A., Gagliardi, S., Buchanan, S., Archibald, S., Gálvez Ciani, A., 2018. Agroecology in Canada: towards an integration of agroecological practice, movement, and science. Sustainability 10, 3299. https://doi.org/10.3390/su10093299.

Jacobs, C., Donaldson, C., Ives, J.T., Keeshig, K., Day, T., Febria, C., 2021. Weaving indigenous and Western science knowledges through a land-based field course at Bkejwanong territory (Laurentian Great Lakes). Case Stud. Environ. 5. https://doi.org/10.1525/cse.2021.1422042.

Jones, H.P., Jones, P.C., Barbier, E.B., Blackburn, R.C., Benayas, J.M.R., Holl, K.D., McCrackin, M., Meli, P., Montoya, D., Mateos, D.M., 2018. Restoration and repair of Earth's damaged ecosystems. Proc. R. Soc. B 285. https://doi.org/10.1098/rspb.2017.2577. 20172577.

Joosse, P.J., Baker, D.B., 2011. Context for re-evaluating agricultural source phosphorus loadings to the Great Lakes. Can. J. Soil Sci. 91, 317–327. https://doi.org/10.4141/cjss10005.

Kimmerer, R.W., 2013. Braiding Sweetgrass: Indigenous Wisdom, Scientific Knowledge and the Teachings of Plants. Milkweed Editions.

Kramer, J., 2020. Women in science may suffer lasting career damage from COVID-19. Sci. Am.

Lam, W.V., Macrae, M.L., English, M.C., O'Halloran, I.P., Wang, Y.T., 2016. Effects of tillage practices on phosphorus transport in tile drain effluent under sandy loam agricultural soils in Ontario, Canada. J. Gt. Lakes Res. 42, 1260–1270. https://doi.org/10.1016/j.jglr.2015.12.015.

Larivière, V., Ni, C., Gingras, Y., Cronin, B., Sugimoto, C.R., 2013. Bibliometrics: global gender disparities in science. Nature 504, 211–213. https://doi.org/10.1038/504211a.

Latulippe, N., Klenk, N., 2020. Making room and moving over: knowledge co-production, indigenous knowledge sovereignty and the politics of global environmental change decision-making. Curr. Opin. Environ. Sustain 42, 7–14. https://doi.org/10.1016/j.cosust.2019.10.010.

Maas, B., Grogan, K.E., Chirango, Y., Harris, N., Liévano-Latorre, L.F., McGuire, K.L., Moore, A.C., Ocampo-Ariza, C., Palta, M.M., Perfecto, I., Primack, R.B., Rowell, K., Sales, L., Santos-Silva, R., Silva, R.A., Sterling, E.J., Vieira, R.R.S., Wyborn, C., Toomey, A., 2020. Academic leaders must support inclusive scientific communities during COVID-19. Nat. Ecol. Evol. 4, 997–998. https://doi.org/10.1038/s41559-020-1233-3.

Macrae, M., Jarvie, H., Brouwer, R., Gunn, G., Reid, K., Joosse, P., King, K., Kleinman, P., Smith, D., Williams, M., Zwonitzer, M., 2021. One size does not fit all: toward regional conservation practice guidance to reduce phosphorus loss risk in the Lake Erie watershed. J. Environ. Qual. 50, 529–546. https://doi.org/10.1002/jeq2.20218.

McElwee, P., Fernández-Llamazares, Á., Aumeeruddy-Thomas, Y., Babai, D., Bates, P., Galvin, K., Guèze, M., Liu, J., Molnár, Z., Ngo, H.T., Reyes-García, V., Chowdhury, R.R., Samakov, A., Shrestha, U.B., Díaz, S., Brondízio, E.S., 2020. Working with indigenous and local knowledge (ILK) in large-scale ecological assessments: reviewing the experience of the IPBES global assessment. J. Appl. Ecol. 57, 1666–1676. https://doi.org/10.1111/1365-2664.13705.

Moreno-Mateos, D., Barbier, E.B., Jones, P.C., Jones, H.P., Aronson, J., López-López, J.A., McCrackin, M.L., Meli, P., Montoya, D., Rey Benayas, J.M., 2017. Anthropogenic ecosystem disturbance and the recovery debt. Nat. Commun. 8, 14163. https://doi.org/10.1038/ncomms14163.

Norström, A.V., Cvitanovic, C., Löf, M.F., West, S., Wyborn, C., Balvanera, P., Bednarek, A.T., Bennett, E.M., Biggs, R., de Bremond, A., Campbell, B.M., Canadell, J.G., Carpenter, S.R., Folke, C., Fulton, E.A., Gaffney, O., Gelcich, S., Jouffray, J.-B., Leach, M., Tissier, M.L., Martín-López, B., Louder, E., Loutre, M.-F., Meadow, A.M., Nagendra, H., Payne, D., Peterson, G.D., Reyers, B., Scholes, R., Speranza, C.I., Spierenburg, M., Stafford-Smith, M., Tengö, M., van der Hel, S., van Putten, I., Österblom, H., 2020. Principles for knowledge co-production in sustainability research. Nat. Sustain., 1–9. https://doi.org/10.1038/s41893-019-0448-2.

Norton, D.A., Young, L.M., Byrom, A.E., Clarkson, B.D., Lyver, P.O., McGlone, M.S., Waipara, N.W., 2016. How do we restore New Zealand's biological heritage by 2050? Ecol. Manage. Restor. 17, 170–179. https://doi.org/10.1111/emr.12230.

Paine, R.T., 1966. Food web complexity and species diversity. Am. Nat. 100, 65–75.

Palmer, M.A., 2008. Reforming watershed restoration: science in need of application and applications in need of science. Estuaries Coasts 32, 1–17. https://doi.org/10.1007/s12237-008-9129-5.

Palmer, M.A., 2009. Reforming watershed restoration: science in need of application and applications in need of science. Estuaries Coasts 32, 1–17. https://doi.org/10.1007/s12237-008-9129-5.

Palmer, M.A., Kramer, J.G., Boyd, J., Hawthorne, D., 2016. Practices for facilitating interdisciplinary synthetic research: the National Socio-Environmental Synthesis Center (SESYNC). Curr. Opin. Environ. Sustain. Sci. 19, 111–122. https://doi.org/10.1016/j.cosust.2016.01.002.

Powell, K., 2018a. Should we steer clear of the winner-takes-all approach? Nature 553, 367–369. https://doi.org/10.1038/d41586-018-00482-y.

Powell, K., 2018b. These labs are remarkably diverse—here's why they're winning at science. Nature 558, 19–22. https://doi.org/10.1038/d41586-018-05316-5.

Prokopy, L.S., Floress, K., Klotthor-Weinkauf, D., Baumgart-Getz, A., 2008. Determinants of agricultural best management practice adoption: evidence from the literature. J. Soil Water Conserv. 63, 300–311. https://doi.org/10.2489/jswc.63.5.300.

Rayne, A., Byrnes, G., Collier-Robinson, L., Hollows, J., McIntosh, A., Ramsden, M., Rupene, M., Tamati-Elliffe, P., Thoms, C., Steeves, T.E., 2020. Centring indigenous knowledge systems to re-imagine conservation translocations. People Nat. 2, 512–526. https://doi.org/10.1002/pan3.10126.

Reid, A.J., Eckert, L.E., Lane, J.-F., Young, N., Hinch, S.G., Darimont, C.T., Cooke, S.J., Ban, N.C., Marshall, A., 2021. "Two-Eyed Seeing": an indigenous framework to transform fisheries research and management. Fish Fish. 22, 243–261. https://doi.org/10.1111/faf.12516.

Rzhetsky, A., Foster, J.G., Foster, I.T., Evans, J.A., 2015. Choosing experiments to accelerate collective discovery. PNAS 112, 14569–14574. https://doi.org/10.1073/pnas.1509757112.

Sayles, J.S., Baggio, J.A., 2017. Social–ecological network analysis of scale mismatches in estuary watershed restoration. Proc. Natl. Acad. Sci. U. S. A. 114, E1776–E1785. https://doi.org/10.1073/pnas.1604405114.

Schuster, R., Germain, R., Bennett, J., Reo, N., Secord, D., Arcese, P., 2018. Biodiversity on Indigenous lands equals that in protected areas. bioRxiv. https://doi.org/10.1101/321935. 321935.

Schuster, R., Germain, R.R., Bennett, J.R., Reo, N.J., Arcese, P., 2019. Vertebrate biodiversity on indigenous-managed lands in Australia, Brazil, and Canada equals that in protected areas. Environ. Sci. Policy 101, 1–6. https://doi.org/10.1016/j.envsci.2019.07.002.

Sterner, R.W., Ostrom, P., Ostrom, N.E., Klump, J.V., Steinman, A.D., Dreelin, E.A., Vander Zanden, M.J., Fisk, A.T., 2017. Grand challenges for research in the Laurentian Great Lakes. Limnol. Oceanogr. 62, 2510–2523. https://doi.org/10.1002/lno.10585.

Suding, K., Higgs, E., Palmer, M., Callicott, J.B., Anderson, C.B., Baker, M., Gutrich, J.J., Hondula, K.L., LaFevor, M.C., Larson, B.M.H., Randall, A., Ruhl, J.B., Schwartz, K.Z.S., 2015. Committing to ecological restoration. Science 348, 638–640. https://doi.org/10.1126/science.aaa4216.

Truth and Reconciliation Commission of Canada, 2015. Truth and Reconciliation Commission of Canada: Calls to Action.

Tuck, E., Yang, K.W., 2012. Decolonization is not a metaphor. Decolonization: indigeneity. Educ. Soc. 1.

Turner, R.K., 1988. Pluralism in environmental economics: a survey of the sustainable economic development debate. J. Agric. Econ. 39, 352–359. https://doi.org/10.1111/j.1477-9552.1988.tb00594.x.

UN Environment Program, 2019. UN Decade on Ecosystem Restoration 2021–2030 [WWW Document]. UN Decade Restor. URL http://www.decadeonrestoration.org/partners. (accessed 9.27.20).

Universities Canada, 2015. New Principles on Indigenous Education.

Van Esbroeck, C.J., Macrae, M.L., Brunke, R.I., McKague, K., 2016. Annual and seasonal phosphorus export in surface runoff and tile drainage from agricultural fields with cold temperate climates. J. Gt. Lakes Res. 42, 1271–1280. https://doi.org/10.1016/j.jglr.2015.12.014.

van Goethem, K., Damphousse, L., Stammler, K.L., Carroll, E., Febria, C.M., n.d. Ecological impacts of management practices in agricultural drain networks: a literature synthesis. Can. Water Resour. J. Rev. Can. Ressour. Hydr. (In revision).

Weller, L.A., 2022. Carbon Characterization in Agricultural Landscapes (MSc Thesis). University of Windsor, Windsor, ON.

White, B.E., McIntosh, A.R., Febria, C.M., Warburton, H.J., 2021. The potential role of biotic interactions in stream restoration. Restor. Ecol. https://doi.org/10.1111/rec.13396, e13396.

Wilson, R.S., Beetstra, M.A., Reutter, J.M., Hesse, G., Fussell, K.M.D., Johnson, L.T., King, K.W., LaBarge, G.A., Martin, J.F., Winslow, C., 2019. Commentary: achieving phosphorus reduction targets for Lake Erie. J. Gt. Lakes Res. 45, 4–11. https://doi.org/10.1016/j.jglr.2018.11.004.

Wong, C., Ballegooyen, K., Ignace, L., Johnson, M.J., Swanson, H., 2020. Towards reconciliation: 10 calls to action to natural scientists working in Canada. Facets 5, 769–783.

Wood, B.A., Blair, H.T., Gray, D.I., Kemp, P.D., Kenyon, P.R., Morris, S.T., Sewell, A.M., 2014. Agricultural science in the wild: a social network analysis of farmer knowledge exchange. PLoS One 9, e105203. https://doi.org/10.1371/journal.pone.0105203.

CHAPTER NINE

Disrupting the governance of social-ecological rigidity traps: Can pluralism foster change towards sustainability?

Pablo F. Méndez[a,*], David Fajardo-Ortiz[b,c], and Jennifer M. Holzer[d]

[a]Institute of Innovation and Knowledge Management, INGENIO (CSIC-UPV), Universitat Politècnica de València, Valencia, Spain
[b]German Centre for Higher Education Research and Science Studies, Hannover, Germany
[c]KU Leuven, Leuven, Belgium
[d]Environmental Sustainability Research Centre, Brock University, St. Catharines, ON, Canada
*Corresponding author: e-mail address: pfmendez@ingenio.upv.es

Contents

Advances in Ecological Research, Volume 66
ISSN 0065-2504
https://doi.org/10.1016/bs.aecr.2022.04.011

Abstract

Processes of change towards sustainability are urgently needed to ensure that humans continue to live within safe biophysical limits. Governance and politics are inherently implicated in any effort to foster successful change processes. Pluralism has been argued as a key ingredient for increasing the likelihood of robust and viable processes of change. Here, we explore pluralism vis-a-vis resource and ecosystem governance from a social-ecological systems (SES) perspective, considering both plurality of values and methodological pluralism. We focus on (1) understanding situations of limited pluralism and factors contributing to it being perpetuated, and (2) exploring how "latent pluralism" could be drawn out to foster change towards sustainability. Our particular interest lies in understanding the role of discursive-institutional dynamics—discourses and how they play out via institutions—underlying governance. To these ends, we synthesize and build on existing case study research in the Doñana region (Guadalquivir estuary, SW Spain), an estuary-delta SES characterized by a rigidity trap, a situation of inflexibility that makes the SES more vulnerable. Using the illustrative example of a failed hydraulic megaproject, we use a novel analytical approach blending institutional analysis with discursive-institutional dynamics. We show how tight relationships between institutional structures and dominant discourses in the Doñana region, underpinned by a hegemonic hydraulic-mission discourse, played a prominent role in curtailing pluralism in Doñana's governance contexts over time. We also show how latent pluralism exists in the region, and reflect upon current ideas and discourses that might help loosen existing governance stalemates and foster change towards sustainability.

1. Introduction

Change towards sustainability (Fischer-Kowalski and Haberl, 2007; Leach et al., 2012; Schot and Kanger, 2018; Scoones et al., 2020) is now more urgent than ever to maintain a habitable and healthy environment for human and non-human life (Folke et al., 2021). Governance and politics are inherently implicated in any effort to foster such a change, irrespective of the focal system (e.g., socio-technical, social-ecological) or disciplinary stance (e.g., transitions, transformations, pathways) (Patterson et al., 2017). Existing governance configurations can create conditions ripe for change (Chaffin et al., 2016; Kemp et al., 2007; Olsson et al., 2006; Plummer et al., 2020), or may need to be transformed to enable those conditions to emerge (Meadowcroft, 2007; Rijke et al., 2013; Schlüter et al., 2021). However, despite critical research around the capacity of governance to foster change towards sustainability, the understanding of such a role remains underdeveloped in the literature (Berkes, 2007; Kellner, 2021; Muiderman et al., 2022; Muñoz-Erickson et al., 2016; Orach and Schlüter, 2021; Patterson et al., 2017; Rocha et al., 2022). In this article, we contribute to fill that general gap with a focus on social-ecological systems (SES) traps, the situation where social and ecological feedbacks mutually reinforce each other to maintain or push the system towards an undesirable state (Carpenter and Brock, 2008; Holling et al., 2002). Through a case of an entrapped SES, we explore how greater pluralism, which has been found to be instrumental for ensuring the preservation of ecological functions critical for sustainability (Gavin et al., 2015; Pascual and Howe, 2018), may help steer governance towards more desirable trajectories.

1.1 Social-ecological systems traps and main research aims

SES traps emerge due to the persistence of self-reinforcing dynamics among social and ecological systems maintained through strong feedback loops, producing negative outcomes at system level (Cumming, 2017). We are particularly interested in the interrelationship between the rigidity trap (Enqvist et al., 2016; Stedman, 2016) and the lock-in trap (Allison and Hobbs, 2004). Rigidity traps are characterized by the recurrent application of command-and-control approaches (e.g., Holling and Meffe, 1996) for resource and ecosystem management. Command-and-control leads to: (1) diminished SES resilience to shocks, (2) increased socioeconomic dependence on fewer resources, and (3) the erosion of a system's potential to adapt to change (Cox, 2016; Gunderson et al., 1995; Holling and Meffe, 1996). In a lock-in trap, continued degradation of the resource base turns into a syndrome consisting of low capacity for change, high connectedness among structural variables, and a near-total removal of natural capital (Allison and Hobbs, 2004). This degradation results in low resource yields, economic returns, social welfare, critical ecological functions and overall potential to escape undesirable situations (Allison and Hobbs, 2004). Rigidity traps, therefore, can be seen as more desirable, having higher potential for change, though if pushed enough against certain social or biophysical thresholds, they can enter a self-reinforcing dynamic leading to a (less desirable) lock-in trap (based on Gunderson et al., 2017; Reyers et al., 2018; Dornelles et al., 2020). In the past decade, the explanatory focus of SES traps research has changed from deficiencies in human adaptive capacity to analysing the influence of historical, institutional, discursive and/or power factors regarding the emergence and persistence of traps (e.g., Baker et al., 2018; Enqvist et al., 2016; Graziano et al., 2019; Laborde et al., 2016; Méndez et al., 2019).

In this paper, we demonstrate a novel approach that integrates those factors inherent in the above traps by blending different institutional analysis paradigms. We build on existing case study research carried out in the Doñana region, an estuary-delta SES (Guadalquivir estuary, SW Spain). The Doñana SES is characterized by a rigidity trap, including among other features: pervasive command-and-control approaches to water resources management and wetland conservation; high degradation of key SES functions but some degree of well-preserved natural capital; lack of demonstrated learning from management failures and environmental crises; and strong path dependence (Méndez et al., 2012, 2019, 2022). The Doñana case study research ("Doñana") aims to understand and explain the evolution of the governance configuration at the core of the rigidity trap, relying on an adaptive inference protocol (based on Holling and Allen, 2002; see details in Méndez et al., 2019). Following previous research insights, our more specific aims here were:

- To understand and explain how a situation of limited pluralism emerged in governance and was perpetuated, leading to an undesirable rigidity trap.
- To explore how 'latent pluralism' could foster change towards sustainability.[a]

Next, we will explain our focus on limited and latent pluralism within the scope of values pluralism. The findings of our analyses will elucidate patterns in governance and its influencing discursive-institutional mechanisms and will provide insights into how latent pluralism may be nurtured to trigger change towards sustainability in SES.

1.2 Pluralism of values and methodological pluralism

Despite well-developed knowledge about the meso-level (e.g., aggregated institutional arrangements) and micro-level (e.g., strategic interactions among actors) mechanisms orientating Doñana's current path towards a rigidity trap (see, e.g., Méndez et al., 2019, 2022), we still lack understanding about the process of "rigidification". Without understanding how this process occurred, we are ill-prepared to avoid lock-in traps or advise on change towards more sustainable states. In understanding this historical process, we are particularly interested in values pluralism. We also conceptualize our novel approach as constituting methodological pluralism, i.e., plurality of methods or frameworks.

[a] Through this research we also seek to inform the development of an environmental a nature policy analysis task for the building of a major European Research Infrastructure: 'European Long-Term Ecosystem, Critical Zone and Socio-Ecological Research Infrastructure' (eLTER RI).

Concerning the plurality of values, the Doñana SES rigidity trap emerged historically, among other factors, from recurrent and increasing levels of top-down command-and-control as a dominant approach to water resource use and wetland conservation. Here, we explore whether and how that phenomenon precluded alternative paradigms and values, and so resulted in 'limited pluralism' in governance settings. Limited pluralism is defined as a top-down authoritarian tendency to control and curtail participation, without fully suppressing it (based on Owen and Bindman, 2019). Paradoxically, as we show below, Doñana is also characterized by 'latent pluralism'. Latent pluralism refers to the "pluralism" of potential possibilities, of the different ways values and knowledge could influence decision-making if they were to develop in the future under the right conditions (based on Szalai, 2005). Here, we explore whether and how the pluralism latent in Doñana could be expanded to disrupt the rigidity trap and foster change towards sustainability. We do this in the discussion by reflecting on current literature about how to orient policy to advance transformative change (Ghosh et al., 2021). As pluralism of both values and knowledge has been found to be instrumental for promoting sustainability in SES (Avriel-Avni and Dick, 2019; Gavin et al., 2015; Lang et al., 2012; Pascual and Howe, 2018; Wyborn et al., 2019), we employ this concept as our main focus throughout the article. Pluralism is also capable of producing more ambitious visions of the future in which SES tensions are seen as opportunities for change, rather than merely as technical problems to be solved through managerial governance (Gillard et al., 2016). For example, the diverse frameworks employed by actors to make sense of change processes can be used to figure out compromises and propose workable solutions (Feola, 2015).

Regarding methodological pluralism, we delve into institutional-analysis approaches used in previous research (see Section 3) and reflect on them at an ontological level, rather than adding new analytic or framework layers like in previous iterations. In this way, we seek to operate at a higher level of analysis and further refine understanding and explanations of discursive-institutional dynamics in limiting pluralism and rigidifying governance in the Doñana SES.

1.3 An example within the case study: A hydraulic megaproject

To undertake our illustrative exercise, we use a hydraulic megaproject launched in 2000 to improve the maritime access to Seville's inland port

(IMAP megaproject) through the Guadalquivir estuary. The shallow dredging of the estuary to allow for light-ship navigation and maritime trading remained a relatively uncontested planning measure until 2000. The IMAP megaproject proposed deeper dredging, regardless of the negative impacts identified on estuary and wetland ecosystems and local socioeconomic sectors. This brought about civil contestation and discord among the main actors involved (Vargas and Paneque, 2015). Although some measures were implemented, the megaproject faced constant challenges from the outset and was finally cancelled by the Spanish Supreme Court in 2019. Important lessons can be drawn from this failed megaproject. The IMAP megaproject posed a high risk of creating negative effects difficult to reverse at system level, thus increasing the likelihood of a lock-in trap. Despite deep-dredging being cancelled, the megaproject uncovered the inherent failure of the main governance actors involved to coordinate to achieve more sustainable outcomes in the region.

The article is organized as follows. Section 2 describes the methodological pluralism underlying the development of our novel approach. The approach itself is presented in Section 3. Section 4 describes how we operationalized our approach to analyze the IMAP megaproject. Section 5 combines results and discussion, and Section 6 offers our concluding remarks.

2. The pluralist harmonizing movement in institutional analysis

Institutional analysis is a key tool for public policy process research, encompassing a rich diversity of theoretical frameworks and logics. Policy research analyses complex interactions involving public policy with a focus on change and stasis over time involving individual and collective action (Schlager and Weible, 2013). Two decades ago, Campbell and Pedersen (2001) called for a 'second movement' towards "a more constructive dialogue among [institutional analysis] paradigms". By paradigms, they referred to the application of neoinstitutionalist strands of political science for the understanding of 'institutions' and 'discourses' in policy studies. Although based on different theories and logics, blending these strands can provide a more robust and exhaustive approach to tackling pressing problems (sensu Howlett et al., 2016). In this article we respond to Campbell and Pedersen's call by applying a blend of neoinstitutionalist strands to enhance understanding and explanatory power of SES studies (see similar efforts in e.g., Boettcher and Kim, 2022; Haller, 2019; Lubell, 2015; O'Riordan and

Jordan, 1999; Ostrom, 2010; Vatn and Vedeld, 2012; Young et al., 2008). For Hotimsky et al. (2006), different neoinstitutionalist strands provide a balance in SES research and governance that would not otherwise be achieved. Therefore, a consolidation exercise is crucial to increase our comprehension of the diverse dimensions of SES interactions. Here we contribute to bridging this gap by combining different neoinstitutionalist strands through a replicable analytical approach. By working at an ontological level, we critically analyse institutions and discourses and their relationships with human behaviour, decisions and actions in the SES context (sensu Hotimsky et al., 2006). Our approach primarily builds on Schmidt's proposal for methodological pluralism (2008, 2010) by combining neoinstitutional approaches for both theoretical- and solution-oriented purposes.

3. A neoinstitutionalist approach

Our approach uses concepts from three neoinstitutionalist strands: rational-choice, historical, and discursive. We summarize and compare differences among their theories and logics to identify potential areas of overlap or interaction below.

3.1 Comparing three neoinstitutionalist approaches

Rational choice institutionalism (RI) assumes that actors are fully rational, self-interested agents seeking to maximize their own utility. These agents pursue their self-interest by expressing preferences and goals in the institutional context, which is conceived as a structure composed of external incentives and disincentives (Hall and Taylor, 1996; Schmidt, 2008; Shepsle, 1989). Having evolved from neoclassical economics, RI considers institutions as entities that provide rationally motivated actors with a degree of certainty about others' behaviour in transactions (e.g., negotiations, information acquisition, enforcement rules) or the provision of public goods (e.g., knowledge, flood control, transport networks) (North, 1990; Ostrom, 1990). Actors employ information rationally and their behaviour is not driven by impersonal historical forces, but by strategic calculations or internalized rules (Shepsle, 1989).

Historical institutionalism (HI) relies heavily on two ideas (termed "logics", as articulated by Schmidt, 2008): logic of path dependence' and contingency (the seeming unpredictability of historical events and sequences that resulted in current institutional structures). For HI, institutions are sets of regularized practices with rules or norms that structure action and outcomes, thus situating power asymmetries within historical dynamics (Hall and Thelen, 2006). Variations in political, governance or institutional settings shape actors' capacities for action, policy making, and institution building (Thelen, 1999). Despite actors operating within given social, psychological or cultural structures that conditions their behaviour, they are also seen as having the ability to interpret situations and make choices within a given structure (Hall and Taylor, 1996). Institutions are seen as shaping actors' behaviour, rather than just a rational product of actors' decisions (as in RI above) (Bell, 2002). Since HI focuses on path dependence, rationality is present "only insofar as institutions are the intended consequences of actors' choices", so using HI as a theory lends itself to mechanistic explanations or historical determinism (versus RI's economic determinism) (Schmidt, 2008).

Both RI and HI have difficulties explaining the causes of crises, institutional change, and human agency, and both require adding elements from other analytic approaches to produce successful explanations (Schmidt, 2008). Institutions act as constraints, whether as incentives structuring action (RI) or paths shaping action (HI), hence both of these approaches conform strongly to a rule-following logic (Schmidt, 2008). This begs the question: how do we explain institutional change if everyone follows rules? The discursive strand helps explain how institutional change occurs by focusing on the relationships between institutions (rules) and discourses (ideas) (Schmidt, 2008).

Discursive institutionalism (DI) synthesizes several theoretical concepts concerned with the communicative and coordinative processes of discourse argumentation (Schmidt, 2008). DI gives explanatory power to discourse, defined here as encompassing "not only the substantive content of ideas but also the interactive processes by which ideas are conveyed" (Schmidt, 2008). Discourses are used by actors when generating, deliberating, and/or legitimizing ideas as part of political or policy processes, all of which are influenced by their institutional context (i.e., rules, legislation, norms) (Schmidt, 2008). However, for DI, institutions are 'carriers of ideas' (Williams, 2019). This interpretation is what makes for a more dynamic approach to institutional change/stasis, because here the actors' interests are seen as subjective (versus objective as in RI), but they are also seen as real and observable empirical entities (Schmidt, 2008). The focus on the interactive communication processes involved in discourse is what differentiates DI from the other approaches. DI goes beyond the other strands' use of ideas, believing that ideas can largely explain institutional dynamics

(Schmidt, 2010). The discourse-focused approach of DI (vs the rational actor focus of RI and HI) operates on the assumption that interactions largely determine political, policy and institutional change/stasis dynamics. Inspired by this approach, we conducted the illustrative exercise of this study guided by the concept of "discourse inertia" (Williams, 2019), due to its plasticity to combine insights and logics from different neoinstitutionalist approaches.

3.2 Discourse inertia

Discourse inertia is defined as "the persistent promotion and dominance of a perception of reality" through the "continuous reproduction of certain storylines and elements of a discourse" (Williams, 2019). According to Williams (2019), dominant or hegemonic discourses are often institutionalized, and are often fraught with inertia due to path-dependent tendencies (as in HI) and powerful entrenched interests (as in RI). However, despite those tendencies, dominant discourses are also dynamic, for they must constantly be conveyed, reinforced, and defended. This can lead to certain narratives to be taken for granted, self-reinforced, and thus prone to continuity as they become entrenched within institutions, policy and political practices. Discourse inertia can help us to understand the persistence of certain institutions or policy pathways, despite evidence of their ineffectiveness or the existence of more suitable alternatives. Finally, discourse inertia may aid reflection on reasons for the persistence of dominant discourses and resistance to change. This has been visualized by the term 'discursive-institutional spiral', which illustrates how actors' ideas, discourses and institutions interact and co-evolve (Fig. 1 and Williams, 2019).

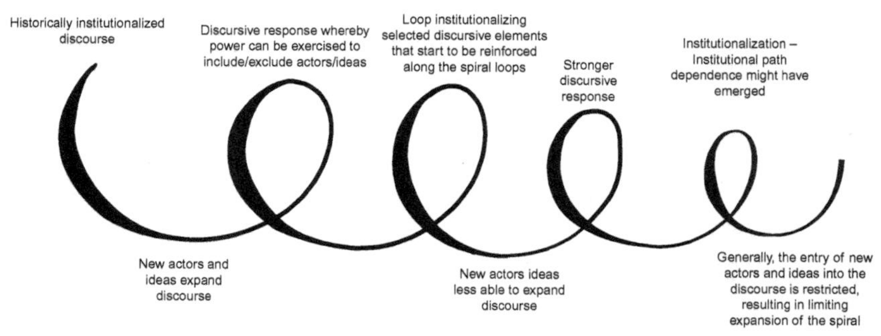

Fig. 1 Discursive-institutional spiral. *Modified from Williams (2019). Spiral art: https:// www.csaimages.com/ (not for commercial use).*

4. Methods

To operationalize our illustrative exercise, we performed a thematic analysis and then reanalysed Doñana's historical pattern, as we describe below (see Fig. 2 for a detailed workflow).

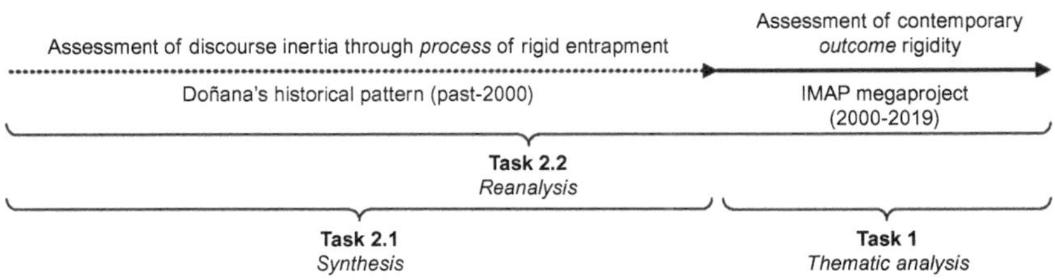

Fig. 2 Workflow of operationalization methods.

4.1 Task 1: Thematic analysis

In task 1 we performed a *thematic analysis* of the IMAP megaproject to analyse command-and-control traits in Doñana in the present-day. We collected data from online newspaper articles about the IMAP megaproject and triangulated these data with a previous study of the IMAP megaproject that used a variety of sources, such as planning documents and research articles (Méndez et al., 2022). The objective was to match evidence from the newspaper articles against a series of categories from command-and-control theory (synthesized from Cox, 2016), through deductive thematic analysis (Nowell et al., 2017). We used these categories as "sensitizing" concepts (Kauffman and Wiering, 2021) to filter content and identify the most relevant information reflecting the command-and-control categories. We also developed a *historical profile* of the IMAP megaproject (2000–2019) tracing key events using secondary sources.

4.2 Task 2: Reanalysis of Doñana's historical pattern

Task 2 was carried out through two sub-tasks. First, we produced a *synthesis* of results from previous research iterations of the Doñana case (henceforward "*Doñana's historical pattern*", Task 2.1). This task used resilience concepts such as traps theory to contextualize the IMAP megaproject in its broader SES context. Then we reanalysed *Doñana's historical pattern,* reflecting on results obtained through the thematic analysis of the IMAP megaproject (task 1). Relying on the concept of *discourse inertia* and informed by the *discursive-institutional spiral,* our aims were to trace the historical persistence of command-and-control governance and to reflect on discursive-institutional dynamics that played out. We compared those dynamics to theoretical insights and logics from our *neo-institutionalist approach* (Section 3). Combining tasks 1 and 2 allowed us to assess contemporary rigidity in Doñana focusing on the IMAP megaproject and the historical processes of rigid entrapment (Fig. 2).

5. Results and discussion

In Section 5.1, we briefly describe the IMAP megaproject historical profile (task 1) and contextualize the megaproject within Doñana's historical pattern (task 2.1; see full thematic analysis results in Appendix A; see full synthesis pattern in Appendix B). Then, we show how Doñana is characterized by a wealth of alternative viewpoints, which we characterize here as "latent pluralism" (Section 5.2.1). Finally, through our reanalysis of Doñana's historical pattern, we expound upon how and why novel ideas quickly lose momentum in Doñana, resulting in limited pluralism and missed opportunities to foster change (task 2.2, Sections 5.2.2–5.2.4) and conclude with a discussion of how pluralism could be expanded and mainstreamed in Doñana (Section 5.3).

5.1 Contextualizing the IMAP megaproject

At present, the inland Port of Seville and the Guadalquivir estuary's navigable canal (90 km) are both managed by the Seville Port Authority (see geographical location in Fig. 3).

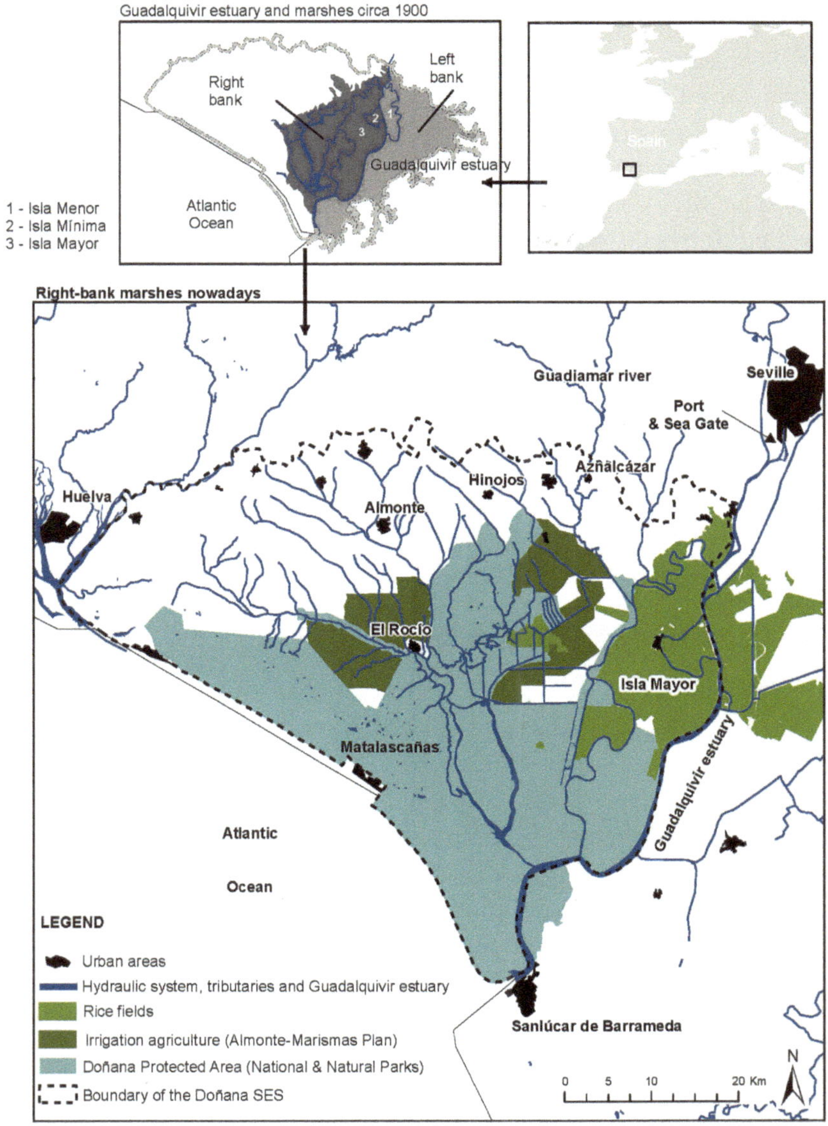

Fig. 3 Geographical location of the Doñana's SES and Guadalquivir estuary. *Modified from Méndez, P.F., Amezaga, J.M., Santamaría, L., 2019. Explaining path-dependent rigidity traps: increasing returns, power, discourses, and entrepreneurship intertwined in social-ecological systems. Ecol. Soc. 24 (2), 30. doi:10.5751/ES-10898-240230, under a Creative Commons CC BY-NC 4.0 International Licence.*

The IMAP megaproject was announced in 2000 as part of the Port Authority's 2020 Strategic Plan (see historical profile in Fig. 4). The megaproject rested on the assumption that making the port more accessible and improving its connection to the Trans-European Transport Network, thus increasing

Date	Event	Organization
2000	Announcement of IMAP megaproject by Port Authority	Port Authority (Spanish Ministry of Public Works)
2003	Creation of supportive platform "Sevilla por su Puerto"	Local association
	Environmental Impact Assessment (EIA), issuing approval decision for IMAP megaproject	Spanish Ministry of Environment[†]
	Prescriptive report rejecting EIA approval decision and requesting scientific study to assess megaproject's impact on turbidity, salinity and biodiversity	Doñana's Participation Council
	2003 EIA approval decision made contingent on conditions from rejection report of Doñana's Participation Council	Spanish Ministry of Environment
	Creation of opposing platform "Por un Río Vivo"	Local association
2004	Agreement to create Estuary Scientific Commission as required by new conditions for 2003 EIA approval decision	Doñana's Participation Council (plenary session with Port Authority)
2005	Creation of Estuary Scientific Commission, consisting of same experts as Scientific Commission of Doñana 2005 project	Interorganizational
	Commissioning of in-depth scientific study by Estuary Scientific Commission	Spanish National Research Council (CSIC) and universities
2006	Beginning of Sea Gate Construction	Port Authority
	Ratification of 2000-2020 Master Plan of Seville's port, including IMAP megaproject	State Ports (Spanish Ministry of Public Works)
2007	Presentation of project for in-depth study about the estuary (CSIC study)	CSIC
2010	Publication of CSIC study's results	CSIC
	Negative statement rejecting deep dredging measure	Estuary Scientific Commission
	Guadalquivir Water Plan draft, leaving out deep dredging measure	Guadalquivir River Authority (Spanish Ministry of Environment)
	Sea Gate comes into operation	Port Authority
2011	Letter from Secretary of State for Climate Change to Port Authority rejecting deep dredging and binding a new EIA approval decision to compliance with requirements from CSIC study	Spanish Ministry of Environment
2012	Formal complaint against deep dredging to the European Commission	WWF
	Opening of infringement proceeding against Spain, requesting to make binding the Estuary Scientific Commission negative statement	European Commission
2013	Passing of Royal Decree (355/2013), approving Guadalquivir River Water Plan, including deep dredging measure as complementary and not referencing new knowledge from CSIC study	Spanish Government
	Urgent request to Spanish Government to forbid deep dredging measure, requesting to make binding Estuary Scientific Commission's negative statement	UNESCO
2014	Issuing of Statement on potential infringement of Habitats Directive (92/43/CEE) by deep dredging measure, responding to complaint from WWF	European Commission
	Rejection of Port Authority's Project for the Improvement of the Riverbanks	Doñana's Participation Council
	Resignation and replacement of Port Authority's president	Port Authority
	Binding of deep dredging approval to compliance with new 2011 EIS requirements, based on CSIC study	Andalusian Regional Government
2015	Cancelation of deep dredging as complementary measure in Guadalquivir River Water Plan, responding to appeal from WWF	Spanish Supreme Court
2016	Passing of Royal Decree (1/2016), approving new Guadalquivir River Water Plan, fully incorporating deep dredging	Spanish Government
	Urgent request to Spanish Government to reject deep dredging - Doñana Protected Area on the verge of being included in UNESCO's black list	UNESCO
	Commitment to not authorize IMAP megaproject	Spanish Government
2019	Final sentence cancelling deep dredging measure in 2016 Guadalquivir River Water Plan, responding to appeal from WWF	Spanish Supreme Court
	Appointment of new president for the Port Authority	Port Authority

[†] Currently Ministry for Ecological Transition and Demographic Challenge

Fig. 4 Historical profile of the IMAP megaproject (based on our thematic analysis and key secondary sources, e.g., Donadei, 2020; Sancho-Royo and Del Moral, 2015; Vargas, 2014; Vargas and Paneque, 2015).

freight traffic from (deep-draft) high-tonnage ships, would be key for local and provincial socioeconomic development (e.g., generation of thousands of jobs, increasing cruise tourism) (Donadei, 2020; Vargas and Paneque, 2015). The megaproject's main measures included (1) deep dredging of the estuary's navigable canal to increase its depth (from 6.5 to 8 m on average) and width (from 60 to 90 m on average), (2) the construction of a higher-capacity lock (named Sea Gate), (3) deep dredging maintenance for 20 years, and (4) auxiliary infrastructure. Since 1985, the Port Authority has performed maintenance shallow dredging for light-ship traffic on a regular basis (García and Marín, 2006).

Our thematic analysis of the IMAP megaproject (Appendix A) shows ubiquitous command-and-control traits in Doñana's governance configuration for water resources use and wetland conservation. These traits were inferred across the six categories synthesized from Cox (2016), although with different degrees of actualization in reality. According to previous research, Doñana's rigid governance configuration evolved historically through a path-dependent dynamic and the persistence of a command-and-control governance approach. At the system level, this path-dependent historical pattern led to the emergence of a rigidity trap, which is the current state in the Doñana SES today (see description and definitions in Fig. 5).

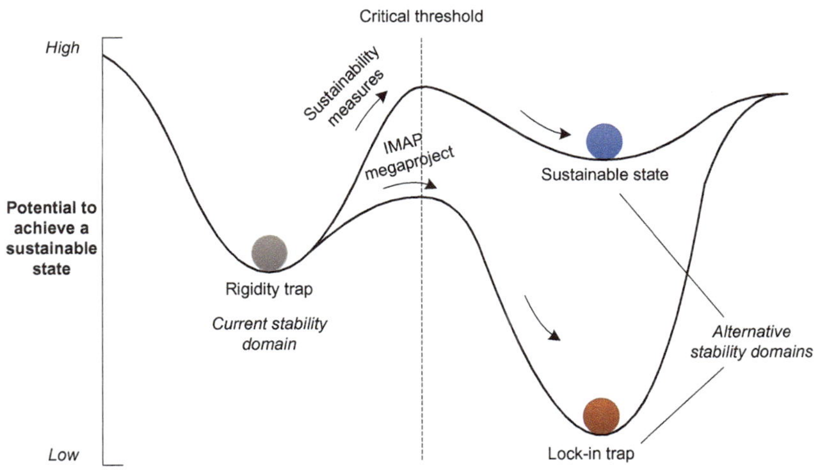

Explanation: the relatively fast development of irrigation and rice agriculture between the 1950s-1980s led to a process of natural capital destruction that could have driven the Doñana region directly into a lock-in trap. Instead, a rigidity trap with higher potential for change materialized due to the creation of the Doñana National Park (1969) thanks to A reactive protection-for-conservation race (see main text). Conceptually, the Doñana case presumes that, if the current rigidity trap is pushed enough against certain biophysical thresholds, the whole SES might potentially fall into a lock-in trap. In this conception, the IMAP megaproject appears as a short-lived disturbance posing a high risk of a regime shift towards a lock-in trap, which would generate insurmountable shifting costs hindering a more sustainable state.

Fig. 5 Doñana's current stability landscape. Based on Walker et al. (2004).

In this context, the IMAP megaproject can be characterized as a disturbance that might have been the trigger that pushed the region towards a lock-in trap (Fig. 5). Recent scholarship shows that short-lived anthropogenic disturbances might induce collapse into irreversible SES functions or states, suppressing the potential of change towards more favourable situations (Haider et al., 2018; Phelan et al., 2013; Standish et al., 2014).

5.2 Doñana's historical pattern reanalysed

5.2.1 Alternative views: Latent pluralism in Doñana

Our thematic analysis shows the presence of diverse paradigms and views proposed by various actors as alternatives to deep dredging and associated development in the estuary and broader Doñana SES. One example of an alternative to the command-and-control paradigm that gave rise to the IMAP megaproject is the protection-for-conservation model, to which direct or indirect reference was made in most articles we reviewed. The ecological restoration of the estuary was also put forward as an option in the context of the Spanish Supreme Court's ruling (Fig. 4), suggested as a countermeasure after the cancellation of deep dredging. References were also made to sustainable development, which influenced narratives in the region since the 1990s. Some alternative views were backward-looking, promoting a return to an idealized form of "*equilibrium that broke in the 20th century.*" This refers to enhanced navigability through meander cuts, the modification of the marshlands and floodplains for agricultural development and the pollution of the estuary (see e.g., ED_050 in Appendix A). Clean renewable energy was also put forward as an alternative to gas storage in the region; for example, by using the Estuary Scientific Commission to generate knowledge "*about cumulative impacts to dispel doubts about the complex risks*" entailed by these type of projects (EPRESS_041). As these examples illustrate, "working with nature" is another alternative paradigm that came up repeatedly in our analysis.

Vast experience and know-how has been accumulated over recent decades in Doñana, particularly through numerous participatory and transdisciplinary initiatives that have been carried out, resulting in a wealth of latent pluralism (e.g., the sustainable development plans put forward in the 1990s; Castells et al., 1992). Efforts to promote transdisciplinarity and enhance stakeholder engagement continued throughout the late 1990s and the 2000s, such as the hydro-ecological restoration projects launched as a response to Los Frailes' mining accident (Doñana 2005 and Guadiamar Green Corridor) (see e.g., García and Marín, 2006). These projects were relatively successful at coupling research with policy through action research

(Escalera, 2003; Méndez et al., 2012) and collaborative adaptive management (Santamaría et al., 2006). Research underpinning these projects focused on a wide range of topics, such as: the study of cultural and ecosystem services (García-Llorente et al., 2011), the conservation-vs.-development paradigm in and around the protected areas (Martín-López et al., 2011), the institutional arrangements in place (Gómez-Baggethun and Kelemen, 2008), and the contribution of traditional ecological knowledge to community resilience (Gómez-Baggethun et al., 2012). In parallel to such research efforts, the Doñana Long-term Socio-Ecological research (LTSER) platform was established in 2007 as part of efforts to formalize lasting infrastructures for integrated ecosystem research at the European level (Díaz-Delgado et al., 2016). These efforts have evolved during the last decade and the insights, know-how and experience gained are currently informing a major European effort to build integrated, continental infrastructure for social-ecological research (this project is known as "eLTER RI"). Doñana's participation in this project constitutes evidence of latent pluralism Doñana, as it is being showcased as one among several flagship cases where components of long-term social-ecological, transdisciplinary research are present—albeit still in need of an overarching strategy to ensure its relevance and further promotion into policy and practice (Holzer et al., 2019). How and why has this latent pluralism been limited in Doñana and which are the risks entailed by that limitation? What can we learn from the IMAP megaproject?

5.2.2 *Was deep dredging irrational behaviour? Yes and no—And "history matters"*

From the perspective of rational choice institutionalism, which considers institutions as incentive/disincentive structures, the Port Authority and coalition of supportive actors, enabled by the Water Authority, behaved rationally between 2000 and 2010. They considered expert knowledge supporting the IMAP project's purported socioeconomic benefits as relevant and valid, despite the fact that this knowledge was discredited or considered incomplete by opposing actors. The Port Authority and supportive actors acted strategically and in self-interest, disregarding the potential reduction of overall socioeconomic welfare or the suppression of SES functions. According to this perspective, the key problem between 2000 and 2010 was of formal regulatory nature: the 2003 megaproject's EIA was incomplete and was amended ex post by a CSIC study (Fig. 4). This study showed the estuary to be in poor environmental condition due to high organic matter content and turbidity, biochemical pollution and high biogeochemical alteration, and as a non-stratified water body limiting biodiversity, with reduced habitat and potential for fishing activities (CSIC, 2010). Two key conclusions from the study were that: (1) cumulative impacts from the IMAP megaproject may have worsened the situation, both environmentally and socioeconomically, and (2) resource optimization for all sectors was not possible, especially in dry years in the current climate change scenario (CSIC, 2010). If this was the case, then why did the Port Authority persist—seemingly irrationally—to implement deep dredging after the CSIC study? To explain this illogical behaviour, we draw on insights gained through the Doñana's historical pattern.

The development of irrigation and rice agriculture in Doñana (past-1980) appears unexpected, if assessed as rational behaviour. At the outset of the development process (1920s–1940s), local actors seemed to behave irrationally; rice and irrigation agriculture was an unsuitable economic choice to maximize earnings in a harsh saline environment—at least not without heavy public investment to dam and impolder the area (Appendix B). Three factors can make this choice seem more rational:

(A) The discourse, which we refer to as "political-level hydraulic mission discourse" in Section 5.2.4, embedded in institutions (e.g., legislation, regulations, incentives), in which actors speak about mobilizing top-down resources and that will yield ever higher economic profit to local actors.

(B) Local organizational (e.g., companies) and individual actors operating on the ground were responsive to the discursive and institutional signals and, backed by public funding and after several agronomic experimental failures, successfully implemented irrigation and rice agriculture against unfavourable environmental conditions.

(C) Coalition of local and international actors engaged in a protection-for-conservation race for preserving the last remnants of terrestrial, coastal and wetland ecosystems, acting in a reactive way to the proposed agricultural development.

The actions of actors from factors B and C clashed and inadvertently contributed to rigidifying governance, leading the Doñana SES towards its current rigidity trap. The whole process was path-dependent because there were at least two alternative paths as described above (for detail see Méndez et al., 2019). At the time of creation of Doñana National Park in 1969, these two alternative paths both had a significant chance of happening. The sustainability trajectory was (and still is) linked to a better and coordinated integration with nature, and the lock-in trap was (and still is) linked to failure to preserve key ecological functions due to hydraulic and agricultural modifications (see Allison and Hobbs, 2004). Takeaways from Doñana's historical pattern are that factors A, B, and C above continue to operate nowadays, and that the alternative trajectories described above are both still possible today; however, large interventions such as the IMAP megaproject pose a high risk of pushing Doñana into a lock-in trap, thus hindering change processes towards sustainability (Fig. 5). Next, we elaborate on these last points.

5.2.3 The rigidifying power of clashing discourses

According to our thematic analysis, the IMAP megaproject was dominated by (1) an economic development discourse underpinned by modernization and neoliberal ideas, and (2) a protection-for-conservation discourse. Over the course of the megaproject, the Port Authority advocated the economic development discourse, legitimized through a mandate by State Ports (the umbrella state body for all Spanish port authorities) to develop infrastructures and promote maritime trading activities. Based on contested and discredited economic assessments, the Port Authority overstated the IMAP megaproject's potential for high returns on investment (Sancho-Royo and Del Moral, 2014; Sancho-Royo and Del Moral, 2015; Vargas and Paneque, 2015) and ultimately built the Sea Gate using public funds (Fig. 4).

Prior to the 2010 CSIC study and a decisive action of the Spanish Supreme Court to cancel the project, the IMAP megaproject had been hampered by the protection-for-conservation discourse, used by the conservation nonprofit organization WWF and the quasi-governmental Doñana Participation Council. At critical points, both organizations were supported by regional, national, and international organizations or governments (Fig. 4). The Doñana Participation Council played a key role as moderator of environmental governance conflicts in the region, amending the defective EIA, and paving the way for the Estuary Scientific Commission and the CSIC study (Fig. 4). WWF was (and is) a leading conservation organization in Doñana; importantly, they made formal appeals to national and international legislative bodies over the incorporation of deep dredging as a measure in water planning (Fig. 4; see also "lock-in factors" in Appendix A).

The economic development and the protection-for-conservation discourses were both embedded in the values, beliefs and cultural attributes of different groups of local actors. The struggle between supporters and opponents of the IMAP megaproject was dynamic in terms of legal interactions and power/knowledge conflicts. The situation resulted in a tug-of-war with no clear winner, which, in turn, meant a step back to a stable but fragile situation among actors and a deadlock regarding the future of the estuary. We might conclude that both supportive and opposing actors fit the rational model—they all behaved rationally and strategically to achieve their own objectives while responding to their respective economic development and protection-for-conservation rule settings. However, this clashing of discourses and rational behaviour still does not fully explain the persistence of the plan for deep dredging, nor the inability of actors to break the deadlock after the CSIC study was released. Why didn't a new counter-narrative emerge from the supportive actors? Why did it take legal power to cancel deep dredging? We argue that factors A (the dominant discourse in Doñana), B (the successful implementation of irrigation agriculture in unfavourable conditions) and C (the coalition advocating for conservation and opposing agricultural development) complement this explanation underlying the deadlock situation. Beneath all of this history, it was the discourses—the ideas and how they played out in institutions, plans, and on the ground—that shaped Doñana's discursive-institutional spiral, to which we now turn to explain the rigidification process of governance in Doñana (Fig. 6).

5.2.4 Doñana's discursive-institutional spiral

The Guadalquivir estuary deadlock situation is based on the *hydraulic mission discourse* in Doñana. Prevalent worldwide, the hydraulic mission discourse is based on narratives and actions of 'hydrocracies' (Williams, 2019), elite groups of policy and decision makers who consider water resource development a state duty to be implemented through technical and institutional efficiency—despite evidence of negative effects on the environment and society (see e.g., Molle et al., 2009; Verhoeven, 2019).

In Spain the hydraulic mission discourse (factor A above) is deeply embedded in water planning and has steered geographical transformation since before the 18th century (Fig. 6). It mobilizes large public investments in hydraulic infrastructure and irrigation agriculture, based on a narrative of modernization and returns on investments to augment national wealth. With little concern for hydrological and ecological conditions, the discourse operates using the paradigm of nature as an object to be dominated and shaped by humans (Saurí and Del Moral, 2001; Swyngedouw, 2015; Vargas and Paneque, 2015).

The hydraulic mission discourse was promoted by key figures in "regenerationism", an intellectual movement seeking to rationalize Spain's

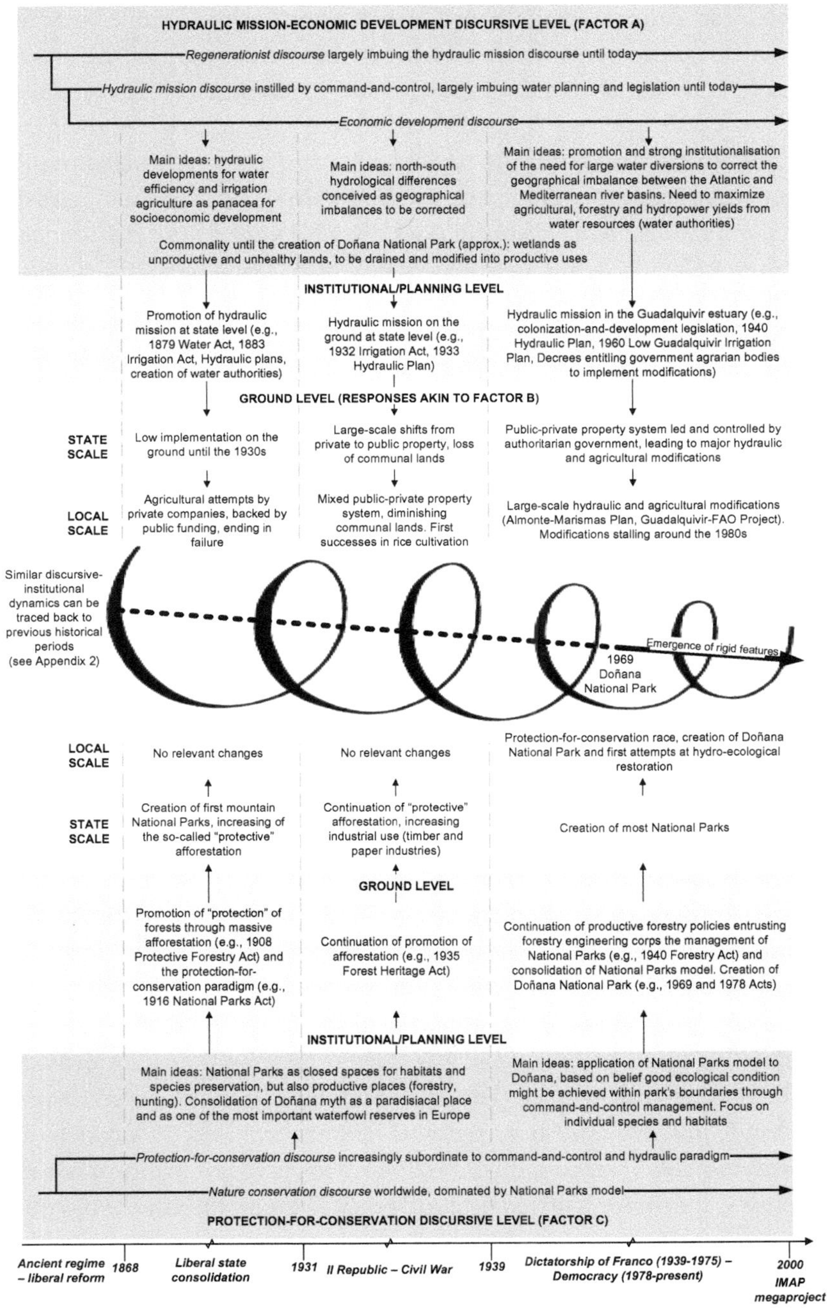

Fig. 6 Application of the discursive-institutional spiral to the Doñana case (based on empirical evidence from Appendix B).

decline as a global political and economic power at the end of the 19th century (Font and Subirats, 2010; Swyngedouw, 1999). The hydraulic mission discourse gathered momentum at the beginning of the 20th century and became more deeply embedded at multiple levels (institutional, planning, local) during the 1930s–1940s (Fig. 6). In the Guadalquivir estuary and Doñana region, the hydraulic mission discourse started in the 1920s with irrigation and rice agriculture experiments, incentivized by policy and legislation. These rules and laws persisted through the 1930s–1970s, giving rise to a massive hydraulic infrastructure, funded by the state through the Guadalquivir River Authority. This took place in synergy with a legislative and planning framework that promoted the colonization of large "unproductive" areas (Fig. 6). The *discourse inertia* of the hydraulic mission discourse can be inferred from Doñana's historical pattern. Such discourse was codified in rules and embedded in actors' thinking, constraining the possibilities of what could be viewed as rational actions. This in turn limited the possibilities for change.

The economic development discourse is impervious to alternative pathways based in different paradigms because it is deeply imbued with the ideas, narratives and perception of reality imposed by the hydraulic mission discourse (Fig. 6). The IMAP megaproject would not likely have been undertaken without the hydraulic mission discourse to persistently structure the cognition, perception of reality and, hence, the rationality of the supportive actors. This project also required the River Authority's passive attitude, neglect of governance duties and lack of compliance with environmental policy guidelines and planning regulations.

Through this alignment of events and factors, the hydraulic mission discourse continued to influence Doñana's recent history, reinforcing path dependence. This is evidenced in particular by: (1) a high pay-off for further investment in the deep dredging measure, due to public funding and construction of the Sea Gate; (2) expectation of higher efficiency and returns from the enhancement and increased use of the estuary's navigable canal (moving from shallow dredging and light ship traffic to deep dredging and heavy traffic); (3) increasing prevalence of the navigable canal operation, i.e., adaptive expectations (based on North, 1990). The assumptions made by the hydraulic mission discourse can thus be used predict the economic policy options and outcomes available in Doñana for certain powerful actors; options that did not (and do not) seem to be the most 'rational choices' at system level. These options could have been realized if it wasn't for the constant opposition from groups operating under the protection-for-conservation discourse.

Not surprisingly, the protection-for-conservation discourse in Doñana is subordinate to the command-and-control approach and the hydraulic paradigm (Fig. 6). During the 1970s, the protection-for-conservation institutions and organizational structures were co-opted by engineering corps through top-down command-and-control, although not without opposition from some pro-conservation engineers (see Appendix B). Meanwhile, the 'hydrocracy' continued throughout the 1970s–1980s, promoting intensive modifications for irrigation and rice agriculture, heavily transforming the natural physical features that regulated the hydrological regime of the protected wetlands (Fig. 6).

Agriculture and protected-area developments, driven by two opposing discourses, reduced the original estuary's floodplain marshland-wetland by ~80%. This created a hard geographical and operational boundary limiting the remaining semi-pristine nature of the inland Doñana SES (Fig. 3). The IMAP megaproject and its contentious deep dredging measure, justified through the hydraulic mission discourse, has essentially created another rigid boundary on the eastern side of the Doñana SES (Fig. 3). A different set of menaces and systemic risks might therefore materialize in the absence of key governance actors such as the Doñana Participation Council and WWF, who act as guardians of the remaining natural capital. The IMAP megaproject jeopardized the fragile governance stability currently existing in the region, triggering a highly conflictive situation pointing to lose-lose prospects for the actors involved, despite their powerful discourses.

Strikingly, discourse inertia can be found in the last strides of the IMAP megaproject. In 2017, after the first backlash from the Spanish Supreme Court (Fig. 4), the Port Authority shifted to a narrative promoting scientific knowledge production about the environmental sustainability of the estuary as a *"permanent and priority goal"* (president of the Port Authority, PUB_003; Appendix A). However, in the same communication, the Port Authority advocated some type of dredging measure based on "*better techniques and better integration*". Today, the new strategy of the Port Authority aims "*for the optimization of navigation in the Guadalquivir [estuary]*" (ibidem), while simultaneously "*working with nature*". Since the Port Authority stated that "*the sentence of the Supreme Court do not disrupt our plans for the port*" (Diario de Sevilla, 2019), it is yet to be seen whether this new strategy will materialize to protect the integrity of the estuary, or whether just the working-with-nature paradigm will be co-opted again.

5.3 Escaping the rigidity trap: Expanding latent pluralism

Our analysis of Doñana's rigid discursive-institutional spiral shows that discourse inertia is the mechanism responsible for Doñana's gradual curtailment of pluralism, resulting in rigid entrapment of the SES. How can this mechanism be disrupted to escape the rigidity trap? Specifically, how can latent pluralism be expanded to foster change towards sustainability? We answer by drawing from the sustainability transitions literature, specifically from recent efforts to make this body of theory more suitable for evaluation purposes (Ghosh et al., 2021; Molas-Gallart et al., 2021). Below, we focus on the initial steps of a potential process of change.

Broadly speaking, spaces for governance experimentation need to be created to nurture latent pluralism. Participation by a range of engaged actors is needed, from across the political spectrum and many walks of life, creating enough diversity to create conditions ripe for minimizing polarization and decision paralysis (Hull, 2009). Such diversity encourages a wide variety of viewpoints, grounded in different knowledge sources. To successfully inform and shape planning policy and decision making, these experimentation spaces also need to be process orientated rather than technical results orientated. These spaces must be protected ("shielded"; Ghosh et al., 2021). Examples of protecting these participatory spaces might include: aligning policy/decision makers and other actors through mutually agreed environmental goals and ways forward that cut across diverse governance levels and SES spatiotemporal scales ("deepen shielding"; Ghosh et al., 2021); voluntary agreements between actors with similar conceptualizations of nature conservation (e.g., Nature-Based Solutions, Nature as other, human/non-human relations); or new regulations to enhance pluralism in formal environmental decision making (e.g., land-use planning, Environmental Impact Assessment) (based on Raven et al., 2010; Smith et al., 2014; Spash, 2022). "Learning" is another critical component to be considered in these spaces and must encompass both first-order learning (e.g., changes in routines or protocols based on new monitoring information) and second-order learning (e.g., changes in core values and beliefs or different ways of approaching problems based on existing or new transdisciplinary knowledge) (Ghosh et al., 2021). In the long term, second-order learning based on confronting core assumptions and biases from actors may be more transformative, due to it challenging well-established rules (Van Poeck et al., 2020). "Deepening of learning" (Ghosh et al., 2021) can happen if actors engage in processes where their thinking can be stretched in collaborative spaces that enhance reflexivity and mutual trust (Raven et al., 2019). These processes can ultimately result in "deepening of networks", drawing together actors with different power identities regarding resources and decisions (Ghosh et al., 2021). This can happen via interventions that are moderated by intermediaries who facilitate mutual connections and reconcile narratives (Kivimaa et al., 2019; Smith and Raven, 2012). Finally, the actors' expectations must be navigated to prevent disappointment and minimize ambiguity. This can be achieved by creating collaborative, inclusive visions, which can lead to more robust, shared expectations (Ghosh et al., 2021). We argue that these initial steps—involving broadening participation and deepening learning and networking—would contribute to setting a new governance baseline more apt to nurture changes towards sustainability in Doñana that would be widely accepted. If that new governance baseline was achieved, conditions would be ripe for expanding and mainstreaming latent pluralism. This could happen by "replicating" innovative governance experiments to resolve issues or working at broader scales to "institutionalize" such experiments through new regulations and the development of new cultural meanings ("upscaling") about the critical role of water resources for human and non-human well-being (Ghosh et al., 2021).

This whole process must start with a crack that breaks the dominant hydraulic discourse and command-and-control paradigm—which presently influence institutions and discourse and, in turn, give rise to a governance configuration unfit to advance SES sustainability. Greater reflexivity must be promoted in governance; the transformative potential of pluralism can be hindered by the stance, still common among scientists, to behave as if they are neutral parties rather than as engaged in a mission-driven activity (Crouzat et al., 2018). Joint learning and the creation of a common vision for change can be a precondition to overcome these type of challenges (Chaffin et al., 2016). At the same time, we must recognize that most perspectives (if not all) are not ideologically neutral. In the absence of equitable power structures to resolve disputes about how nature should be valued and shaped, certain perspectives dominate policy and decision making, pushing change processes to unfavourable conditions and trajectories (Garrett et al., 2021; Jacobs et al., 2020; Leonardsson et al., 2021).

6. Conclusions

Governance plays a crucial role in fostering and steering change towards sustainability. Values pluralism constitutes a key ingredient for increasing the likelihood of robust and viable change processes. Our case study shows that a tight relationship has historically existed between institutional structures and dominant discourses in Doñana. Current ecosystem and water resources governance is characterized by a deeply entrenched contest between economic-development and protection-for-conservation discourses, imbued with hydraulic mission ideas and ways of thinking. At present, Doñana's governance configuration prevents the further deterioration of water resources and ecosystems through actions taken as part of a rich legal legacy, reactive judicial rulings, integrated knowledge co-production, and local nature safeguards and watchful international actors. These features are instrumental in averting the systemic risks posed by high-impact interventions promoted by elite actors with narrow interests who have historically been disinclined to work towards broad socioeconomic welfare.

Our study identified several mechanisms (e.g., rationality restricted by discourses, path dependence mechanisms, discourse inertia) by which values pluralism can be systematically inhibited. Tightly interwoven dominant discourses and institutions can be impervious to new paradigms and knowledge even if they reach the agenda at high level. Novel ideas and narratives can be rapidly cancelled out by dominant discourses. This might reflect underlying coordination failures among governance actors to integrate new discourses into planning and policy. Such failures can be compounded when multiple, powerful fronts (legal, planning, economic, conservation) act at odds with each other, risking inaction. Also, when critical issues such as modernizing irrigation, enhancing navigability, and protecting nature all tend to be issues high on the political agenda, attempts to steer the discourse in a new direction can result in a deadlock that hinders new policy directions.

Finally, we have shown how latent pluralism might be expanded and mainstreamed to escape undesirable SES traps by reflecting on state-of-the-art concepts that can be leveraged towards sustainability transformations. Governance experimentation should be nurtured through the engagement of diverse actors and the protection of innovations set in motion by new networks emerging from these spaces. First- and second-order learning, accounting for power asymmetries, and mitigating actor expectations are key ingredients that contribute to reorienting discourse and changing governance—making these transformed discourses and governance configurations more likely to advance sustainability initiatives that are widely accepted by multiple actors. The main limitation of the research presented here stems from its single-case study design, which restricts its generalization to other cases of SES traps. Another limitation stems from the lack of integration of an analysis of power, a key factor to understanding outcomes in SES traps (Boonstra, 2016)—although power has been considered elsewhere (Méndez et al., 2022). Future research should conduct comparative case studies to understand the influence of contextual factors such as different histories, cultures, and power dynamics. This work must also consider the potential social inequalities associated with deep processes of change that, in turn, will be affected by complex and uncertain social-ecological dynamics and major impacts resulting from global environmental and climate change.

Acknowledgements

The work of Pablo F. Méndez has been funded by the European Commission through the Horizon 2020 Framework Programme for Research and Innovation under Grant Agreement No. 871128 to eLTER PLUS Project. Jennifer M. Holzer gratefully acknowledges the support of the Natural Sciences and Engineering Research Council of Canada (NSERC), (ref. NSERC NETGP 523374-18) [Cette recherche a été financée par le Conseil de recherches en sciences naturelles et en génie du Canada (CRSNG) (numéro de référence NSERC NETGP 523374-18)]. The authors greatly appreciate the comments made by two anonymous reviewers and Jill Tellier, who reviewed initial drafts of the manuscript and helped to strengthen it. In loving memory: Antonio Jesús Fernández Méndez.

Appendix A. Illustrative extracts of results from the thematic analysis (task 1)

A.1 Technical details

We searched for online articles from trusted newspapers between 2000 (first available news), until June 1, 2021 (data collection date), using the built-in search engines of the newspapers and "dragado del Guadalquivir" ("Guadalquivir dredging") as search string. The latter term has been (and still is) consistently used in the media to refer to the IMAP megaproject, due to its emphasis on deep dredging, the megaproject's most controversial measure. We synthesized command-and-control categories from Cox (2016), using a codebook that we then transported to a coding table matching the final list of analysed newspaper articles against four types of categories:

Table A1 Synthesized categories from command-and-control theory.

Sequential categories	Category
Enabling path	Nature of the problem
	Centralized authority and local disempowerment
Technical fix	Analytical simplification
	Capitalization and control
Outcome	Lack of fit due to social-ecological suppression
	Lock-in factors

Based on Cox, M., 2016. The pathology of command and control: a formal synthesis. Ecol. Soc. 21. doi:10.5751/ES-08698-210333.

(1) descriptive and standing categories; (2) six conceptual command-and-control categories (Table A1) with a description of each category; (3) one broad category to capture alternative paradigms; (4) a final set of open questions to allow for inductive coding. For each article, we assessed whether the command-and-control themes or any alternative paradigm could be inferred from the article's content. We also provided our interpretation and/or exemplification with speakers. Finally, we responded to our final set of questions for capturing new recurring regularities sharing meaning that could be abstracted to themes and novel concepts.

A.2 Extract of results

Below, we offer some illustrative random extracts from our thematic analysis. We link general themes elicited during analysis to illustrative newspaper articles (e.g., EM_052 for "El Mundo" article number 52, Table A2). To gather data that was not captured by our search, we consulted ad hoc some new references and referenced them using the newspaper's name (e.g., Diario de Sevilla, 2019; Table A2).

A.2.1 Nature of the problem

Deep dredging appeared in numerous instances portrayed alongside other developments at basin and sea level (e.g., metalliferous mining, underground gas storage, maritime crude oil transport, groundwater extraction), as posing environmental hazards, nature conservation threats and intricate conflicts in the social realm (EPRESS_039; EPRESS_041). At regional level, water and some agricultural uses are affected by illegal activities and conversion of uses, due to inaction by competent authorities to implement approved plans,

Table A2 Random subset of articles used for our illustrative extracts of results.

ID	Article
EPRESS_039	España se compromete ante la UNESCO a no dragar el Guadalquivir para favorecer a Doñana
EPRESS_019	La plataforma del Puerto teme que el dragado siga la suerte del metro y "no haya dinero"
EPRESS_041	Salvemos Doñana pide al Consejo de Participación que actúe ante la "amenaza" del gas
EPRESS_069	El Estado remite un informe a la Unesco en el que expresa su convicción de evitar un deterioro de Doñana
ED_050	Doñana crece 14.000 hectáreas cercada por las amenazas
EPRESS_113	El Parlamento andaluz pide un "gran pacto" por Doñana y que el Gobierno paralice el proyecto de gas
EP_036	Manuel Fernández deja la presidencia del puerto de Sevilla tras 16 años
ED_063	La Unesco retira su ultimátum: Doñana no entrará en su lista negra de patrimonio en peligro por ahora
PUB_003	El Puerto de Sevilla suspende el dragado del Guadalquivir que amenazaba Doñana
ED_068	El Tribunal Supremo anula la inclusión del dragado del Guadalquivir en el Plan Hidrológico
DS_021	La Unesco pide a España que no autorice el dragado del Guadalquivir
EM_051	Zoido: 'La zona franca es viable con o sin dragado'
EM_031	Sevilla vs. Resto del Mundo
EM_052	La sentencia sobre el dragado
EM_004	Ecologistas critican que ni se solucionan ni se evitan las amenazas ni riesgos a Doñana
DS_015	La Sevilla de las ocasiones perdidas
EPRESS_126	Doñana cumple 50 años entre amenazas como el agua, proyectos de infraestructuras y en medio de la emergencia climática
DS_079	Beltrán Pérez apuesta por el dragado del Guadalquivir y el crecimiento demográfico
ABC, May 15, 2021	El Puerto de Sevilla involucra a los arroceros en la mejora de la navegabilidad del río
Diario de Sevilla, July 10, 2019	"La sentencia del Supremo no trastoca nuestros planes en el Puerto"

concomitant with passivity from legal authorities to penalize illegal water and agricultural uses (EM_004; EPRESS_126). These threats, conflicts and illegalities are amplified through a classical discourse conveying the economic development vs nature conservation divide, often representing Doñana as a precious place of universal value, in critical condition and under "attack" (EPRESS_126; EM_004). Arguments made by the Port Authority and key actors from its supportive coalition ("Sevilla por su Puerto"), show how deep dredging was seen as a panacea to bring about socioeconomic progress:

– "*The future of the port is [deep] dredging, in optimal conditions for the environment.*" Unattributed comment (EPRESS_019). By "environment", supporters of the project normally referred to estuarine and protected wetland ecosystems, and to downstream irrigation agriculture, a sector quite against deep dredging due to the hypothetical increase in salinity of estuarine waters (EPRESS_019, PUB_003).
– "*Dredging or dying*", "*it is absolutely essential to ensure navigability in the river [to allow for] modern ships [and be] competitive*", "*[if not] we will jeopardize a fundamental pillar for employment and logistic activity in Seville, because logistics and transport are the key to the future*", "*[the role of] politicians and administration is to help. The only thing lacking is the project for the protection of the riverbanks. They must say how to implement every projected measure.*" President of Sevillian Employers Confederation (EPRESS_019).

A.2.2 Centralized authority and local disempowerment

Deep dredging was envisaged merely a territorial conflict underpinned by a complex political puzzle of alliances and confrontations (EM_31), evidencing the existence of multiple centres of concentrated power divided for and against deep dredging. The potential failure of deep dredging was portrayed as another hindrance to the long-term historical endeavour to make the Guadalquivir estuary navigable in competition with more accessible seaports (e.g., Cadiz's port), within a broader landscape of failing urban engineering projects preventing Seville to "*stand out in the Spanish [economic] landscape*" at present (DS_015). Our main inference regarding local disempowerment from several articles focusing on deep dredging (often in the context of other developments in the region) was not so much about how disempowerment was brought about during the lifecycle of the IMAP megaproject (and other major developments) (EPRESS_126; ED_063). Disempowerment was evident, otherwise, from the lack of clear and institutionalized mechanisms for stakeholder and public participation. It was also noticeable the almost lack of reference to public participation from key international organisms during the megaproject lifecycle such as UNESCO, evidencing the low position of that issue in the political and diplomatic agenda.

A.2.3 Analytical simplification

Analytical simplification could be inferred directly from the arguments of members of the Port Authority or its supportive actor coalition (see above). Also, it could be indirectly inferred from the tendency in the region to subject large development projects (e.g., metalliferous mining, underground gas storage) to more sophisticated and integrated technical, scientific assessments of cumulative impacts after mandatory but incomplete Environmental (or Strategic) Impact Assessment, often as a reaction to formal complaints from nature guardians (e.g., WWF, Doñana Participation Council) (EPRESS_113; ED_063; see also Vargas and Paneque, 2015). Analytical simplification was also exemplified in the propensity of the Port Authority and supportive actors to propose technical fixes as panaceas to resolve the problems created by the deep dredging engineering solution (see "lock-in factors"). The turn towards "*optimization*" of the Port Authority in 2017 (see Section 5.2.4 in main text) was also exemplary, for it indirectly recognized the need to consider richer knowledge inputs a priori, even if it was by commissioning ad-hoc studies to continue justifying a purportedly harmless intervention in the resource system to maximize outputs. And overarching all this was the argument of the Spanish Supreme Court (see "lock-in factors"), a reproach to malpractice upon considering deep dredging as a *"complementary measure"* unfettered from mandatory assessments of ecological impacts.

A.2.4 Capitalization and control

The introduction of substantial physical capitalization and mechanization to exercise control over the target water resource system, thus effectuating the analytical simplifications, was indirectly inferred from the expectations on high returns on investments from actors supporting the IMAP megaproject and directly from the megaproject's costs (DS_026). The IMAP megaproject was considered key for socioeconomic development at city and provincial level, based on the arguments that it would suppose the generation of 15,000–17,000 jobs and the increasing of cruise tourism (ibidem) and, as mentioned above, "*higher profit margins*" for the Port's companies. The alleged benefits of the IMAP megaproject have been convincingly argued as unreliable, based on old forecasts no longer valid (Sancho-Royo and Del Moral, 2014; Sancho-Royo and Del Moral, 2015; Vargas and Paneque, 2015). Other key arguments in favour of the megaproject were the improvement of navigability in the estuary allowing for its connection to the Trans-European Transport Network, as well as the direct access to the European Regional Development Fund for financing the megaproject's development costs (€250 million approx.). Indeed, the fund partially covered the construction of the Sea Gate (€63 million from a total of €163 million) and was expected to cover 80% of deep dredging costs (€37.7 million, plus €31.1 for waste management) (ibidem; see also Diario de Sevilla, 2019; El Confidencial, 2015; Vargas and Paneque, 2015).

A.2.5 Lack of fit due to social-ecological suppression

Inferences about this category were about the disregard by the IMAP megaproject supportive actors for key socioeconomic and ecological functions in the Doñana SES, which could have been suppressed if deep dredging would have been implemented. Since this was not the case, we contextualized our inferences within key insights from other sources informing the risk of large-scale interventions in the estuary (see Méndez et al., 2022). Concern for large environmental and socioeconomic impacts were latent in multiple articles (ED_068; EM_052).

A.2.6 Lock-in factors

From data rich articles, it stood out the argumentation of the Spanish Supreme Court to justify the cancellation of deep dredging, a Kafkaesque process that lasted 4 years and indicates lock-in among actors. In 2015, despite the modifications of the IMAP megaproject by the Port Authority to comply with the environmental requirements of the 2003 EIS and CSIC study, the European Commission issued a statement on the project's potential infringement of the EU Habitats Directive (EM_052). The same year, in response to an appeal from WWF, the Spanish Supreme Court ruled the removal of deep dredging as a 'complementary' measure from the 2013 Water Plan, prepared by the Guadalquivir

River Authority (ibidem). That qualification turned deep dredging a binding measure but freed from justifying compliance with water quality standards (ibidem). The megaproject's supportive actor coalition reacted to the Court's ruling by considering the qualification of 'complementary' as a "*procedural defect*", rectifiable during the next planning phase starting in 2016 (ibidem). The Court's ruling was directed at the Spanish government, not at the Guadalquivir River Authority (water planning authority), nor at the Port Authority (promoter of the megaproject), both of which are state bodies. Defended by the State's Attorney, the Spanish government adduced that deep dredging was "mere [planning] forecast" without providing proper justification, thus downplaying the potential consequences of the measure (ibidem). What perhaps was a "*procedural defect*" was the lack of justification for a harmful measure like deep dredging in the Water Plans, something that was required by both the Water Framework Directive and the Spanish Water Planning Regulation (Royal Decree 907/2007). In addition to water planning, there were procedural defects also along the IMAP megaproject EIA process and our evidence indicated that this seemed to constitute a ubiquitous pattern for other major developments in the region (e.g., underground gas storage, metalliferous mining, groundwater exploitation) (EV_EI_023). The Port Authority and its supportive actor coalition seemed to firmly believe that it was possible to fully balance out the negative environmental and socioeconomic impacts of deep dredging through further engineering and technical fixes. Standing out in this regard is the general argument of "*fixing the estuary*" to allow for deeper dredging, including proposals such as the protection of the riverbanks from the erosion caused by future ship wakes of higher-tonnage vessels, through measures that are normally implemented in emergency situations (e.g., "*geotubes*") (EP_036; EM_051).

Appendix B. Doñana's historical pattern

See our *synthesis* of results from previous research iterations of the Doñana case (*"Doñana's historical pattern"*) in Table B1.

Table B1 Doñana's historical pattern (HP = historical period; for information sources, see Méndez et al., 2012, 2019).

HP	Institutional events	Relevant insights
Ancient regime	Local nobility entitled by royal governments or local agreements to use highly valuable resources and to regulate the use of the less valuable ones by the civilian population. Unregulated hunting 1778—Free Trade Decree promoting trade at state level. Seville and Cádiz (and their ports) lose their monopoly 1794—Royal Order for enhancing navigability in the Guadalquivir estuary (Merlina's meander cut) 1798—First Disentitlements for the expropriation of clergy's assets	Property system based on communal lands or municipal properties for agricultural activities immersed in larger crown and nobility properties Low water consumption regimes, scattered planning and projects, rudimentary infrastructure, and marginal control of water Self-sufficient economy mainly based on traditional knowledge with limited environmental and biophysical impacts Increasing State public deficit Beginning of high-level promotion of navigation in the Guadalquivir estuary (navigability tests, mythology revival), after Seville's loss of trade monopoly and under the Enlightenment's Spirit of Reform Small-scale disentitlements (nationalization-for-privatization) of low-productivity lands for increasing their productivity First public works for enhancing navigation and river defence. Marginal agricultural development
Liberal revolution (1808–1833)	1805—First technical planning for large-scale modifications (Sanlúcar province) 1814—Royal Order constituting the Guadalquivir Navigation Royal Company, preceded by navigability tests 1818—Transaction of large marshland areas (Isla Menor) to the Guadalquivir Navigation Royal Company 1819/1849—Two acts promoting irrigation agriculture among smallholders through tax exemption 1833—Forestry Bylaws establishing the first protection system for forests	Resource exploitation ideas emerge in a context of state deficit-driven impositions, availability of private capital, and enlightened-absolutist socioeconomic trends promoting navigation and economic development in the Guadalquivir estuary Local key actors and entrepreneurs effectively promoting their views of the entire estuary and marshlands as an under-exploited region with enormous potential (navigation, agriculture, colonization, mining) Continuation of works for enhancing navigability (meander cuts) granted to companies participated by private shareholders First attempts for drainage and agricultural development, failing due to lack of adequate technology

Continued

Table B1 Doñana's historical pattern (HP = historical period; for information sources, see Méndez et al., 2012, 2019).—cont'd

HP	Institutional events	Relevant insights
Liberal reform (1833–1868)	1851–1852—Royal Orders for further enhancing navigability 1855—Madoz's Disentitlement nationalizing "low-yield lands" 1859–1862—State regulations entitling Forestry Engineering Corps to catalogue "excludable areas" from the disentitlement process 1865—Creation of the Hydrographic Demarcations 1866—Water Act declaring water of public domain	Agrarian Technical Reform at state level and increasing interventionism at regional and local levels Increasing sale and/or public auction of the so-called "dead hands" (clergy and nobility) properties (marshlands, scrublands and forests) and beginning of extensive forestry operations across the State for agricultural and paper-industry development Large policy and regulatory changes steadily start to signal the "unhealthy/unproductive" nature of marshlands and high returns on investment by providing the lands surrounding the marshlands with exchange value and potential for diversification, easy access/direct transfer of common-property rights, and financial incentives to develop infrastructure High investment costs and long-term recouping (land adaptation to irrigation and crop modification) slow down technology implementation Increasing application of innovative technologies (e.g., canalization and impoldering) for the enhancement of agricultural productivity, based on middle-class and corporate private capital Further enhancement of Guadalquivir's navigability through meander cuts and bottom dredging First middle-class expeditions (naturalists, ornithologists and hunters) to the Doñana wetlands for naturalistic and hunting purposes (A. Chapman and W.J. Buck), promoting Doñana as one of the most important waterfowl reserves in Europe

Liberal state consolidation (1868–1931)	1870—Royal Order promoting drainage engineering projects (left-bank). Most of the projects were finished in 1928, due to inadequate planning 1870—Canals and Reservoirs Act, granting infrastructure operation to public-works companies 1877—Public Works Act, conceiving Hydraulic Works Plans as a centralization instrument and providing the foundations secondary legislation incentivizing private investments	End of laissez-faire policies Projects justified on the need to fight the harmful effects of swampy areas Emergence of 'regenerationism' (from the Spanish 'Regeneracionismo'), an intellectual-scientific movement aimed at understanding the political, scientific, social and economic causes of Spain's political and economic decline, largely in response to the loss of Spanish overseas territories
Liberal state consolidation (1868–1931)	1877—Act for the enhancement and afforestation of public forests excluded from the disentitlement process 1877—Royal Order, granting the first large-scale drainage works (Gallega marsh, failed and reinstated in 1910; Almonte marsh, not executed) 1878—Royal Order (leprosy RO), wetlands as a source of disease and unhealthy areas 1879—Water Act, giving priority to irrigation agriculture (over navigation) through public intervention at the state level. Repealed more than one century afterward (1985). Regenerationist	1879 Water Act marked by regenerationist ideas. Declaration of surface water as a public good. Distinction between continental (e.g., lakes; public property) and coastal waters (e.g., marshes; private property). Focus on water allocation: drinking water ("agua de boca"), agricultural and industrial (agricultural mills and sawmills) Key regenerationism actors describe regional differences in water availability as the main shortcoming for the development of irrigation agriculture, and promote hydraulic-forestry policy and planning as a panacea for socioeconomic development Sustained conflicts between Civil and Forestry engineers, due the poor discrimination of their respective responsibilities Beginning of the forestry protection-for-conservation policies and regulations in Spain. Forestry Engineering Corps entitled to catalogue areas to be protected from colonization and exploitation. Increasing links between nature (forests) conservation and forestry policies

Continued

Table B1 Doñana's historical pattern (HP = historical period; for information sources, see Méndez et al., 2012, 2019).—cont'd

HP	Institutional events	Relevant insights
	1883—Irrigation Act (Gamazo Act), continuing the legislative trend of 1866 and 1879 Water Acts. Regenerationist	1902 Hydraulic Plan included novel knowledge about Spanish geography and hydrography. Elaborated by civil engineers coordinated by Rafael Gasset, several times minister of public works. Based upon more than 300 reports elaborated by the Hydraulic Works Divisions. Planned conversion into irrigable lands of more than 1,460,000 ha (6% of the total Spanish territory). Provisional character, but in force until the creation of the Water Authorities at the end of the 1920s.
	1899—General Plan (draft) of Reservoirs and Irrigation Canals, 1902 Hydraulic Plan	
	1901—Creation of the Hydrographic-Forestry Demarcations, precursors of the water authorities	Continued through 1933 Hydraulic Plan Main perspectives of Hydraulic Plans defended in National Irrigation Congresses (1913, 1918, 1921, 1927, 1934) and paralleled by colonization strategies.
	1902—Hydraulic Plan. Revised in 1909, 1916, 1919 and 1922, reducing the initial planned irrigable surface	Regenerationist ideas absorbed at a political level through the Civil Engineering Corps and Rafael Gasset. Promotion of the main arguments of Joaquín Costa, the key ideologist of the hydraulic mission discourse in both the media and the parliament
	1907—Inner Colonization and Repopulation Act, halting emigration and promoting the cultivation of low-yield lands	In the Guadalquivir estuary, several floods ravaged privately funded dikes and impoldered lands and colonial towns between 1892 and the 1920s, due to deficient publicly funded hydraulic infrastructure. Dikes were however rebuilt, heightened, and reinforced in the entire area (e.g., by the Guadalquivir Marshes Company for rice cultivation)
	1908—Protective Forestry Act promoting afforestation of public and private forests	
	1911—Irrigation of Large Areas Act (Gasset Act) promoting large public works. Legal coverage to 1902 Hydraulic Plan. Synergy with the hydro-electric sector, which continued until the II Republic	Projected economic targets for irrigated lands (e.g., cultivation of cereals, commercial crops, forage) were not met, due to the lack of adequate infrastructure for drainage and reducing soil salinity, compounded by fragmentary planning and uncoordinated work
	1916—National Parks Act	
	1917 to date—Battery of regulations implementing the park-based model	Emergence of protection-for-conservation paradigm with the passing of the National Parks Act of 1916. Top-down implementation of a National Parks system

	1918—Act for the Drainage of Lagoons, Marshes and Marshy Areas as unproductive lands (Cambó Act) 1924—Royal Order of Ministry of Promotion granting private financial group drainage and exploitation rights (Hinojos marshes, ~10,000 ha) 1926—Royal Decree for the creation of the modern water authorities 1927—Creation of the Guadalquivir River Authority 1929—General Plan of Public Works for the Guadalquivir estuary, operative until the 1970s. Emphasis on the need to regulate and streamline irrigation activities through reservoir control 1930—Plan for the Expansion of Irrigation in the Low Guadalquivir (122,750 ha)	taking elements and techniques from other Western countries (such as the USA). Conservation paradigm co-opted by the powerful forestry- and civil-engineering corps, which systematically excluded from protection marshlands and wetlands, considered as unproductive and unhealthy lands Empowerment of regenerationist policies in the primary sector. Included an extensive typology of marshlands and wetlands. Minimum surface liable to be modified: 100 ha. Large diversity of financial instruments, tax reductions and transfer rights for both public and private actors. Not mandatory to justify investments upon cleaning and purification purposes Several attempts by private companies between 1921 and 1937 to modify the Guadalquivir estuary for farming and agricultural purposes, backed by international investment funds. Attempts were mostly driven by speculation and, despite small successes, ended in broad financial and technical failure due to low profitability of crops in saline soils, fragmentary planning and lack of coordination, and underestimation of damages produced by seasonal natural flooding and the harshness of the saline soils High GDP in Spain during the 1920s
II Republic (1931–1939) Civil War (1936–1939)	1932—Agrarian Reform Act, banning landowners from expelling land tenants from rented lands. Obligation to cultivate, to avoid confiscation. Creation of the Institute of Agrarian Reform 1932—Irrigation Act, increasing the State responsibilities in the development of secondary and complementary public works,	Interventionist character of 1932 Irrigation Act and the 1933 Hydraulic Plan. Hydraulic Plan elaborated by the General Directorate of Hydraulic Works, through the creation of a Centre for Hydrographic Studies Need for large water diversions ("trasvases") becomes institutionalized. The state financially assumes the construction of primary and secondary irrigation hydraulic infrastructure and main role in the colonization-and-development strategies 'Regenerationist' emphasis on the need to establish a hydrological balance between the Atlantic and the Mediterranean basins, based on the argument of the higher

Continued

Table B1 Doñana's historical pattern (HP = historical period; for information sources, see Méndez et al., 2012, 2019).—cont'd

HP	Institutional events	Relevant insights
	to realize the functioning of the areas already modified into irrigable lands 1933—Hydraulic Plan (plan of Lorenzo Pardo) 1935—Forest Heritage Act promoting afforestation for production purposes 1937–1939—First small-scale successes in rice cultivation in "Isla Mayor" and "Isla Mínima" (right bank) by Rafael Beca & Co, Agricultural Industries PLC	agricultural productivity and economic potential of the former, and the lack of planned works implemented in the latter Water Authorities entrusted with the management and maximization of the agricultural, forestry, and hydropower yield of water resources at basin level, bringing together the widest range of regional stakeholders, but promoting the national interest to the detriment of local Republican attempts for increasing public property in Doñana by declaring the marshes of public interest Beginning of an intensive shift in jurisdiction in the private-property system towards nationalization During the 1920s, until 1933, just about 12% of the planned primary infrastructure and irrigated land surface carried out, funded through the emission of public debt instruments justified as investments with long-term returns based on taxable income
Dictatorship of Franco (1939–1975)	1939–1949—Battery of regulations providing further legal coverage to the colonization-and-development strategy 1939—Colonization of Large Areas Act. Creation of the National Institute of Colonization 1940—Hydraulic Plan (plan of Alfonso Peña), operative until 1963. Synthesized the main principles of the plan of 1933 1940—Forestry Act entitling Forestry	Franco's government abandons agrarian reformist aspects of the 1933 Hydraulic Plan but undertook the major hydraulic modifications of the 20th century Increasing improvement of the Spanish economy (high GDP) during the 1940s Construction of the Low Guadalquivir Canal between 1940–1951, declared of National interest, to irrigate the left-bank marshes (65,000 ha). Malfunctioned for water transport and distribution. It currently forms part of a more functional irrigation system Major property shifts from communal lands and public property to a mixed public-private system led and controlled by the engineers of the new authoritarian government.

	Engineering Corps for leading National Parks planning and management 1940—Declaration of the Guadalquivir marshes as being of National Interest by the National Institute of Colonization	Further promotion of drainage-canalization-colonization strategy established during the previous decades, including the development of extensive eucalyptus afforestation aimed at providing raw materials for industry. Process fueled by synergies with other economic sectors (e.g., navigation, forestry), an all-time records Spanish GDP and a autarchic system
	1940s—First large-scale successes in rice cultivation by Rafael Beca & Co, Agricultural Industries PLC	First large-scale projects in Doñana, backed by the 1911 Gasset Act. Canalization and diversion by Rafael Beca's company of two of the main channels inflowing into the right-bank marshes: Guadiamar channel and Brazo de la Torre. Works carried out by private contract and publicly funded
	1941—Restitution of Forest Heritage Act declaring Huelva province (including Doñana) region of interest for forestry 1949—Colonization and Property Distribution of Irrigable Areas Act, prompting the modification of vast areas. Modification decided, financed and executed by the state. Continuing 1939 and 1946 Acts	New institutions (e.g., Forestry Heritage Act of 1941, Forestry Act of 1957) turn the incipient the regime for nature-conservation and forest-resources management increasingly bureaucratic and rigid, and focused on maximizing economic returns. Massive afforestation projects promoted at the State level, for production (Forest Heritage/Ministry of Agriculture) and for conservation (Directorate of Forestry/ Ministry of Agriculture)
	1957—Forestry Act entrusting the State Ministry of Agriculture with the stewardship of nature conservation and promoting production-driven planning and technologies	First scientific expeditions of key national and international scientists (e.g., J.A. Valverde, F. Bernis), naturalists (e.g., G. Mountfort) and ornithologists (e.g., Roger T. Peterson) to the Doñana marshes during the 1950s–1960s, supported by private owners (e.g., M. González) who constituted key stakeholders in the protection-for-conservation race. Support provided by international organizations through key knowledge and funding (e.g., the International Union for Conservation of Nature, IUCN, and the International Waterfowl and Wetlands Research Bureau, IWRB). The lobbying and fund-raising campaign reaches a peak with the constitution of the World Wildlife Fund for Nature (WWF)

Continued

Table B1 Doñana's historical pattern (HP = historical period; for information sources, see Méndez et al., 2012, 2019).—cont'd

HP	Institutional events	Relevant insights
Dictatorship of Franco (1939–1975)	1960—Irrigation Plan of the Low Guadalquivir (55,000 ha). Construction of a drainage network in the left-bank	Backed by the FAO-Guadalquivir project and the experiences of the left-bank, realization of a second hydrological study affecting 23,589 ha, declared of National Interest in 1971
Transition to democracy (1975–1978)	1969—Doñana Act creating the Doñana National Park 1971—Decree dictating the merger of the Forest Heritage and the Directorate of Forestry within the Nature Conservation Institute (ICONA) 1972 and 1974—Two Decrees entitling the Institute for Agrarian Reform and Development (IRYDA) to implement large modification projects 1975—Nature Reserves Act increasing the decision power of local administrations, landowners and the ICONA, and favouring the agricultural projects of the IRYDA 1978—Spanish Constitution replacing the authoritarian government by a democratic government 1978—Doñana Act increasing the Park's surface (50,270 ha)	Drainage and modification of 35,000 ha of marshlands for irrigation agriculture (Almonte-Marismas Plan and Guadalquivir-FAO Project), preceded by the discovery of the Almonte-Marismas aquifer The struggle was finally dominated by the establishment engineering lobby, reinforced by the merger of state forestry bodies in 1971, giving rise to the Nature Conservation Institute (ICONA) (Fig. 6). In 1971, the Forestry Directorate and the Forestry Heritage merge to become the Nature Conservation Institute (ICONA) By 1974, J.A. Valverde, the most significant champion of the Doñana's protection-for-conservation race, is replaced by an old-school engineer as director of the National Park 1978 National Park Act establishes novel operational guidelines for conservation inside and outside the National Park showing, for the first time, de jure public interest in the conservation of its aquatic ecosystems. The new institution laid the foundations for the consolidation of the protection-for-conservation discourse and the renewal of the old water resource use institutions regime However, new measures proposed by the 1978 Act co-opted by the engineering lobby, which dominated the public sector and still endorsed the previous management-for-productivity, command-and-control paradigm 1978 National Park Act thus consolidates dominance of engineering lobby in management positions. Direction of National Park and the Biological Reserve respectively assigned to ICONA (Ministry of Agriculture) and the Doñana Biological Station (Spanish Research Council CSIC, Ministry of Education)

Democracy (1978–?)	1981—First Doñana Hydraulic Regeneration Plan aimed at the restoration of the main tributaries to the marsh (Guadiamar and Travieso Channels) and the modification of the Guadalquivir River levee, allowing to control the marsh inflow and discharges by means of sluice gates 1983—Water Act, subordinating large water diversions to specific legislation and to future Water Plans, and creating the opportunity for an open water market. Inclusion of underground water in the Hydraulic Public Domain 1989—Nature Reserves Inventory Act creating the Doñana Natural Park (54,250 ha) buffering the core National Park and the Andalusian Nature Reserves Network 1993—First Doñana Sustainable Development Plan 1998—"Los Frailes" mining accident (known as "Aznalcóllar" mining accident) 1999—Doñana Nature Reserve Bill transferring jurisdiction for the management of Doñana National Park to the Andalusian regional government	Top-down drastic reductions in the irrigable surfaces established in the plans of the prior decades, which become more intensively used due to technological advances As a response to the mining accident at Los Frailes the Spanish government and the Andalusian government launch the Doñana 2005 and Guadiamar Green Corridor restoration projects, respectively The Andalusian regional government wins a political and judicial process to claim jurisdiction over the Doñana National Park from the Spanish government (Doñana Nature Reserve Act)

References

Allison, H.E., Hobbs, R.J., 2004. Resilience, adaptive capacity, and the "lock-in trap" of the Western Australian agricultural region. Ecol. Soc. 9 (1), 3. http://www.ecologyandsociety.org/vol9/iss1/art3/.

Avriel-Avni, N., Dick, J., 2019. Differing perceptions of socio-ecological systems: insights for future transdisciplinary research. In: Advances in Ecological Research. Elsevier, pp. 153–190, https://doi.org/10.1016/bs.aecr.2019.03.001.

Baker, D., Murray, G., Agyare, A.K., 2018. Governance and the making and breaking of social-ecological traps. Ecol. Soc. 23 (1), 38. https://doi.org/10.5751/ES-09992-230138.

Bell, S., 2002. The limits of rational choice: new institutionalism in the test bed of central banking politics in Australia. Political Stud. 50, 477–496. https://doi.org/10.1111/1467-9248.00380.

Berkes, F., 2007. Community-based conservation in a globalized world. Proc. Natl. Acad. Sci. 104, 15188–15193. https://doi.org/10.1073/pnas.0702098104.

Boettcher, M., Kim, R.E., 2022. Arguments and architectures: discursive and institutional structures shaping global climate engineering governance. Environ. Sci. Pol. 128, 121–131. https://doi.org/10.1016/j.envsci.2021.11.015.

Boonstra, W.J., 2016. Conceptualizing power to study social-ecological interactions. Ecol. Soc. 21 (1), 21. https://doi.org/10.5751/ES-07966-210121.

Campbell, J.L., Pedersen, O., 2001. Introduction. In: Campbell, J.L., Pedersen, O. (Eds.), The Rise of Neoliberalism and Institutional Analysis. Princeton University Press, Princeton, NJ, pp. 1–24.

Carpenter, S., Brock, W., 2008. Adaptive capacity and traps. Ecol. Soc. 13 (2), 40. https://doi.org/10.5751/ES-02716-130240.

Castells, M., Cruz, J., Custodio, E., García-Novo, F., de Gaudemar, J.P., et al., 1992. Dictamen sobre estrategias para el desarrollo socioeconómico sostenible del entorno de Doñana. Comisión Internacional de Expertos sobre el Desarrollo del Entorno de Doñana, Junta de Andalucía, Seville, Spain. http://www.juntadeandalucia.es/servicios/publicaciones/detalle/41400.html.

Chaffin, B.C., Garmestani, A.S., Gunderson, L.H., Benson, M.H., Angeler, D.G., Arnold, et al., 2016. Transformative environmental governance. Annu. Rev. Environ. Resour. 41, 399–423. https://doi.org/10.1146/annurev-environ-110615-085817.

Cox, M., 2016. The pathology of command and control: a formal synthesis. Ecol. Soc. 21. https://doi.org/10.5751/ES-08698-210333.

Crouzat, E., Arpin, I., Brunet, L., Colloff, M.J., Turkelboom, F., Lavorel, S., 2018. Researchers must be aware of their roles at the interface of ecosystem services science and policy. Ambio 47, 97–105. https://doi.org/10.1007/s13280-017-0939-1.

CSIC, 2010. Methodological Proposal to Diagnose and Predict the Consequences of Human Activities in the Guadalquivir Estuary. Institute of Marine Sciences of Andalusia, Cadiz, Spain.

Cumming, G.S., 2017. A review of social dilemmas and social-ecological traps in conservation and natural resource management. Conserv. Lett., 1–15. https://doi.org/10.1111/conl.12376.

Diario de Sevilla, 2019. La sentencia del Supremo no trastoca nuestros planes en el Puerto. [Last accessed: 9th February, 2022]. https://www.diariodesevilla.es/sevilla/sentencia-Supremo-no-trastoca-planes-Puerto_0_1371762995.html.

Díaz-Delgado, R., Carro, F., Quirós, F., Osuna, A., Baena, M., 2016. Contribución del seguimiento ecológico a largo plazo a la investigación y la gestión en la plataforma LTSER—Doñana. Ecosistemas 25 (1), 9–18. https://doi.org/10.7818/ECOS.2016.25-1.03.

Donadei, M., 2020. Coaliciones y conflictos alrededor de las cuestiones socioambientales: Hacia visiones integradas e integradoras. El caso del estuario del Guadalquivir. Ciudades 23, 115–136. https://doi.org/10.24197/ciudades.23.2020.115-136.

Dornelles, A.Z., Boyd, E., Nunes, R.J., Asquith, M., Boonstra, W.J., et al., 2020. Towards a bridging concept for undesirable resilience in social-ecological systems. Glob. Sustain. 3. https://doi.org/10.1017/sus.2020.15.

El Confidencial, 2015. Dragado del Guadalquivir: entre los 15.000 empleos de Sevilla y el peligro para Doñana. [Last accessed: 9th February, 2022]. https://www.elconfidencial.com/espana/andalucia/2015-03-29/dragado-del-guadalquivir-entre-los-15-000-empleos-para-sevilla-y-el-peligro-para-donana_746474/.

Enqvist, J., Tengö, M., Boonstra, W.J., 2016. Against the current: rewiring rigidity trap dynamics in urban water governance through civic engagement. Sustain. Sci. 11, 919–933. https://doi.org/10.1007/s11625-016-0377-1.

Escalera, J., 2003. Investigación participativa para el desarrollo sostenible en el Corredor Verde del río de Guadiamar. In: Arenas, J.M., Martínez, F.R., Mora, A., Montes, C., Borja, F. (Eds.), Ciencia y restauración del río Guadiamar. Consejería de Medio Ambiente, Junta de Andalucía, Sevilla, pp. 528–537.

Feola, G., 2015. Societal transformation in response to global environmental change: a review of emerging concepts. Ambio 44, 376–390. https://doi.org/10.1007/s13280-014-0582-z.

Fischer-Kowalski, M., Haberl, H. (Eds.), 2007. Socioecological Transitions and Global Change: Trajectories of Social Metabolism and Land Use, Advances in Ecological Economics. Edward Elgar, Cheltenham, UK; Northampton, MA.

Folke, C., Polasky, S., Rockström, J., Galaz, V., Westley, F., et al., 2021. Our future in the Anthropocene biosphere. Ambio 50, 834–869. https://doi.org/10.1007/s13280-021-01544-8.

Font, N., Subirats, J., 2010. Water management in Spain: the role of policy entrepreneurs in shaping change. Ecol. Soc. 15. https://doi.org/10.5751/ES-03344-150225.

García, F., Marín, C., 2006. Doñana, Water and Biosphere. Doñana 2005 Project. Guadalquivir River Authority, Ministry of Environment, Madrid, España. https://portals.iucn.org/library/node/46194.

García-Llorente, M., Martín-López, B., Díaz, S., Montes, C., 2011. Can ecosystem properties be fully translated into service values? An economic valuation of aquatic plant services. Ecol. Appl. 21 (8), 3083–3103. https://doi.org/10.1890/10-1744.1.

Garrett, R.D., Cammelli, F., Ferreira, J., Levy, S.A., Valentim, J., Vieira, I., 2021. Forests and sustainable development in the Brazilian Amazon: history, trends, and future prospects. Annu. Rev. Environ. Resour. 46, 625–652. https://doi.org/10.1146/annurev-environ-012220-010228.

Gavin, M.C., McCarter, J., Mead, A., Berkes, F., Stepp, J.R., Peterson, D., Tang, R., 2015. Defining biocultural approaches to conservation. Trends Ecol. Evol. 30, 140–145. https://doi.org/10.1016/j.tree.2014.12.005.

Ghosh, B., Kivimaa, P., Ramirez, M., Schot, J., Torrens, J., 2021. Transformative outcomes: assessing and reorienting experimentation with transformative innovation policy. Sci. Public Policy 00, 1–18. https://doi.org/10.1093/scipol/scab045.

Gillard, R., Gouldson, A., Paavola, J., Van Alstine, J., 2016. Transformational responses to climate change: beyond a systems perspective of social change in mitigation and adaptation. WIREs Clim. Chang. 7, 251–265. https://doi.org/10.1002/wcc.384.

Gómez-Baggethun, E., Kelemen, E., 2008. Linking institutional change and the flows of ecosystem services. Case studies in Spain and Hungary. In: Kuvánková-Oravská, T., Chobotova, V., Jilkova, J., Sauer, P. (Eds.), Institutional Analysis of Sustainability Problems. Slovak Academy of Sciences, Bratislava, Slovakia, pp. 118–145.

Gómez-Baggethun, E., Reyes-García, V., Olsson, P., Montes, C., 2012. Traditional ecological knowledge and community resilience to environmental extremes: a case study in Doñana, SW Spain. Glob. Environ. Chang. 22 (1), 640–650. https://doi.org/10.1016/j.gloenvcha.2012.02.005.
Graziano, M., de Groot, G., Pilato, L., Sánchez, M., Izaguirre, I., Pizarro, H., 2019. Fostering urban transformations in Latin America: lessons around the ecological management of an urban stream in coproduction with a social movement (Buenos Aires, Argentina). Ecol. Soc. 24 (4), 13. https://doi.org/10.5751/ES-11226-240413.
Gunderson, L.H., Holling, C.S., Light, S.S., 1995. Barriers and Bridges to the Renewal of Ecosystems and Institutions. Columbia University Press, New York, NY, USA.
Gunderson, L., Cosens, B., Chaffin, B., et al., 2017. Regime shifts and panarchies in regional scale social-ecological water systems. Ecol. Soc. 22 (1), 31. https://doi.org/10.5751/ES-08879-220131.
Haider, L.J., Boonstra, W.J., Peterson, G.D., Schlüter, M., 2018. Traps and sustainable development in rural areas: a review. World Dev. 101, 311–321. https://doi.org/10.1016/j.worlddev.2017.05.038.
Hall, P.A., Taylor, R., 1996. Political science and the three new institutionalisms. Political Stud. 44, 936–957. https://doi.org/10.1111/j.1467-9248.1996.tb00343.x.
Hall, P., Thelen, K., 2006. Institutional change in varieties of capitalism. In: Paper prepared for presentation to the Europeanists Conference (Chicago: March).
Haller, T., 2019. Towards a new institutional political ecology: how to marry external effects, institutional change and the role of power and ideology in commons studies. In: Haller, T., Breu, T., de Moor, T., Rohr, C., Znoj, H. (Eds.), The Commons in a Glocal World: Global Connections and Local Responses. Routledge, Taylor & Francis Group, pp. 90–120.
Holling, C.S., Allen, C.R., 2002. Adaptive inference for distinguishing credible from incredible patterns in nature. Ecosystems 5, 319–328. https://doi.org/10.1007/s10021-001-0076-2.
Holling, C.S., Meffe, G.K., 1996. Command and control and the pathology of natural resource management. Conserv. Biol. 10 (2), 328–337. https://doi.org/10.1046/j.1523-1739.1996.10020328.x.
Holling, C.S., Gunderson, L.H., Peterson, G.D., 2002. Sustainability and panarchies. In: Gunderson, L.H., Holling, C.S. (Eds.), Panarchy: Understanding Transformations in Human and Natural Systems. Island Press, Washington, DC, USA, pp. 63–102.
Holzer, J.M., Adamescu, C.M., Cazacu, C., Díaz-Delgado, R., Dick, J., Méndez, P.F., Santamaría, L., Orenstein, D.E., 2019. Evaluating transdisciplinary science to open research-implementation spaces in European social-ecological systems. Biol. Conserv. 238, 108228. https://doi.org/10.1016/j.biocon.2019.108228.
Hotimsky, S., Cobb, R., Bond, A., 2006. Contracts or scripts? A critical review of the application of institutional theories to the study of environmental change. E&S 11, art41. https://doi.org/10.5751/ES-01718-110141.
Howlett, M., McConnell, A., Perl, A., 2016. Weaving the fabric of public policies: comparing and integrating contemporary frameworks for the study of policy processes. J. Comp. Policy Anal. Res. Pract. 18, 273–289. https://doi.org/10.1080/13876988.2015.1082261.
Hull, R.B., 2009. Environmental pluralism. In: Callicott, J.B., Frodeman, R. (Eds.), Encyclopedia of Environmental Ethics and Philosophy. Macmillan Reference USA, Detroit, pp. 384–387.
Jacobs, S., Zafra-Calvo, N., Gonzalez-Jimenez, D., Guibrunet, L., Benessaiah, K., et al., 2020. Use your power for good: plural valuation of nature—the Oaxaca statement. Glob. Sustain. 3. https://doi.org/10.1017/sus.2020.2.

Kaufmann, M., Wiering, M., 2021. The role of discourses in understanding institutional stability and change—an analysis of Dutch flood risk governance. J. Environ. Policy Plan. 0, 1–20. https://doi.org/10.1080/1523908X.2021.1935222.
Kellner, E., 2021. The controversial debate on the role of water reservoirs in reducing water scarcity. WIREs Water 8, e1514. https://doi.org/10.1002/wat2.1514.
Kemp, R., Rotmans, J., Loorbach, D., 2007. Assessing the Dutch energy transition policy: how does it deal with dilemmas of managing transitions? J. Environ. Policy Plan. 9, 315–331. https://doi.org/10.1080/15239080701622816.
Kivimaa, P., Boon, W., Hyysalo, S., Klerkx, L., 2019. Towards a typology of intermediaries in sustainability transitions: a systematic review and a research agenda. In: Research Policy, New Frontiers in Science, Technology and Innovation Research from SPRU's 50th Anniversary Conference. 48, pp. 1062–1075, https://doi.org/10.1016/j.respol.2018.10.006.
Laborde, S., Fernández, A., Phang, S.C., Hamilton, I.M., Henry, N., et al., 2016. Social-ecological feedbacks lead to unsustainable lock-in in an inland fishery. Glob. Environ. Chang. 41, 13–25. https://doi.org/10.1016/j.gloenvcha.2016.08.004.
Lang, D.J., Wiek, A., Bergmann, M., Stauffacher, M., Martens, P., et al., 2012. Transdisciplinary research in sustainability science: practice, principles, and challenges. Sustain. Sci. 7, 25–43. https://doi.org/10.1007/s11625-011-0149-x.
Leach, M., Rockström, J., Raskin, P., Scoones, I., Stirling, A., et al., 2012. Transforming innovation for sustainability. Ecol. Soc. 17. https://doi.org/10.5751/ES-04933-170211.
Leonardsson, H., Kronsell, A., Andersson, E., Burman, A., Blanes, R., Da Costa, K., Hasselskog, M., Stepanova, O., Öjendal, J., 2021. Achieving peaceful climate change adaptation through transformative governance. World Dev. 147, 105656. https://doi.org/10.1016/j.worlddev.2021.105656.
Lubell, M., 2015. Collaborative partnerships in complex institutional systems. Curr. Opin. Environ. Sustain. 12, 41–47. https://doi.org/10.1016/j.cosust.2014.08.011.
Martín-López, B., García-Llorente, M., Palomo, I., Montes, C., 2011. The conservation against development paradigm in protected areas: valuation of ecosystem services in the Doñana social-ecological system (southwestern Spain). Ecol. Econ. 70, 1481–1491. https://doi.org/10.1016/j.ecolecon.2011.03.009.
Meadowcroft, J., 2007. Who is in charge here? Governance for sustainable development in a complex world. J. Environ. Policy Plan. 9, 299–314. https://doi.org/10.1080/15239080701631544.
Méndez, P.F., Isendahl, N., Amezaga, J.M., Santamaría, L., 2012. Facilitating transitional processes in rigid institutional regimes for water management and wetland conservation: experience from the Guadalquivir estuary. Ecol. Soc. 17 (1), 26. https://doi.org/10.5751/ES-04494-170126.
Méndez, P.F., Amezaga, J.M., Santamaría, L., 2019. Explaining path-dependent rigidity traps: increasing returns, power, discourses, and entrepreneurship intertwined in social-ecological systems. Ecol. Soc. 24 (2), 30. https://doi.org/10.5751/ES-10898-240230.
Méndez, P.F., Clement, F., Palau-Salvador, G., Díaz-Delgado, R., Villamayor-Tomas, S., 2022. Understanding the Governance of Sustainability Pathways: Hydraulic Megaprojects, Social-Ecological Traps, and Power in Networks of Action Situations. Preprint, https://doi.org/10.31235/osf.io/b3p7h.
Molas-Gallart, J., Boni, A., Giachi, S., Schot, J., 2021. A formative approach to the evaluation of transformative innovation policies. Res. Eval. 1–12. https://doi.org/10.1093/reseval/rvab016.
Molle, F., Mollinga, P.P., Wester, P., 2009. Hydraulic bureaucracies and the hydraulic mission: flows of water, flows of power. Water Alternat. 2 (3), 328–349.

Muiderman, K., Zurek, M., Vervoort, J., Gupta, A., Hasnain, S., Driessen, P., 2022. The anticipatory governance of sustainability transformations: hybrid approaches and dominant perspectives. Glob. Environ. Chang. 73, 102452. https://doi.org/10.1016/j.gloenvcha.2021.102452.

Muñoz-Erickson, T.A., Campbell, L.K., Childers, D.L., Grove, J.M., Iwaniec, D.M., Pickett, S.T.A., Romolini, M., Svendsen, E.S., 2016. Demystifying governance and its role for transitions in urban social–ecological systems. Ecosphere 7, e01564. https://doi.org/10.1002/ecs2.1564.

North, D.C., 1990. Institutions, Institutional Change and Economic Performance. Cambridge University Press, Cambridge, MA.

Nowell, L.S., Norris, J.M., White, D.E., Moules, N.J., 2017. Thematic analysis: striving to meet the trustworthiness criteria. Int J Qual Methods 16, 1–13. https://doi.org/10.1177/1609406917733847.

Olsson, P., Gunderson, L., Carpenter, S., Ryan, P., Lebel, L., Folke, C., Holling, C.S., 2006. Shooting the rapids: navigating transitions to adaptive governance of social-ecological systems. Ecol. Soc. 11. https://doi.org/10.5751/ES-01595-110118.

Orach, K., Schlüter, M., 2021. Understanding the dynamics of fish politics: the role of diverse actor interactions in transformations towards co-management. Environ. Sci. Pol. 124, 195–205. https://doi.org/10.1016/j.envsci.2021.06.010.

O'Riordan, T., Jordan, A., 1999. Institutions, climate change and cultural theory: towards a common analytical framework. Glob. Environ. Chang. 9, 81–93. https://doi.org/10.1016/S0959-3780(98)00030-2.

Ostrom, E., 1990. Governing the Commons: The Evolution of Institutions for Collective Action, first ed. Cambridge University Press, https://doi.org/10.1017/CBO97805118077.

Ostrom, E., 2010. Institutional analysis and development: elements of the framework in historical perspective. In: Crothers, C. (Ed.), Historical Developments and Theoretical Approaches in Sociology. Encyclopaedia of Life Support Systems (EOLSS). EOLSS Publishers, Oxford, UK.

Owen, C., Bindman, E., 2019. Civic participation in a hybrid regime: limited pluralism in policymaking and delivery in contemporary Russia. Gov. Oppos. 54, 98–120. https://doi.org/10.1017/gov.2017.13.

Pascual, U., Howe, C., 2018. Seeing the wood for the trees: exploring the evolution of frameworks of ecosystem services for human wellbeing. In: Schreckenberg, K., Mace, G., Poudyal, M. (Eds.), Ecosystem Services and Poverty Alleviation: Trade-Offs and Governance. Routledge, Oxford, pp. 3–21.

Patterson, J., Schulz, K., Vervoort, J., van der Hel, S., Widerberg, O., Adler, C., Hurlbert, M., Anderton, K., Sethi, M., Barau, A., 2017. Exploring the governance and politics of transformations towards sustainability. Environ. Innov. Soc. Trans. 24, 1–16. https://doi.org/10.1016/j.eist.2016.09.001.

Phelan, L., Henderson-Sellers, A., Taplin, R., 2013. The political economy of addressing the climate crisis in the earth system: undermining perverse resilience. New Political Econ. 18, 198–226. https://doi.org/10.1080/13563467.2012.678820.

Plummer, R., Baird, J., Farhad, S., Witkowski, S., 2020. How do biosphere stewards actively shape trajectories of social-ecological change? J. Environ. Manag. 261, 110139. https://doi.org/10.1016/j.jenvman.2020.110139.

Raven, R., Van Den Bosch, S., Weterings, R., 2010. Transitions and strategic niche management: towards a competence kit for practitioners. Int. J. Technol. Manag. 51, 57–74.

Raven, R., Sengers, F., Spaeth, P., Xie, L., Cheshmehzangi, A., de Jong, M., 2019. Urban experimentation and institutional arrangements. Eur. Plan. Stud. 27, 258–281. https://doi.org/10.1080/09654313.2017.1393047.

Reyers, B., Folke, C., Moore, M.-L., Biggs, R., Galaz, V., 2018. Social-ecological systems insights for navigating the dynamics of the Anthropocene. Annu. Rev. Environ. Resour. 43, 267–289. https://doi.org/10.1146/annurev-environ-110615-085349.

Rijke, J., Farrelly, M., Brown, R., Zevenbergen, C., 2013. Configuring transformative governance to enhance resilient urban water systems. Environ. Sci. Pol. 25, 62–72. https://doi.org/10.1016/j.envsci.2012.09.012.

Rocha, J., Lanyon, C., Peterson, G., 2022. Upscaling the resilience assessment through comparative analysis. Glob. Environ. Chang. 72. https://doi.org/10.1016/j.gloenvcha.2021.102419, 102419.

Sancho-Royo, F., Del Moral, L., 2014. Estuario del Guadalquivir. Observatorio de las Políticas del Agua (OPPA), Evaluación del Primer Ciclo de Evaluación. Fundación Nueva Cultura del Agua. Available online http://www.conama.org/conama/download/files/conama2014/STs%202014/1996969113_asociado_Informe%20OPPA.pdf.

Sancho-Royo, F., Del Moral, L., 2015. Dragado del estuario del Guadalquivir: Plan Hidrológico, sentencia del Tribunal Supremo y cambio de rumbo., https://doi.org/10.13140/RG.2.1.4782.6802.

Santamaría, L., Green, A., Díaz-Delgado, R., Bravo, M.A., Castellanos, E.M., 2006. Caracoles—a new laboratory for science and wetland restoration. In: García, F., Marín, C. (Eds.), Doñana, Water and Biosphere. Madrid, Doñana 2005 Project: Guadalquivir River Authority. Ministry of Environment, pp. 325–327.

Saurí, D., Del Moral, L., 2001. Recent developments in Spanish water policy. Alternatives and conflicts at the end of the hydraulic age. Geoforum 32, 351–362. https://doi.org/10.1016/S0016-7185(00)00048-8.

Schlager, E., Weible, C.M., 2013. New theories of the policy process. Policy Stud. J. 41, 3. https://doi.org/10.1111/psj.12030.

Schlüter, M., Lindkvist, E., Basurto, X., 2021. The interplay between top-down interventions and bottom-up self-organization shapes opportunities for transforming self-governance in small-scale fisheries. Mar. Policy 128, 104485. https://doi.org/10.1016/j.marpol.2021.104485.

Schmidt, V.A., 2008. Discursive institutionalism: the explanatory power of ideas and discourse. Annu. Rev. Polit. Sci. 11, 303–326. https://doi.org/10.1146/annurev.polisci.11.060606.135342.

Schmidt, V.A., 2010. Taking ideas and discourse seriously: explaining change through discursive institutionalism as the fourth 'new institutionalism'. Eur. Polit. Sci. Rev. 2, 1. https://doi.org/10.1017/S175577390999021X.

Schot, J., Kanger, L., 2018. Deep transitions: emergence, acceleration, stabilization and directionality. Res. Policy 47, 1045–1059. https://doi.org/10.1016/j.respol.2018.03.009.

Scoones, I., Stirling, A., Abrol, D., Atela, J., Charli-Joseph, L., et al., 2020. Transformations to sustainability: combining structural, systemic and enabling approaches. Curr. Opin. Environ. Sustain. 42, 65–75. https://doi.org/10.1016/j.cosust.2019.12.004.

Shepsle, K.A., 1989. Studying institutions: some lessons from the rational choice approach. J. Theor. Polit. 1, 131–147. https://doi.org/10.1177/0951692889001002002.

Smith, A., Raven, R., 2012. What is protective space? Reconsidering niches in transitions to sustainability. Res. Policy Spec. Sec. Sustain. Trans. 41, 1025–1036. https://doi.org/10.1016/j.respol.2011.12.012.

Smith, A., Kern, F., Raven, R., Verhees, B., 2014. Spaces for sustainable innovation: solar photovoltaic electricity in the UK. Technol. Forecast. Soc. Chang. 81, 115–130. https://doi.org/10.1016/j.techfore.2013.02.001.

Spash, C.L., 2022. Three decades of environmental values: some personal reflections. Environ. Val. 31, 1–14. https://doi.org/10.3197/096327122X16386102423930.

Standish, R.J., Hobbs, R.J., Mayfield, M.M., et al., 2014. Resilience in ecology: abstraction, distraction, or where the action is? Biol. Conserv. 177, 43–51. https://doi.org/10.1016/j.biocon.2014.06.008.
Stedman, R.C., 2016. Subjectivity and social-ecological systems: a rigidity trap (and sense of place as a way out). Sustain. Sci. 11, 891–901. https://doi.org/10.1007/s11625-016-0388-y.
Swyngedouw, E., 1999. Modernity and hybridity: nature, regeneracionism, and the production of the Spanish waterscape, 1890–1930. Ann. Assoc. Am. Geogr. 89 (3), 443–465. https://doi.org/10.1111/0004-5608.00157.
Swyngedouw, E., 2015. Liquid Power. Contested Hydro-Modernities in Twentieth-Century Spain. MIT Press, Cambridge.
Szalai, E., 2005. Socialism: An Analysis of its Past and Future. Central European University Press, Budapest.
Thelen, K., 1999. Historical institutionalism in comparative politics. Annu. Rev. Polit. Sci. 2, 369–404.
Van Poeck, K., Östman, L., Block, T., 2020. Opening up the black box oflearning-by-doing in sustainability transitions. Environ. Innov. Soc. Trans. 34, 298–310. https://doi.org/10.1016/j.eist.2018.12.006.
Vargas, J., 2014. Consecuencias del megaproyecto de dragado de profundización en el sistema socio-ecológico del estuario del Guadalquivir, España. Agua y Territorio 4, 23–34.
Vargas, J., Paneque, P., 2015. Major hydraulic projects, coalitions and conflict. Seville's harbour and the dredging of the Guadalquivir (Spain). Water 7 (12), 6736–6749. https://doi.org/10.3390/w7126658.
Vatn, A., Vedeld, P., 2012. Fit, interplay, and scale: a diagnosis. E&S 17, art12. https://doi.org/10.5751/ES-05022-170412.
Verhoeven, H., 2019. The mirage of supply-side development: the hydraulic mission and the politics of agriculture and water in the Nile Basin. Oxf. Res. Encyclopedia Environ. Sci. https://doi.org/10.1093/acrefore/9780199389414.013.645.
Walker, B., Holling, C.S., Carpenter, S., Kinzig, A., 2004. Resilience, adaptability and transformability in social–ecological systems. Ecol. Soc. 9. https://doi.org/10.5751/ES-00650-090205.
Williams, J.M., 2019. Discourse inertia and the governance of transboundary rivers in Asia. Earth Syst. Govern. 3, 100041. https://doi.org/10.1016/j.esg.2019.100041.
Wyborn, C., Datta, A., Montana, J., Ryan, M., Leith, P., Chaffin, B., Miller, C., van Kerkhoff, L., 2019. Co-producing sustainability: reordering the governance of science, policy, and practice. Annu. Rev. Environ. Resour. 44, 319–346. https://doi.org/10.1146/annurev-environ-101718-033103.
Young, O.R., King, L.A., Schroeder, H. (Eds.), 2008. Institutions and Environmental Change: Principal Findings, Applications, and Research Frontiers. MIT Press, Cambridge, Mass.

Further reading

Del Moral, L., 1991. La obra hidráulica en la cuenca del Bajo Guadalquivir (siglos XVIII-XX). Gestión del agua y organización del territorio. Publicaciones de la Universidad de Sevilla, Sevilla, España, https://doi.org/10.12795/9788447220908.
Fernández-Delgado, C., 2017. Doñana natural space: the uncertain future of a crown jewel in Europe's protected areas. Case Stud. Environ. 1 (1), 1–12. https://doi.org/10.1525/cse.2017.000570.
Green, A.J., Alcorlo, P., Peeters, E.T., et al., 2017. Creating a safe operating space for wetlands in a changing climate. Front. Ecol. Environ. 15, 99–107. https://doi.org/10.1002/fee.1459.

Ostrom, E., 2007. A diagnostic approach for going beyond panaceas. PNAS 104, 15181–15187. https://doi.org/10.1073/pnas.0702288104.

Paredes, I., Ramírez, F., Aragonés, D., Bravo, M.Á., Forero, M.G., Green, A.J., 2021. Ongoing anthropogenic eutrophication of the catchment area threatens the Doñana world heritage site (South-West Spain). Wetl. Ecol. Manag. https://doi.org/10.1007/s11273-020-09766-5.

Parra-Supervía, F., 1990. La política de espacios naturales: una historia ambigua. Ciudad y territorio 83, 67–76. https://recyt.fecyt.es/index.php/CyTET/article/view/83652.

Stirling, A., 2014. Emancipating Transformations: From Controlling 'the Transition' to Culturing Plural Radical Progress, STEPS Working Paper 64. STEPS Centre, Brighton, UK.

CHAPTER TEN

Governance to manage the complexity of nature's contributions to people co-production

Roman Isaac[a,*], Jana Kachler[b,c,d], Klara J. Winkler[e], Eerika Albrecht[f], María R. Felipe-Lucia[b,d], and Berta Martín-López[a]

[a]Social-Ecological Systems Institute, Leuphana University of Lüneburg, Lüneburg, Germany
[b]Department of Ecosystem Services, Helmholtz Centre for Environmental Research—UFZ, Leipzig, Germany
[c]Faculty of Biological Sciences, Friedrich-Schiller-University Jena, Jena, Germany
[d]Department of Ecosystem Services, German Centre for Integrative Biodiversity Research (iDiv), Halle-Jena-Leipzig, Germany
[e]Department of Natural Resource Sciences, McGill University, Ste-Anne-de-Bellevue, Canada
[f]LYY Institute for Natural Resources, Environment and Society, University of Eastern Finland, Kuopio, Joensuu, Finland
*Corresponding author: e-mail address: isaac@leuphana.de

Contents

Advances in Ecological Research, Volume 66
ISSN 0065-2504
https://doi.org/10.1016/bs.aecr.2022.04.009

Abstract

Nature's contributions to people (NCP) support people's quality of life and are generated by an interplay of both natural and anthropogenic capitals, known as NCP co-production, which also includes ecosystem services. A governance system of formal and informal institutions on different yet interlinked spatial, administrative, and temporal scales influences the management of these capitals underpinning the co-production of NCP. While the field of NCP governance is well established, the role of governance on natural and anthropogenic capitals in NCP co-production is under-researched, which might undermine its integration into decision-making. Here, we analyse how governance of anthropogenic capitals influencing NCP co-production has been researched in the scientific literature. We conducted two systematic literature reviews of publications on ecosystem service governance and ecosystem service and NCP co-production. We distinguish different modes of governance directed at NCP co-production following the classification described by Primmer et al. (2015)—hierarchical, scientific-technical, (adaptive)-collaborative governance and the governing of strategic behaviour. These governance modes resemble the pluralism in ecosystem services governance as they involve various actors and governance approaches. We tie these governance modes to the anthropogenic capitals (human, social, physical, or financial) involved in NCP co-production at the respective administrative, spatial, and temporal scales. Our results show that the literature refers to a variety of combinations of governance modes and anthropogenic capitals involved in NCP co-production. Anthropogenic capitals are mostly studied in multiple governance modes, with material NCP being the most commonly NCP investigated. Our results highlight a major research gap in the NCP co-production literature, which currently neglects the role of governance when assessing the interplay of anthropogenic and natural capitals.

1. Introduction

Within its conceptual framework, the Intergovernmental Science-Policy Platform on Biodiversity and Ecosystem Services (IPBES) explicitly recognises the possibility of achieving a good quality of life through the access to nature's contributions to people (NCP), which do not only depend on the state and functioning of natural elements but also on anthropogenic assets or capitals (Díaz et al., 2015). This idea was formerly conceived in the ecosystem services literature with the concept of ecosystem services co-production (Palomo et al., 2016). Lavorel et al. (2020) define co-produced NCP as "the benefits [provided] from nature to people through multiple, long-term and dynamic socio-ecological interactions and the associated interplay of natural and human-derived capitals" (p.3). Recent research has assessed the contributions of anthropogenic capitals to the provision of ecosystem services and NCP (see, e.g. Bruley et al., 2021; Fischer and Eastwood, 2016; Gómez-Baggethun et al., 2013; Lavorel et al., 2020; Palliwoda et al., 2021; Raymond et al., 2010; Torralba et al., 2018). For example, Lavorel et al. (2020) highlight that a lack of operationalisation regarding the co-production concept exists and apply it to the case of climate change adaptation across several study sites. Bruley et al. (2021) conducted a comprehensive study on NCP co-production in a multifunctional landscape in the French Alps, concluding that governance, which might also be conceptualised as social capital, is relevant for the co-production of NCP.

In addition, the IPBES conceptual framework accentuates the reciprocal relationships between institutions and governance with NCP, anthropogenic assets, and ultimately people's quality of life (Díaz et al., 2015). In fact, the IPBES conceptual framework places institutions at its core to emphasise their key role in mediating the provision of NCP as well as the access and control over NCP (Díaz et al., 2015; Martín-López et al., 2019). Governance, however, goes beyond the role of institutions. Thus, this review follows a framework proposed by Primmer et al. (2015) who differentiate between hierarchical governance—which includes institutions—, scientific-technical governance, (adaptive)-collaborative governance, and the governing of strategic behaviour. These diverse modes of governance not only reflect the plurality of governing ecosystem services, but the plurality of actors involved in this governance. While hierarchical governance is predominantly based on formal institutions, for example in form of national legislation, (adaptive) collaborative and scientific-technical governance is based on involving various non-state actors due to their expertise and stakes in governing ecosystem functions and services (Primmer et al., 2015). Méndez et al. (in this special issue) aim to understand when the pluralism of actors and governance modes is limited in achieving its goals and explore how strengthening pluralism can contribute to more sustainable trajectories.

While research on ecosystem services has increased exponentially over the last two decades (Droste et al., 2018), ecosystem services governance research has not grown correspondingly (Winkler et al., 2021). Nonetheless, ecosystem services governance has gained increasing attention in the science-policy interface over the last years and is now an established field of research (Sattler et al., 2018). Despite this, few empirical studies refer explicitly to the involvement of governance in the co-production of NCP (see e.g. Auer et al., 2017; Lavorel et al., 2019, 2020; Vialatte et al., 2019). In fact, to our best knowledge, no paper has attempted to review the research status of governance of NCP co-production.

This paper aims to analyse how previous literature on NCP co-production has addressed the different modes of governance proposed by Primmer et al. (2015) and how they are directed at anthropogenic capitals involved in the co-production of NCP. To do so, we conduct a systematic literature review that aims to: (i) show which NCP are studied by research conducted on ecosystem services governance and NCP co-production; (ii) acknowledge the anthropogenic capitals contributing to these NCP; and (iii) identify which modes of governance or combinations of these are directed at anthropogenic capitals involved in the co-production of NCP studied in the literature. Our work builds a bridge between the debate on NCP and ecosystem services governance and the debate focussing on the interplay of natural and anthropogenic capitals in NCP co-production. While the first debate has evolved around the question of how governance and institutions can affect ecosystem services and the way people use and manage them (see e.g. Winkler et al., 2021), the latter debate seeks to understand how humans contribute to the supply of NCP and ecosystem services through the use of different anthropogenic capitals (see e.g. Bruley et al., 2021; Lavorel et al., 2019; Palomo et al., 2016). Thereby, we contribute to a better understanding of how anthropogenic capitals in NCP co-production are governed and lay the foundation for further research to analyse which modes of governance ensure the co-production and use of NCP in a sustainable manner.

2. Theoretical background

2.1 Nature's contributions to people and their co-production

NCP encompass all "contributions, both positive and negative, of living nature (i.e., biodiversity of organisms, ecosystems, and their associated ecological and evolutionary process) to quality of life for people" (Díaz et al., 2018, p. 3). The NCP concept includes two analytical perspectives: a context-specific and a generalising perspective. The context-specific perspective embraces the diversity in perceptions on NCP around the world, highlighting that the same NCP can be perceived differently depending on the personal context. The generalising perspective seeks to provide a context-independent definition for 18 NCP that underpin people's quality of life (Hill et al., 2021). The 18 NCP are classified into three groups: material, regulating and non-material NCP (Díaz et al., 2018).

Unlike the common ecosystem services classifications of provisioning, regulating, and cultural services, the NCP concept acknowledges that the cultural dimension diffuses through all NCP instead of being confined into a single group (Kadykalo et al., 2019). Material NCP include all "physical substances, objects, or materials from nature which people use to consume to live," such as providing of food or building materials (Hill et al., 2021, p. 911). Regulating NCP refer to the way ecosystems "modify environmental conditions experienced by people, and/or sustain and/or regulate the generation of material and non-material contributions" (Díaz et al., 2018, p. 5) like the regulation of freshwater and air quality or the pollination of plants. Last, non-material NCP entail "nature's effects on subjective or psychological aspects underpinning people's quality of life" on both an individual and societal level (Hill et al., 2021, p. 911). This includes the ways nature benefits humans by providing "physical and psychological experiences," "learning and inspiration" and "supporting identities" (Díaz et al., 2018). In this paper, we use the categories of material, non-material and regulating NCP portrayed by the generalising perspective with the aim to have a universal system of classification that aligns with former research on ecosystem services.

While the IPBES conceptual framework highlights that NCP are co-produced by nature and humans through the use of anthropogenic assets (Díaz et al., 2015, 2018), the debate on ecosystem services has long regarded them as a "free gift" solely provided by nature (Spangenberg et al., 2014, p. 40). However, recent literature has considered the dual nature of ecosystem services as being provided by contributions from nature and anthropogenic capitals (see e.g. Bruley et al., 2021; Fischer and Eastwood, 2016; Lavorel et al., 2020; Palomo et al., 2016; Raymond et al., 2010).

In this paper, we use the NCP concept because it is a broader paradigm that encompasses not only the ecosystem services framework but also the multiple ways by which societies relate with nature and contribute different anthropogenic assets to the provision of NCP (see Díaz et al., 2018; Hill et al., 2021). Over the last decade, diverse framings of co-production have evolved in ecosystem services and NCP research. Lele et al. (2013) first noted that co-production involves ecosystem functions and different forms of human agency, technologies, and financial capital. Spangenberg et al. (2014) described how human agency shapes the co-production of ecosystem services later in more detail. Palomo et al. (2016) built on the definition of capitals given by Goodwin (2003) to specify how human, social, physical

BOX 1 Definitions of the anthropogenic capitals involved in NCP co-production.

Human capital refers to people's capabilities, including health, informal knowledge, formal knowledge, skills, motivation, and labour, that can yield a flow of NCP.

Social capital includes all intangible assets associated with interactions between people, including formal and informal networks, trust, and formal and informal institutions required for enhancing the flow of NCP.

Physical capital refers to technological or manufactured assets that contribute to the NCP flow.

Financial capital Virtual mechanisms in the form of savings, credits, and other monetary forms are used for trading, maintaining, or enhancing natural, human, social or physical capitals that contribute to the flow of NCP.

Adapted from Isaac, R., Kachler, J., Martín-López, B., Felipe-Lucia, M.R., 2022. Fieldwork Protocol Effects of land management on the Supply and Distribution of ecosystem services (ESuDis). https://doi.org/10.13140/RG.2.2.24291.84006: S1 and based on Palomo, I., Felipe-Lucia, M.R., Bennett, E.M., Martín-López, B., Pascual, U., 2016. Disentangling the pathways and effects of ecosystem service co-production, in: Advances in Ecological Research. Elsevier, pp. 245–283. https:/doi.org/10.1016/bs.aecr.2015.09.003 and Goodwin, N.R., 2003. Five Kinds of Capital: Useful Concepts for Sustainable Development.

and financial capitals interact with natural capital to provide ecosystem services. In this paper, we use the classification of human, social, physical and financial capitals involved in co-production as explained by Palomo et al. (2016) to determine which capitals are the target of management and governance actions. Box 1 provides a definition of the four types of capitals.

2.2 Governance of nature's contributions to people

Within the IPBES conceptual framework, governance and institutions play an integral role and affect NCP by regulating the access to, control over, and allocation of capitals contributing to the co-production of NCP (Díaz et al., 2015). Ecosystem services governance "refers to the processes by which a range of actors (e.g. government, resource users, environmental groups and private entities) make decisions that influence the use of ecosystem-derived goods and services" (Metzger et al., 2020, p. 2). With the emergence of research on ecosystem services governance, many concepts from the general governance literature have been introduced to the ecosystem service literature. Despite the increased attention on governance in ecosystem services research, Sattler et al. (2018) suggest that ecosystem services literature has only partly succeeded in researching the full complexity of ecosystem services governance.

In an attempt to navigate the genuine complexity of governance, a variety of frameworks have been developed to analyse ecosystem services governance across administrative and spatial scales, beneficiaries, and ecosystems (see e.g. Albert et al., 2017; Fischer and Eastwood, 2016; Knüppe and Pahl-Wostl, 2011; Metzger et al., 2020; Primmer et al., 2015; Vialatte et al., 2019). In some frameworks, the analysis of ecosystem services governance can be specific to a particular ecosystem or set of services. For example, Knüppe and Pahl-Wostl (2011) assess the governance of groundwater to enable scholars to improve their understanding of groundwater-related ecosystem services. Fossey et al. (2020) assess the governance related to soil ecosystem services and how these are integrated into policy-making in the form of land-use planning. Vialatte et al. (2019) assess the multiple ecosystem services provided in agricultural landscapes by considering social and ecological components and the effects of multi-level governance.

To provide a tool for the empirical assessment of ecosystem services governance, Primmer et al. (2015) present a more universal framework that highlights how decision-making actively shapes ecosystems and their services. The framework distinguishes between four modes of governance: scientific-technical, (adaptive) collaborative, hierarchical governance, and the governing of strategic behaviour. Box 2 explains each of the governance modes in detail. The strength of this framework lies in the possibility of applying it across governance scales and in a range of ecosystem contexts. Winkler et al. (2021) applied this framework to understand how ecosystem service research has addressed governance, finding that hierarchical governance is the most prominently studied ecosystem service governance mode and often affects the three other governance modes.

In this manuscript, we apply the framework by Primmer et al. (2015) to the literature on NCP co-production and investigate whether the findings by Winkler et al. (2021) can be confirmed for the governance of capitals in NCP co-production. Building on the findings of Winkler et al. (2021), we differentiate between adaptive and collaborative governance to gain a more nuanced view of how governance has been used in the literature on NCP co-production (see Box 2). Using the framework by Primmer et al. (2015), Box 3 illustrates how governance of the different anthropogenic capitals matters for NCP co-production in an exemplary farming system.

BOX 2 Modes of nature's contributions to people (NCP) governance.

Hierarchical governance concerns regulations that influence NCP and the other modes of governance (Primmer et al., 2015). In contrast to collaborative governance, hierarchical governance is executed by state actors through formal institutions. Formal institutions exist mainly in the form of policies (e.g. frameworks, laws, regulations) that directly affect NCP or the other governance modes. As interdependencies between policies across governance levels exist, hierarchical governance is often studied in multi-level contexts. Therefore, governance of NCP is often linked to scale mismatches (Borgström et al., 2006; Winkler et al., 2021). Scale mismatches occur when ecosystem boundaries do not match with the administrative units' borders. A positive example is the Water Framework Directive, the first European Union directive using a natural unit (water catchments) for governance, rather than administrative units such as counties (Moss, 2012).

Scientific-technical governance refers to decision-making based on scientific and technical arguments (Primmer et al., 2015). Experts, who are not necessarily scientists influence decision-making, play a vital role in this process. The underlying expectation for scientific-technical governance is that effective dissemination of scientific knowledge leads to better governance (Huth and Possingham, 2011; Sutherland et al., 2004). Land-use planning serves as an example for this governance mode as it incorporates scientific arguments when making decisions about potential trade-offs between NCP (Paavola and Hubacek, 2013).

Collaborative governance stresses the importance of collective learning and bottom-up participatory processes that embrace diverse stakeholder perspectives (Primmer et al., 2015). Moreover, collaborative governance is best executed when based on open and transparent processes to collective problem-solving (Newig and Fritsch, 2009). Participatory processes are often considered to contribute positively to ecological outcomes (Bäckstrand et al., 2010).

Adaptive governance tries to increase the responsive capacity to changing circumstances, including changing societal values (Burkhard et al., 2012). Examples are stewardships approaches to ecosystems (Chapin III et al., 2015; Olsson et al., 2008) or approaches that promote local learning (Cork et al., 2006).

Governing of strategic behaviour features the potential of governance arrangements and the interactions between them to be transformative (Primmer et al., 2015). It refers to market-based or commercial approaches. This implies the aim to influence policy-making processes through lobbying with the goal of affecting policies and protecting certain interests (Newig and Fritsch, 2009; Primmer et al., 2015). Besides, governing of strategic behaviour can nudge behaviour in a wanted direction (Primmer, 2011). Payments for ecosystem services are a typical example of governing strategic behaviour often referred to in the literature (see e.g. Börner et al., 2010; Rival, 2010; Van Hecken and Bastiaensen, 2010).

BOX 3 Governance modes affecting capitals in nature's contributions to people (NCP) co-production.

To illustrate how governance affects NCP and the co-production underpinning NCP, we use a fictional farm with grass-fed dairy cows as an example. The various NCP arising at the farm level and the capitals contribution to them as well as the different forms of governance affecting these are described below.

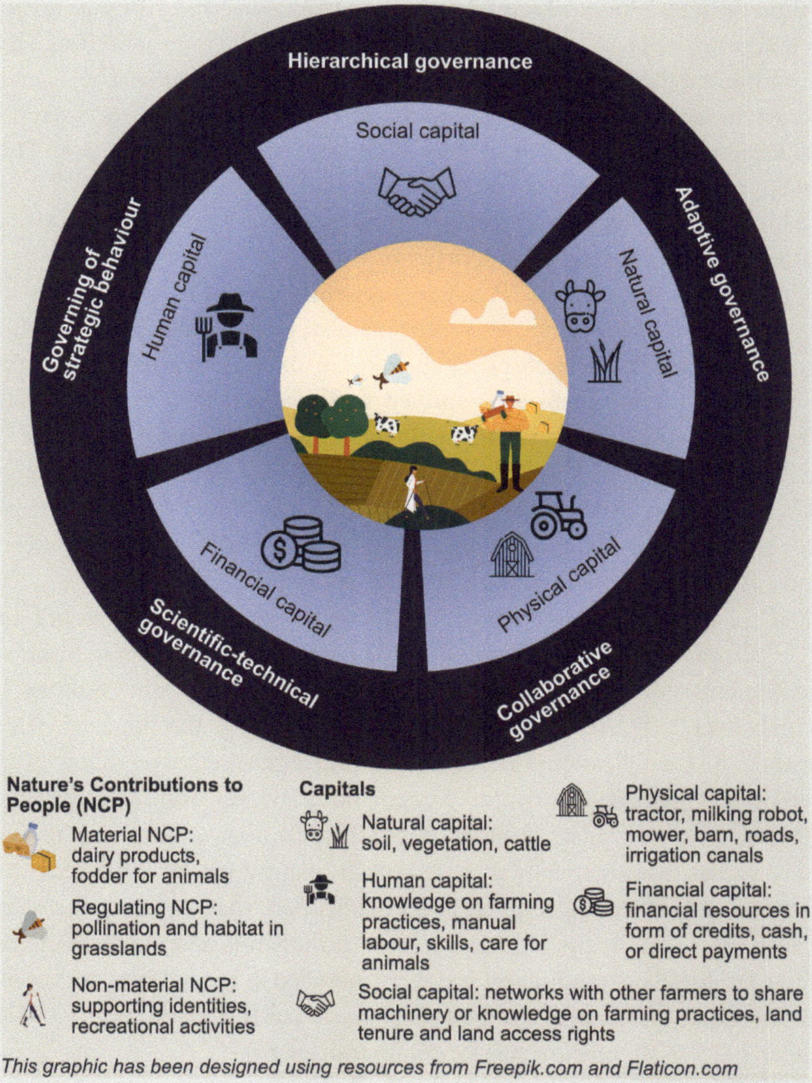

Graphic by Annika R. Schmidt.

3. Methods

3.1 Data collection

We conducted a systematic literature review (Bryman, 2012) to understand how different modes of governance are directed at capitals in NCP co-production. Since co-production of NCP is a relatively new concept, we assumed that little research could be found by only searching for publications on the governance of anthropogenic capitals in NCP co-production. Thus, we combined the results of two searches. The first focuses on publications on ecosystem services and NCP co-production. The second was conducted by Winkler et al. (2021) and deals with ecosystem services governance. Thus, we approached the governance of anthropogenic capitals in NCP co-production from two vantage points. In both cases, we only included articles that were peer-reviewed and written in English.

The first search on NCP and ecosystem services co-production was conducted to capture literature on ecosystem services and natural and anthropogenic capitals in their co-production in an unlimited to 2020 timespan. The search was automatically confined to the years 1945 to 2020 by the *Web of Science*. The search string employed included *'ecosystem services*'* as well as other synonyms (e.g. *'ecosystem good*', 'ecosystem function*', 'nature benefit*', 'provisioning service*', 'regulating service*', 'natures contribution* to people', 'supply*', 'demand*' ,'use*', 'co-production*', 'co-creation*'),* as well as different forms of *'anthropogenic capital'* (e.g. *'human capital*', 'human asset*', 'soci* capital*', 'soci* factor*', 'soci* asset*', 'technological capital*', 'manufactured capital*', 'financial capital*', ' physical capital*'*). For the entire search string see Supplementary material S1 in the online version at https://doi.org/10.1016/bs.aecr.2022.04.009. Based on this search string, the literature search returned 384 publications. For the literature assessment we followed the Preferred Reporting Items for Systematic Reviews and Meta-Analyses (PRISMA) flow diagram (Page et al., 2021). Assessing the abstracts resulted in 170 papers that potentially included NCP co-production. A full-text analysis of these publications resulted in 25 publications that explicitly (i.e. stated by authors) or implicitly (i.e. not stated by authors but covered in the paper) referred to NCP co-production and were included in the full-text coding (see Supplementary material S2 in the online version at https://doi.org/10.1016/bs.aecr.2022.04.009 for the PRISMA flow diagram).

The second search by Winkler et al. (2021) on ecosystem services governance used a search string with the words *'ecosystem service* and* *'governance.'* The search was limited to the years 2006 to 2019, and the year 2020 was added for this review, to capture literature on ecosystem services and governance after the Millenium Ecosystem Assesment's publication (Millennium Ecosystem Assessment (Program), 2005). The selection of publications that had on average at least one citation per year since their publication was a criterion for relevance in the scientific discourse. The selection criteria yielded 157 publications that focused on ecosystem services and governance and were empirical or had an empirical component. Empirical in this case refers to research based on observation and measurement of phenomena. Next, we screened the abstracts and, if uncertain, the full text to see whether the publications explicitly or implicitly dealt with NCP co-production. The application of this final selection criterion resulted in 21 publications on ecosystem services governance which referred to NCP co-production.

After removing one duplicate found in both searches, 45 publications were fully coded (see Supplementary material S3 in the online version at https://doi.org/10.1016/bs.aecr.2022.04.009 for the list of publications reviewed). We followed a deductive code system based on the governance framework proposed by Primmer et al. (2015) and the co-production framework by Palomo et al. (2016). Besides general characteristics for each publication (i.e. the year of publication, location, temporal scale, spatial scale, landscape type, data source and methods for data collection and analysis), we coded information concerning the type of NCP co-production (i.e. explicit or implicit), the different NCP categories (based on Díaz et al., 2018), the anthropogenic capitals involved in NCP co-production (based on Palomo et al., 2016), information about the governance of NCP co-production (i.e., governance level, polycentric governance, governance affecting capitals or NCP, formal governance, informal governance (for definitions please see Supplementary material S4 in the online version at https://doi.org/10.1016/bs.aecr.2022.04.009), and governance modes (based on Primmer et al., 2015). Supplementary material S4 in the online version at https://doi.org/10.1016/bs.aecr.2022.04.009 provides a complete list of the variables used in the analysis, including definitions and references when relevant.

A clear distinction between social capital and governance is difficult to make throughout the literature on NCP co-production given that social capital encompasses formal and informal institutions (see Box 1). We, therefore, counted the involvement of institutions in NCP co-production as both social capital and governance.

3.2 Data analysis

We employed descriptive statistics to determine which anthropogenic capitals were involved in the co-production of the 18 NCP categories and the respective NCP groups and tied these to the governance modes. To see which combinations of governance modes, anthropogenic capitals and NCP groups were referred to most often we counted and ranked the combinations. Finally, to explore the relations between NCP categories, anthropogenic capitals, and governance modes, we conducted a Hierarchical Cluster Analysis (HCA) with the factors obtained in a Multiple Correspondence Analysis (MCA) using *XLSTAT* Version 2020.3.1.2. The MCA was run with the NCP categories, anthropogenic capitals, and governance modes. We also displayed the centroids of governance levels and informal institutions to gain in-depth knowledge of those connections.

4. Results

4.1 General characteristics of publications and geographic distribution of case studies

The number of publications on the governance of NCP co-production reviewed for this paper amounts to 45 publications. Most publications referred to at least one empirical method involving stakeholders (Fig. 1A). Regarding all methods, more than half of the publications used two or more methods (56%; $n=25$). We found that interviews ($n=29$) and surveys ($n=17$) were conducted most often in the reviewed publications. Literature reviews ($n=11$) were only conducted in combination with other methods and as a way to determine research gaps. Moreover, a range of other methods, including observation ($n=7$), ecological data collection ($n=6$) as well as workshops ($n=4$), and focus groups (n=4) were used.

We found that more than half of the publications included case studies on the regional level (62%, $n=28$) (Fig. 1B). The regional level, in this case, describes an entity larger than a village or city (local level), but smaller than the sub-national level (e.g. states, departments). Examples for the regional level are shires, districts, or counties. Only 13 case studies (29%) looked at NCP co-production on the local level, which refers to a municipality, village, or city. Two case studies conducted research at higher administrative levels. Gatzweiler (2006) referred to a case study on the national level, namely Ethiopia, while Hicks and Cinner (2014) took a multi-national approach when comparing case studies from Kenya,

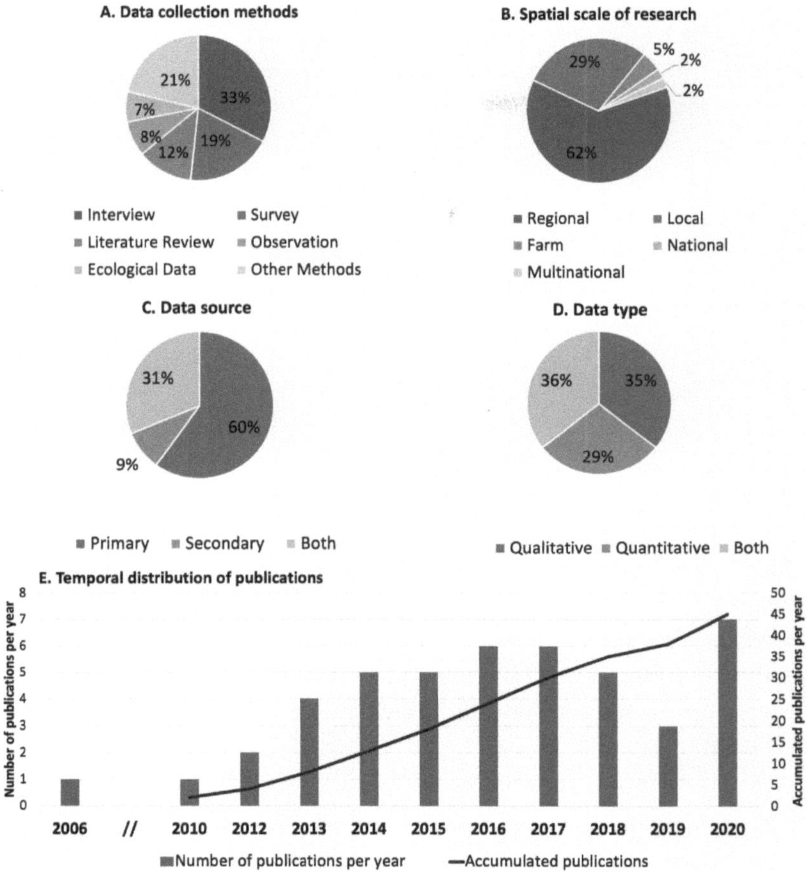

Fig. 1 Distribution of papers on the governance of NCP co-production based on: (A) data collection methods employed (*other methods* include workshops and focus groups), (B) spatial scale at which the research was conducted, (C) source of data, (D) type of data used in publications and (E) temporal distribution of reviewed publications between 2006 and 2020.

Madagascar, Seychelles, and Tanzania. The data used in the reviewed publications were predominantly primary (60%, $n=27$) (Fig. 1C). Some studies used primary as well as secondary data (31%, $n=14$), for example interviews paired with a literature review (e.g. Knüppe et al., 2016). Only 9% used secondary data, which had come from previously conducted case studies ($n=4$). Most publications used qualitative (36%, $n=16$) and a mixture of both types of data (36%, $n=16$). Only 13 publications (28%) used quantitative data (Fig. 1D). We do not see a clear trend with regards to the number of publications per year (Fig. 1E).

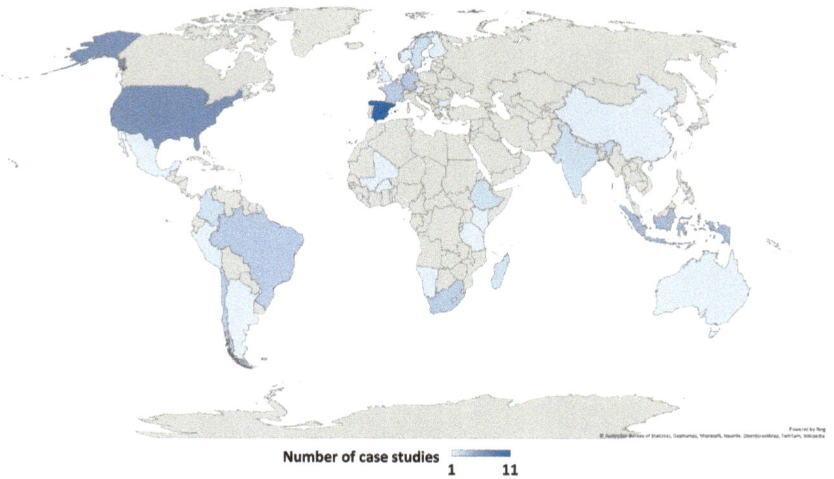

Fig. 2 Geographic distribution of publications on governance and NCP co-production.

Most case studies were located in Spain (*n* = 11) followed by the United States (*n* = 6) and France (*n* = 4) (Fig. 2). Regarding continents, most case studies were placed in Europe (39%, *n* = 26) followed by Africa (22%, n = 14), South America (15%, *n* = 10) and North America (12%, *n* = 8). The least case studies were conducted in Asia (9%, *n* = 6) and Oceania (3%, *n* = 2).

4.2 Anthropogenic capitals in nature's contributions to people co-production

Out of the 45 publications dealing with NCP co-production, only nine (20%) highlight explicit co-production, while 36 (80%) deal with NCP co-production implicitly. Hence, the interplay of natural and anthropogenic capitals in the provision of NCP is recognised by many authors but not defined as NCP co-production. Besides, nine publications (20%) do not refer to any form of governance when studying the co-production of NCP. Thus, no data from these publications can be found in the following.

4.2.1 General findings

We identified a total of 185 co-produced NCP, which are studied with one or more modes of governance. Material NCP are described most often (34%, *n* = 62) in the context of NCP co-production, followed by regulating NCP second most (28%, *n* = 51) and non-material NCP (19%, *n* = 36). A fifth (19%) of interactions between natural and anthropogenic capitals found

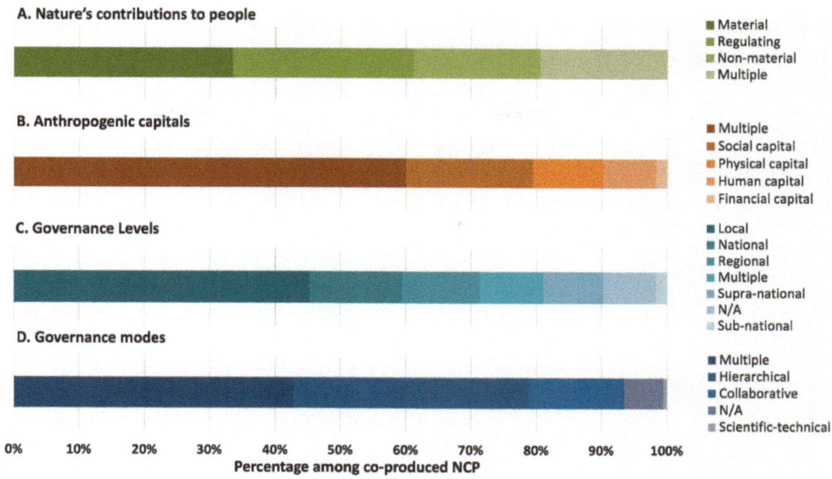

Fig. 3 Percentage of publications referring to the three groups of Nature Contributions to People (NCP), anthropogenic capitals, different governance levels and modes of governance relating to NCP co-production. Some publications did not refer to governance levels and modes and are marked as N/A (Not Applicable).

in the literature cannot be tied to one single NCP but resulted in the co-production of multiple NCP ($n=36$) (Fig. 3A).

More than half of the NCP are co-produced by natural capital and multiple anthropogenic capitals (60%, $n=111$). Co-production of NCP only involving one capital type is based on social capital (19%, $n=36$), physical capital (11%, $n=20$), human capital (8%, $n=15$) and financial capital ($n=3$, 2%) (Fig. 3B). As natural capital is always involved in the co-production of NCP it is not listed in this review.

We found that co-production of NCP is studied at various governance levels. Almost half (45%, $n=84$) of co-produced NCP are studied with governance on the local level and 14% at the national level (n=26). The regional (12%, $n=22$), supra-national (9%, $n=17$) and sub-national (2%, $n=3$) levels play a subordinate role in the governance of NCP co-production. A tenth of co-produced NCP are affected by multi-level governance (10%, $n=18$) while 8% are not related to any specific level ($n=15$) and are represented as N/A (Fig. 3C). For example, local governance is studied by Malinga et al. (2018) who compare NCP in large scale and smallholder farms in South Africa, while Lavorel et al. (2020) focus on the regional level when researching climate adaptation services in the French Alps.

Besides, our review shows that NCP co-production is mostly influenced by combinations of multiple modes of governance (43%, $n=79$). While hierarchical governance affects 36% of co-produced NCP ($n=67$), collaborative governance is only linked to 15% ($n=27$) and scientific-technical governance to only one co-produced NCP (1%). Adaptive governance and the governing of strategic behaviour are only studied in combination with other governance forms and are not studied by themselves. Some co-produced NCP cannot be connected to a specific mode of governance and are labelled as N/A (6%, $n=11$) (Fig. 3D).

4.2.2 Material NCP

Out of the 62 material NCP we found that food and feed is the most studied at 79%. While materials make up 8% of material NCP, energy, and other material NCP only account for 7% and 6%, respectively (Fig. 4A). Formal institutions are the principal anthropogenic capital involved in the co-production of material NCP with 29%. Infrastructure and built capital play an important role as they are linked to 17% of material NCP. Networks (10%), informal knowledge (10%), formal knowledge (8%) and tools and machinery (8%) play a subordinate role in the co-production of material NCP. Financial capital is mostly studied in combination with other capitals and is therefore not referred to on its own in the following (Fig. 5A).

4.2.3 Regulating NCP

Of the 51 regulating NCP we found, freshwater quality (21%), soils and sediments (19%), and freshwater quantity are the most studied. Pollination (14%) and climate (8%) are studied less often. All other regulating NCP make up 20% of this group (Fig. 4B). Formal knowledge and tools and machinery contribute most to the co-production of regulating NCP at 16% each. Networks and other human capitals play a subordinate role at 14% each (Fig. 5B).

4.2.4 Non-material NCP

Psychological and physical experience (61%) are the most studied of the 36 non-material NCP we found. Supporting identities (11%), learning and inspiration (8%) and all other non-material NCP (20%) only account for 39% (Fig. 4C). Non-material NCP are mainly co-produced by informal (34%) and formal institutions (21%) as well as infrastructure and built capital (26%). All other anthropogenic capitals contribute 19% (Fig. 5C).

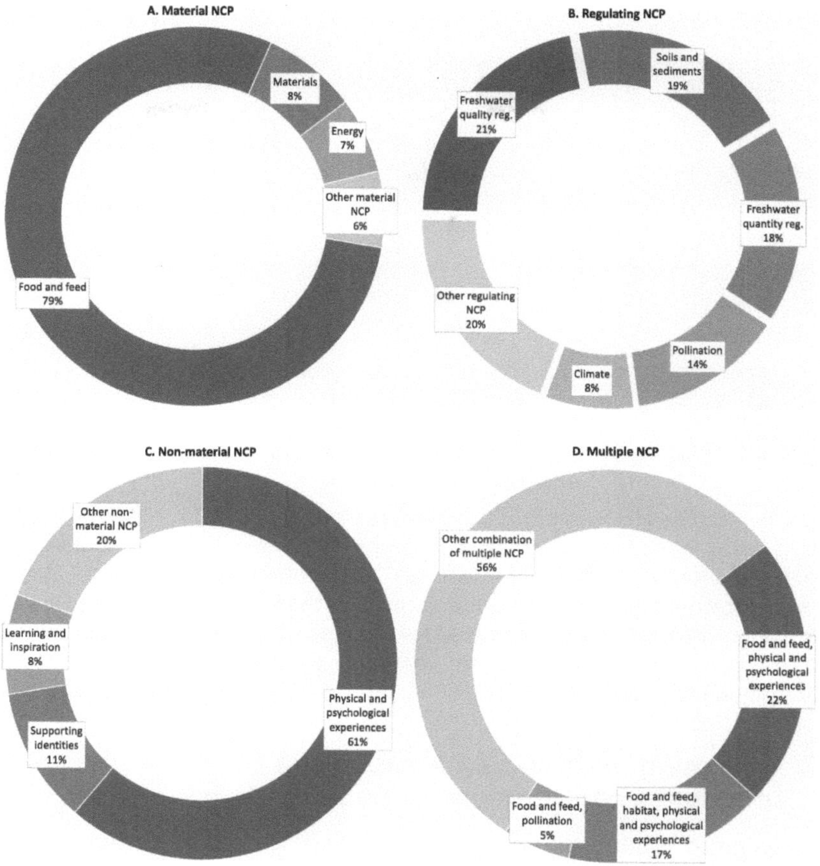

Fig. 4 Nature's contributions to people (NCP) sorted by their group (Abbreviation: reg. = regulation).

4.2.5 Multiple NCP

Within our review, we found 36 combinations of multiple NCP studied together. The material NCP food and feed was most often studied with the non-material NCP psychological and physical experiences. Together they make up 22% of NCP studied in combination. The second-most found combination of multiple NCP includes food and feed, habitat, and psychological and physical experiences (17%), followed by food and feed and pollination (5%). All other combinations of multiple NCP amount to 56% (Fig. 4D). Formal institutions contribute most to the co-production of multiple NCP (28%). Infrastructure and built capital are second (16%) and other human capitals, which include capitals not captured by our human capital definition, come third (15%) (Fig. 5D).

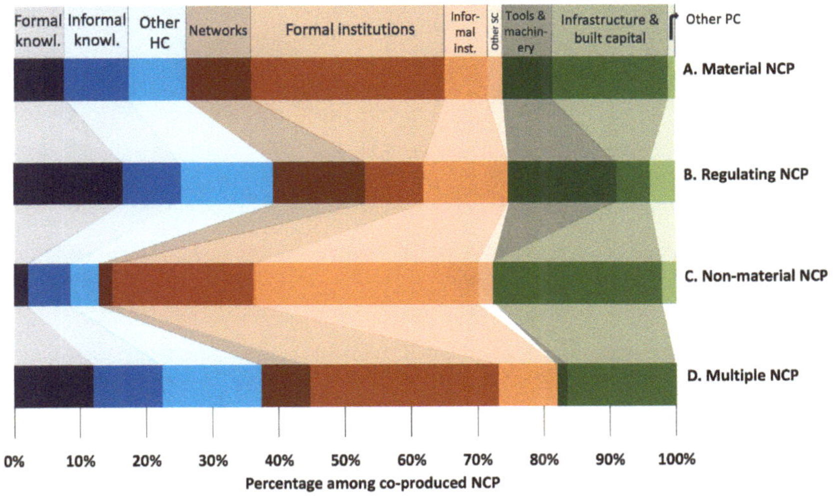

Fig. 5 Percentage of references in the reviewed literature made to anthropogenic capitals involved in nature's contributions to people (NCP) co-production (Abbreviations: knowl. = knowledge, HC = human capital, inst. = institutions, SC = social capital, PC = physical capital). Each anthropogenic capital in the top row is highlighted in a specific colour. Follow the transparent colourway to see how capitals contribute to each NCP group. When a capital does not contribute to a NCP group, the transparent colourway is minimised.

4.3 Governing anthropogenic capitals in the co-production of nature's contributions to people

The literature studies the governance of NCP co-production in multiple ways. However, of the 36 publications we reviewed that included governance of co-produced NCP, only seven publications referred to co-production explicitly (Auer et al., 2017; Felipe-Lucia et al., 2015; Garcia-Llorente et al., 2015; Lavorel et al., 2020; Malinga et al., 2018; Torralba et al., 2018; Vialatte et al., 2019). Despite having in common that they all explicitly refer to co-production, Auer et al. (2017) and Garcia-Llorente et al. (2015) only refer to it on a side-note while the other authors have integrated the concept throughout their publications (Felipe-Lucia et al., 2015; Lavorel et al., 2020; Malinga et al., 2018; Torralba et al., 2018; Vialatte et al., 2019).

The Multiple Correspondence Analysis (MCA) represents the connections between NCP, anthropogenic capitals, and governance modes, as researched in the scientific literature (Fig. 6). The first two factorial axes of the MCA accumulated 66% of the total inertia (Fig. 6A). The first axis (48%) revealed that the governance modes behind the management of the anthropogenic capitals underpinning the provision of NCP varied

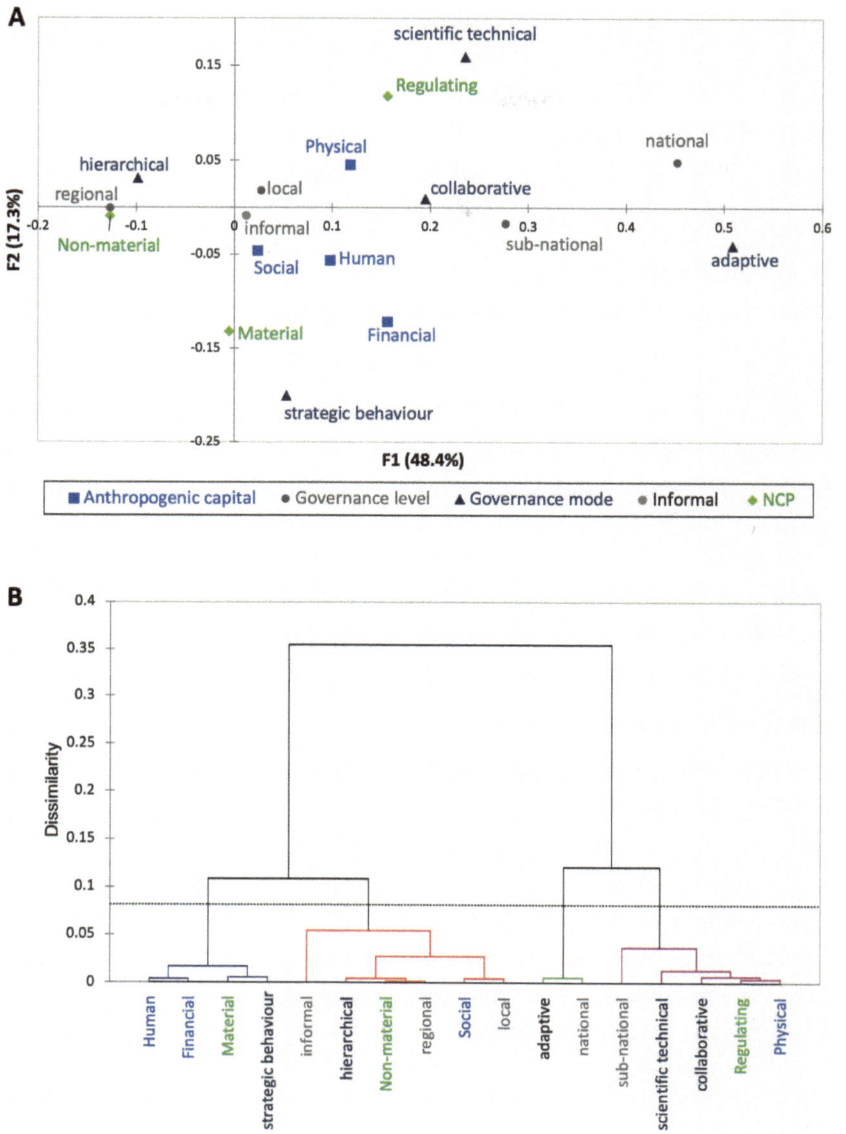

Fig. 6 Multiple correspondence analysis (MCA) and hierarchical cluster analysis (HCA) represent the connections between nature's contributions to people (NCP), anthropogenic capitals and governance modes. (A) MCA biplot, which factors 1 and 2 account for 48.4% and 17.3% of the inertia, respectively. The relative closeness of the variable positions along the axes reflects their tendency to be associated in the scientific literature. (B) Dendrogram of the hierarchical cluster analysis (HCA) performed with the MCA factors. The Bray and Curtis distance and Ward's Method were used as agglomerative techniques.

between non-material (in negative scores) and regulating NCP (in positive scores). Axis 2 (17%) distinguished the scientific literature that studied the governance of NCP co-production of regulating (in positive scores) and material NCP (in negative scores).

The Hierarchical Cluster Analysis (HCA) revealed three clusters representing different ways of researching governance of anthropogenic capitals that underpin NCP provision (Fig. 6B). The first cluster represents a group of literature that researched material NCP as mainly co-produced through an interplay of natural capital with human and financial capital. The governance mode related to this cluster is the governing of strategic behaviour (Fig. 6B). The second cluster represents the co-production of non-material NCP, which mostly involved social capital and hierarchical governance operating at regional and local scales and through informal institutions (Fig. 6B). The third cluster represents the scientific literature that researched regulating NCP. This cluster shows that regulating NCP are mainly co-produced by natural and physical capital and subject to scientific-technical and collaborative governance, operating at the sub-national level (Fig. 6B).

5. Discussion

5.1 Research trends on the governance of NCP co-production

Research on ecosystem services has increased exponentially over the last two decades (Droste et al., 2018; Luederitz et al., 2015). Although not rising exponentially and still underrepresented in ecosystem services literature compared to other research topics (Droste et al., 2018), the governance of ecosystem services has gained growing attention in recent years (Winkler et al., 2021). Similarly, the co-production concept is slowly permeating the literature on ecosystem services and NCP, but has been recently applied in various contexts (see e.g., Bruley et al., 2021; Fischer and Eastwood, 2016; Grosinger et al., 2021; Lavorel et al., 2020). Considering these developments and the results of this review, we can conclude that governance of NCP co-production is currently under-researched and yet to be explored in detail.

The fact that the governance of NCP co-production is a topic that has been traditionally overlooked entails a limitation for this review. Namely, the data availability for our review was relatively small when compared to much larger and more broadly focussed reviews on ecosystem services (see e.g. Chaudhary et al., 2015; Droste et al., 2018) and the governance of ecosystem services (see e.g. Winkler et al., 2021).

5.2 Methods in research on the governance of NCP co-production

Our review's results revealed that research on the governance of NCP co-production is predominantly based on social science methods that gather mainly qualitative data (Fig. 1). For example, Vialatte et al. (2019) used in-depth interviews and focus groups for a socio-cultural valuation among farmers in France. Malinga et al. (2018) used a mixed methods approach including in-depth interviews, expert assessments and participatory mapping paired with biophysical data to quantify NCP in agricultural contexts in South Africa. Similarly, Winkler et al. (2021) highlight that more than half of ecosystem services governance literature uses qualitative data. By contrast, the ecosystem service literature is dominated by ecological and economic methods and quantitative data (see e.g. Fagerholm et al., 2016; Liquete et al., 2013; Martín-López et al., 2019; Nieto-Romero et al., 2014). Liquete et al. (2013), for example, found that marine and coastal ecosystem services are mainly studied using environmental science methods and quantitative data. Similarly, Nieto-Romero et al. (2014) found that most literature on ecosystem services in Mediterranean agroecosystems used economic, environmental or forestry science methods, while Martín-López et al. (2019) point out that biophysical methods dominate research on NCP in mountains.

In line with the findings by Winkler et al. (2021), our review showed that research on the governance of NCP co-production is mostly focused at regional and local scales, which might be due to the availability of data in regionally administered statistical databases. Another potential explanation is that most studies were conducted using social science methods involving local stakeholders (e.g. interviews), which aligns with the fact that more than half of the research on the governance of NCP co-production uses primary data.

5.3 Geographic distribution

Similar to several other reviews on ecosystem services and NCP (Haase et al., 2014; Liquete et al., 2013; Martín-López et al., 2019), our results showed that most case studies are located in Europe, Spain being a hotspot of research on the governance of NCP co-production. In contrast to the reviews by Chaudhary et al. (2015) and Luederitz et al. (2015), who investigated ecosystem services in general and found most case studies were being conducted in the United States, our findings only show six case studies of co-produced NCP in the United States. Similar to Winkler et al. (2021), we found few case studies on the African continent outside of South Africa. While we found some case studies in Colombia, Brazil, China, India and South-East Asia, for example, a research gap exists regarding the governance of NCP co-production in most other countries across Asia and Latin America. Similar to Liquete et al. (2013), Luederitz et al. (2015), and Martín-López et al. (2019), we found no research conducted in Russia. Thus, future studies should aim to close these gaps by conducting further research in smaller countries of the African and Asian continent, across America, particularly in Russia, which appears to be under-researched.

5.4 Governing anthropogenic capitals in NCP co-production

Our results showed that material and regulating NCP are most studied in the NCP co-production governance literature. By contrast, Martín-López et al. (2019) found that most studies on NCP in mountains studied multiple NCP. When looking at multiple NCP, our review showed that food and feed appear in the three most common combinations of NCP (Fig. 4).

Our results also revealed that most research looked at the interplay of natural capital with multiple anthropogenic capitals. This result contrasts with the review by Kachler et al. (under review), which found that literature on ecosystem services co-production mainly focused on material and non-material NCP and human and physical capitals. Besides, our results showed that anthropogenic capitals in NCP co-production are predominantly governed on the local and regional levels. This is comparable to the results obtained by Winkler et al. (2021), who found most NCP governed on the regional and then on the local level. Regarding the modes of governance, our results paint an ambivalent picture compared to the findings by Winkler et al. (2021). In contrast to Winkler et al. (2021), we could not single out one mode of governance as predominant since our findings show that most anthropogenic capitals in NCP co-production are subject to multiple modes of governance. However, in line with Winkler et al. (2021), we found that hierarchical governance is the single most studied governance mode, while collaborative and scientific-technical governance are studied the least. Besides, Winkler et al. (2021) found governing of strategic behaviour to be studied second most often while we found that it is only studied in combination with other forms of governance and thus appears as multiple modes of governance in our findings. Thus, our finding that anthropogenic capitals are predominantly governed by multiple modes of governance suggests that governance pluralism supports the co-production of NCP.

While Droste et al. (2018) found that agriculture is only the sixth most referred topic in ecosystem services research between 2001 and 2016, our analysis showed that most material NCP found in the literature on the governance of NCP co-production are defined as food and feed and thus related to agriculture. This also explains the strong link between material NCP and the capitals involved in their co-production. Human capital in the form of skills, local ecological knowledge and farming specific knowledge, and financial capital in the form of subsidies, direct payments and investments are necessary for viable agricultural production (see e.g. Garrido et al., 2017; Lavorel et al., 2020). As the agricultural sector, particularly within the European Union (EU), is heavily impacted by subsidies and direct payments (Simoncini et al., 2019), it thus comes as no surprise that these capitals behind the co-production of material NCP are influenced by the governing of strategic behaviour at the supra-national level (Fig. 6), which in this case refers to the EU's Common Agricultural Policy.

Similar to other reviews (see e.g. Fagerholm et al., 2016; Haase et al., 2014; Martín-López et al., 2019), we found regulating NCP to be the second most studied group of NCP (Fig. 4). In contrast to Haase et al. (2014) and Fagerholm et al. (2016) but comparable to Martín-López et al. (2019), we identified regulation of freshwater quality as well as regulation of soils and sediments as the most studied regulating NCP. These were predominantly co-produced by physical capital (Fig. 6), for example, in the form of irrigation canals for the regulation of freshwater quantity and quality (Garcia-Llorente et al., 2015) or tillage machinery in the management of soils (Dietze et al., 2019). Anthropogenic capitals underpinning regulating NCP co-production were found to be impacted by scientific-technical and collaborative governance modes (Fig. 6). For example, Garcia-Llorente et al. (2015) highlight that the involvement of local stakeholders predominantly drives collaborative governance on Spanish irrigation systems. Likewise, Hill et al. (2019) stress the importance of local Indigenous knowledge and its transmission through collaborative governance in the management of pollination NCP. Similarly, Hickey and Badry (in this special issue) explore ways of advancing collaborative governance to integrate Indigenous knowledge into decision making better. Besides, Hill et al. (2019) also refer to several national and international policies securing traditional customary laws that support pollination. The integration in legal documents implies a recognition of local Indigenous knowledge in decision-making processes and thus serves as an example on how scientific-technical governance is behind the management of regulating NCP.

Similar to Fagerholm et al. (2016), our results revealed that psychological and physical experiences are the most studied non-material NCP (Fig. 4). Besides, we found that social capital contributes most to non-material NCP and is governed hierarchically on the regional and local level. This finding can be ascribed to promoting tourism and local recreation in protected areas such as national parks (Mannetti et al., 2017) or urban green spaces and infrastructure (Brown et al., 2020).

5.5 Research outlook

The clusters of governance modes affecting anthropogenic capitals in NCP co-production discussed above reveal several knowledge gaps within the literature. First and in line with Fagerholm et al. (2016)[a], we found that the most studied NCP within the literature were material and regulating. However, the anthropogenic capitals contributing to their co-production, namely human, physical, and financial capital, are the least-studied. Second, like Winkler et al. (2021), we found that collaborative and scientific-technical governance are the least-studied modes of governance overall. Nevertheless, these governance modes are most strongly related to the regulating NCP. This is crucial for further research on pluralism in NCP governance as collaborative and scientific-technical governance involve a variety of state and non-state actors, but since regulating NCP affect all parts of society. Hence, future research should focus on better understanding the role of collaborative and scientific-technical governance modes in the management of human, physical and financial capitals behind the co-production of material and regulating NCP. In the light of ever-increasing challenges due to climate change and biodiversity loss, these insights may be crucial to improve adaptive decision making for the management of NCP based on scientific data and inclusive stakeholder engagement.

[a] In contrast to Fagerholm et al. (2016) we did not refer to supporting ecosystem services in our review explicitly as they are already incorporated within the eighteen generalised NCP (Martín-López et al. 2018).

Acknowledgements

This work has been partially funded by the German Research Foundation DFG Priority Program 1374 "Biodiversity Exploratories" (DFG Ref.No.43316337). RI would like to thank Annika Rieke Schmidt for her support and design of Box 3. JK and MFL gratefully acknowledge the support of iDiv funded by the Deutsche Forschungsgemeinschaft (DFG–FZT 118, 202548816). KJW acknowledges the support of the Natural Sciences and Engineering Research Council of Canada (NSERC), [funding reference number NSERC NETGP 523374-18]. La recherche de KJW a été financée par le Conseil de recherches en sciences naturelles et en génie du Canada (CRSNG), [numéro de référence NSERC NETGP 523374-18]."

Disclosure statement

The authors reported no potential conflict of interest.

References

Albert, C., Von Haaren, C., Othengrafen, F., Krätzig, S., Saathoff, W., 2017. Scaling policy conflicts in ecosystem services governance: a framework for spatial analysis. J. Environ. Policy Plan. 19, 574–592. https://doi.org/10.1080/1523908X.2015.1075194.

Auer, A., Maceira, N., Nahuelhual, L., 2017. Agriculturisation and trade-offs between commodity production and cultural ecosystem services: a case study in Balcarce County. J. Rural Stud. 53, 88–101. https://doi.org/10.1016/j.jrurstud.2017.05.013.

Bäckstrand, K., Khan, J., Kronsell, A., Lovbrand, E., 2010. The Promise of New Modes of Environmental Governance, in: Environmental Politics and Deliberative Democracy. Edward Elgar Publishing, p. 13822. doi:10.4337/9781849806411.00009.

Borgström, S.T., Elmqvist, T., Angelstam, P., Alfsen-Norodom, C., 2006. Scale mismatches in Management of Urban Landscapes. Ecol. Soc. 11 (2). https://doi.org/10.5751/ES-01819-110216.

Börner, J., Wunder, S., Wertz-Kanounnikoff, S., Tito, M.R., Pereira, L., Nascimento, N., 2010. Direct conservation payments in the Brazilian Amazon: scope and equity implications. Ecol. Econ. 69, 1272–1282. https://doi.org/10.1016/j.ecolecon.2009.11.003.

Brown, J.A., Larson, K.L., Lerman, S.B., Childers, D.L., Andrade, R., Bateman, H.L., Hall, S.J., Warren, P.S., York, A.M., 2020. Influences of environmental and social factors on perceived bio-cultural services and disservices. Front. Ecol. Evol. 8, 569730. https://doi.org/10.3389/fevo.2020.569730.

Bruley, E., Locatelli, B., Lavorel, S., 2021. Natures contributions to people: coproducing quality of life from multifunctional landscapes. Ecol. Soc. 26, art12. https://doi.org/10.5751/ES-12031-260112.

Bryman, A., 2012. Social Research Methods, fourth ed. Oxford University Press, Oxford, New York.

Burkhard, B., de Groot, R., Costanza, R., Seppelt, R., Jørgensen, S.E., Potschin, M., 2012. Solutions for sustaining natural capital and ecosystem services. Ecol. Indic. 21, 1–6. https://doi.org/10.1016/j.ecolind.2012.03.008.

Chapin III, F.S., Sommerkorn, M., Robards, M.D., Hillmer-Pegram, K., 2015. Ecosystem stewardship: a resilience framework for arctic conservation. Glob. Environ. Chang. 34, 207–217. https://doi.org/10.1016/j.gloenvcha.2015.07.003.

Chaudhary, S., McGregor, A., Houston, D., Chettri, N., 2015. The evolution of ecosystem services: a time series and discourse-centered analysis. Environ. Sci. Policy 54, 25–34. https://doi.org/10.1016/j.envsci.2015.04.025.

Cork, S.J., Peterson, G.D., Bennett, E.M., Petschel-Held, G., Zurek, M., 2006. Synthesis of the storylines. Ecol. Soc. 11 (2). https://doi.org/10.5751/ES-01798-110211.

Díaz, S., Demissew, S., Carabias, J., Joly, C., Lonsdale, M., Ash, N., Larigauderie, A., Adhikari, J.R., Arico, S., Báldi, A., Bartuska, A., Baste, I.A., Bilgin, A., Brondizio, E., Chan, K.M., Figueroa, V.E., Duraiappah, A., Fischer, M., Hill, R., Koetz, T., Leadley, P., Lyver, P., Mace, G.M., Martin-Lopez, B., Okumura, M., Pacheco, D., Pascual, U., Pérez, E.S., Reyers, B., Roth, E., Saito, O., Scholes, R.J., Sharma, N., Tallis, H., Thaman, R., Watson, R., Yahara, T., Hamid, Z.A., Akosim, C., Al-Hafedh, Y., Allahverdiyev, R., Amankwah, E., Asah, S.T., Asfaw, Z., Bartus, G., Brooks, L.A., Caillaux, J., Dalle, G., Darnaedi, D., Driver, A., Erpul, G., Escobar-Eyzaguirre, P., Failler, P., Fouda, A.M.M., Fu, B., Gundimeda, H., Hashimoto, S., Homer, F., Lavorel, S., Lichtenstein, G., Mala, W.A., Mandivenyi, W., Matczak, P., Mbizvo, C., Mehrdadi, M., Metzger, J.P., Mikissa, J.B., Moller, H., Mooney, H.A., Mumby, P., Nagendra, H., Nesshover, C., Oteng-Yeboah, A.A., Pataki, G., Roué, M., Rubis, J., Schultz, M., Smith, P., Sumaila, R., Takeuchi, K., Thomas, S., Verma, M., Yeo-Chang, Y., Zlatanova, D., 2015. The IPBES conceptual framework—connecting nature and people. Curr. Opin. Environ. Sustain. 14, 1–16. https://doi.org/10.1016/j.cosust.2014.11.002.

Díaz, S., Pascual, U., Stenseke, M., Martín-López, B., Watson, R.T., Molnár, Z., Hill, R., Chan, K.M.A., Baste, I.A., Brauman, K.A., Polasky, S., Church, A., Lonsdale, M., Larigauderie, A., Leadley, P.W., van Oudenhoven, A.P.E., van der Plaat, F., Schröter, M., Lavorel, S., Aumeeruddy-Thomas, Y., Bukvareva, E., Davies, K., Demissew, S., Erpul, G., Failler, P., Guerra, C.A., Hewitt, C.L., Keune, H., Lindley, S., Shirayama, Y., 2018. Assessing nature's contributions to people. Science 359, 270–272. https://doi.org/10.1126/science.aap8826.

Dietze, V., Hagemann, N., Jürges, N., Bartke, S., Fürst, C., 2019. Farmers consideration of soil ecosystem services in agricultural management—a case study from Saxony, Germany. Land Use Policy 81, 813–824. https://doi.org/10.1016/j.landusepol.2018.11.003.

Droste, N., D'Amato, D., Goddard, J.J., 2018. Where communities intermingle, diversity grows – the evolution of topics in ecosystem service research. PLoS One 13, e0204749. https://doi.org/10.1371/journal.pone.0204749.

Fagerholm, N., Torralba, M., Burgess, P.J., Plieninger, T., 2016. A systematic map of ecosystem services assessments around European agroforestry. Ecol. Indic. 62, 47–65. https://doi.org/10.1016/j.ecolind.2015.11.016.

Felipe-Lucia, M.R., Martín-López, B., Lavorel, S., Berraquero-Díaz, L., Escalera-Reyes, J., Comín, F.A., 2015. Ecosystem services flows: why Stakeholders' power relationships matter. PLoS One 10, e0132232. https://doi.org/10.1371/journal.pone.0132232.

Fischer, A., Eastwood, A., 2016. Coproduction of ecosystem services as human–nature interactions—an analytical framework. Land Use Policy 52, 41–50. https://doi.org/10.1016/j.landusepol.2015.12.004.

Fossey, M., Angers, D., Bustany, C., Cudennec, C., Durand, P., Gascuel-Odoux, C., Jaffrezic, A., Pérès, G., Besse, C., Walter, C., 2020. A framework to consider soil ecosystem services in territorial planning. Front. Environ. Sci. 8, 28. https://doi.org/10.3389/fenvs.2020.00028.

Garcia-Llorente, M., Iniesta-Arandia, I., Willaarts, B.A., Harrison, P.A., Berry, P., Bayo, M. d.M., Castro, A.J., Montes, C., Martin-Lopez, B., 2015. Biophysical and sociocultural factors underlying spatial trade-offs of ecosystem services in semiarid watersheds. Ecol. Soc. 20. https://doi.org/10.5751/ES-07785-200339.

Garrido, P., Elbakidze, M., Angelstam, P., Plieninger, T., Pulido, F., Moreno, G., 2017. Stakeholder perspectives of wood-pasture ecosystem services: a case study from Iberian dehesas. Land Use Policy 60, 324–333. https://doi.org/10.1016/j.landusepol.2016.10.022.

Gatzweiler, F.W., 2006. Organizing a public ecosystem service economy for sustaining biodiversity. Ecol. Econ. 59, 296–304. https://doi.org/10.1016/j.ecolecon.2005.10.017.

Gómez-Baggethun, E., Kelemen, E., Martín-López, B., Palomo, I., Montes, C., 2013. Scale misfit in ecosystem service governance as a source of environmental conflict. Soc. Nat. Resour. 26, 1202–1216. https://doi.org/10.1080/08941920.2013.820817.

Goodwin, N.R., 2003. Five Kinds of Capital: Useful Concepts for Sustainable Development. Tufts University, Medford, MA, USA.

Grosinger, J., Vallet, A., Palomo, I., Buclet, N., Lavorel, S., 2021. Collective capabilities shape the co-production of nature's contributions to people in the alpine agricultural system of the Maurienne valley, France. Reg. Environ. Change 21, 117. https://doi.org/10.1007/s10113-021-01840-9.

Haase, D., Larondelle, N., Andersson, E., Artmann, M., Borgström, S., Breuste, J., Gomez-Baggethun, E., Gren, Å., Hamstead, Z., Hansen, R., Kabisch, N., Kremer, P., Langemeyer, J., Rall, E.L., McPhearson, T., Pauleit, S., Qureshi, S., Schwarz, N., Voigt, A., Wurster, D., Elmqvist, T., 2014. A quantitative review of urban ecosystem service assessments: concepts, models, and implementation. Ambio 43, 413–433. https://doi.org/10.1007/s13280-014-0504-0.

Hicks, C.C., Cinner, J.E., 2014. Social, institutional, and knowledge mechanisms mediate diverse ecosystem service benefits from coral reefs. Proc. Natl. Acad. Sci. U. S. A. 111, 17791–17796. https://doi.org/10.1073/pnas.1413473111.

Hill, R., Nates-Parra, G., Quezada-Euán, J.J.G., Buchori, D., LeBuhn, G., Maués, M.M., Pert, P.L., Kwapong, P.K., Saeed, S., Breslow, S.J., Carneiro da Cunha, M., Dicks, L.V., Galetto, L., Gikungu, M., Howlett, B.G., Imperatriz-Fonseca, V.L., O'B Lyver, P., Martín-López, B., Oteros-Rozas, E., Potts, S.G., Roué, M., 2019. Biocultural approaches to pollinator conservation. Nat. Sustain 2, 214–222. https://doi.org/10.1038/s41893-019-0244-z.

Hill, R., Díaz, S., Pascual, U., Stenseke, M., Molnár, Z., Van Velden, J., 2021. Nature's contributions to people: weaving plural perspectives. One Earth 4, 910–915. https://doi.org/10.1016/j.oneear.2021.06.009.

Huth, N., Possingham, H.P., 2011. Basic ecological theory can inform habitat restoration for woodland birds: ecological theory informs habitat restoration. J. Appl. Ecol. 48, 293–300. https://doi.org/10.1111/j.1365-2664.2010.01936.x.

Kachler, J., Isaac, R., Martín-López, B., Bonn, A., Felipe-Lucia, M.R., n.d.. Ecosystem services co-production- how much evidence is out there? People and Nature. Submitted for publication.

Kadykalo, A.N., López-Rodriguez, M.D., Ainscough, J., Droste, N., Ryu, H., Ávila-Flores, G., Le Clec'h, S., Muñoz, M.C., Nilsson, L., Rana, S., Sarkar, P., Sevecke, K.J., Harmáčková, Z.V., 2019. Disentangling 'ecosystem services' and 'nature's contributions to people. Ecosyst. People 15, 269–287. https://doi.org/10.1080/26395916.2019.1669713.

Knüppe, K., Pahl-Wostl, C., 2011. A framework for the analysis of governance structures applying to groundwater resources and the requirements for the sustainable management of associated ecosystem services. Water Resour. Manag. 25, 3387–3411. https://doi.org/10.1007/s11269-011-9861-7.

Knüppe, K., Pahl-Wostl, C., Vinke-de Kruijf, J., 2016. Sustainable groundwater management: a comparative study of local policy changes and ecosystem Services in South Africa and Germany: sustainable groundwater management. Environ. Policy Gov. 26, 59–72. https://doi.org/10.1002/eet.1693.

Lavorel, S., Colloff, M.J., Locatelli, B., Gorddard, R., Prober, S.M., Gabillet, M., Devaux, C., Laforgue, D., Peyrache-Gadeau, V., 2019. Mustering the power of ecosystems for adaptation to climate change. Environ. Sci. Policy 92, 87–97. https://doi.org/10.1016/j.envsci.2018.11.010.

Lavorel, S., Locatelli, B., Colloff, M.J., Bruley, E., 2020. Co-producing ecosystem services for adapting to climate change. Philos. Trans. R. Soc. B 375, 20190119. https://doi.org/10.1098/rstb.2019.0119.
Lele, S., Springate-Baginski, O., Lakerveld, R., Deb, D., Dash, P., 2013. Ecosystem services: origins, contributions, pitfalls, and alternatives. Conservat. Soc. 11, 343. https://doi.org/10.4103/0972-4923.125752.
Liquete, C., Piroddi, C., Drakou, E.G., Gurney, L., Katsanevakis, S., Charef, A., Egoh, B., 2013. Current status and future prospects for the Assessment of marine and coastal ecosystem services: a systematic review. PLoS One 8, e67737. https://doi.org/10.1371/journal.pone.0067737.
Luederitz, C., Brink, E., Gralla, F., Hermelingmeier, V., Meyer, M., Niven, L., Panzer, L., Partelow, S., Rau, A.-L., Sasaki, R., Abson, D.J., Lang, D.J., Wamsler, C., von Wehrden, H., 2015. A review of urban ecosystem services: six key challenges for future research. Ecosyst. Serv. 14, 98–112. https://doi.org/10.1016/j.ecoser.2015.05.001.
Malinga, R.H., Jewitt, G.P.W., Lindborg, R., Andersson, E., Gordon, L.J., 2018. On the other side of the ditch: exploring contrasting ecosystem service coproduction between smallholder and commercial agriculture. Ecol. Soc. 23. https://doi.org/10.5751/ES-10380-230409.
Mannetti, L.M., Göttert, T., Zeller, U., Esler, K.J., 2017. Expanding the protected area network in Namibia: an institutional analysis. Ecosyst. Serv. 28, 207–218. https://doi.org/10.1016/j.ecoser.2017.08.008.
Martín-López, B., Church, A., Başak Dessane, E., Berry, P., Chenu, C., Christie, M., Gerino, M., Keune, H., Osipova, E., Oteros-Rozas, E., Paillard, S., Rossberg, A.G., Schröter, M., van Oudenhoven, A.P. E., 2018. Chapter 2: Nature's contributions to people and quality of life. In: Rounsevell, M., Fischer, M., Torre-Marin Rando, A., Mader, A. (Eds.), The IPBES regional assessment report on biodiversity and ecosystem services for Europe and Central Asia. Secretariat of the Intergovernmental Science-Policy Platform on Biodiversity and Ecosystem Services, Bonn, Germany, pp. 57–185.
Martín-López, B., Leister, I., Lorenzo Cruz, P., Palomo, I., Grêt-Regamey, A., Harrison, P.A., Lavorel, S., Locatelli, B., Luque, S., Walz, A., 2019. Nature's contributions to people in mountains: a review. PLoS One 14, e0217847. https://doi.org/10.1371/journal.pone.0217847.
Metzger, J.P., Fidelman, P., Sattler, C., Schröter, B., Maron, M., Eigenbrod, F., Fortin, M., Hohlenwerger, C., Rhodes, J.R., 2020. Connecting governance interventions to ecosystem services provision: a social-ecological network approach. People Nat. https://doi.org/10.1002/pan3.10172.
Millennium Ecosystem Assessment (Program) (Ed.), 2005. Ecosystems and Human Well-Being: Synthesis. Island Press, Washington, DC.
Moss, T., 2012. Spatial fit, from panacea to practice: implementing the EU water framework directive. Ecol. Soc. 17, art2. https://doi.org/10.5751/ES-04821-170302.
Newig, J., Fritsch, O., 2009. Environmental governance: participatory, multi-level - and effective? Environ. Policy Gov. 19, 197–214. https://doi.org/10.1002/eet.509.
Nieto-Romero, M., Oteros-Rozas, E., González, J.A., Martín-López, B., 2014. Exploring the knowledge landscape of ecosystem services assessments in Mediterranean agroecosystems: insights for future research. Environ. Sci. Policy 37, 121–133. https://doi.org/10.1016/j.envsci.2013.09.003.
Olsson, P., Folke, C., Hughes, T.P., 2008. Navigating the transition to ecosystem-based management of the great barrier reef, Australia. Proc. Natl. Acad. Sci. 105, 9489–9494. https://doi.org/10.1073/pnas.0706905105.
Paavola, J., Hubacek, K., 2013. Ecosystem services, governance, and stakeholder participation: an introduction. Ecol. Soc. 18, art42. https://doi.org/10.5751/ES-06019-180442.
Page, M.J., McKenzie, J.E., Bossuyt, P.M., Boutron, I., Hoffmann, T.C., Mulrow, C.D., Shamseer, L., Tetzlaff, J.M., Akl, E.A., Brennan, S.E., Chou, R., Glanville, J.,

Grimshaw, J.M., Hróbjartsson, A., Lalu, M.M., Li, T., Loder, E.W., Mayo-Wilson, E., McDonald, S., McGuinness, L.A., Stewart, L.A., Thomas, J., Tricco, A.C., Welch, V.A., Whiting, P., Moher, D., 2021. The PRISMA 2020 statement: an updated guideline for reporting systematic reviews. BMJ. doi:10.1136/bmj.n71.

Palliwoda, J., Fischer, J., Felipe-Lucia, M.R., Palomo, I., Neugarten, R., Büermann, A., Price, M.F., Torralba, M., Eigenbrod, F., Mitchell, M.G.E., Beckmann, M., Seppelt, R., Schröter, M., 2021. Ecosystem service coproduction across the zones of biosphere reserves in Europe. Ecosyst. People 17, 491–506. https://doi.org/10.1080/26395916.2021.1968501.

Palomo, I., Felipe-Lucia, M.R., Bennett, E.M., Martín-López, B., Pascual, U., 2016. Disentangling the Pathways and Effects of Ecosystem Service Co-Production. In: Advances in Ecological Research. Elsevier, pp. 245–283. doi:10.1016/bs.aecr.2015.09.003.

Primmer, E., 2011. Analysis of institutional adaptation: integration of biodiversity conservation into forestry. J. Clean. Prod. 19, 1822–1832. https://doi.org/10.1016/j.jclepro.2011.04.001.

Primmer, E., Jokinen, P., Blicharska, M., Barton, D.N., Bugter, R., Potschin, M., 2015. Governance of ecosystem services: a framework for empirical analysis. Ecosyst. Serv. 16, 158–166. https://doi.org/10.1016/j.ecoser.2015.05.002.

Raymond, C.M., Fazey, I., Reed, M.S., Stringer, L.C., Robinson, G.M., Evely, A.C., 2010. Integrating local and scientific knowledge for environmental management. J. Environ. Manage. 91, 1766–1777. https://doi.org/10.1016/j.jenvman.2010.03.023.

Rival, L., 2010. Ecuador's Yasuní-ITT initiative: the old and new values of petroleum. Ecol. Econ. 70, 358–365. https://doi.org/10.1016/j.ecolecon.2010.09.007.

Sattler, C., Loft, L., Mann, C., Meyer, C., 2018. Methods in ecosystem services governance analysis: an introduction. Ecosyst. Serv. 34, 155–168. https://doi.org/10.1016/j.ecoser.2018.11.007.

Simoncini, R., Ring, I., Sandström, C., Albert, C., Kasymov, U., Arlettaz, R., 2019. Constraints and opportunities for mainstreaming biodiversity and ecosystem services in the EU's common agricultural policy: insights from the IPBES assessment for Europe and Central Asia. Land Use Policy 88, 104099. https://doi.org/10.1016/j.landusepol.2019.104099.

Spangenberg, J.H., Görg, C., Truong, D.T., Tekken, V., Bustamante, J.V., Settele, J., 2014. Provision of ecosystem services is determined by human agency, not ecosystem functions. Four case studies. Int. J. Biodivers. Sci. Ecosyst. Serv. Manag. 10, 40–53. https://doi.org/10.1080/21513732.2014.884166.

Sutherland, W.J., Pullin, A.S., Dolman, P.M., Knight, T.M., 2004. The need for evidence-based conservation. Trends Ecol. Evol. 19, 305–308. https://doi.org/10.1016/j.tree.2004.03.018.

Torralba, M., Oteros-Rozas, E., Moreno, G., Plieninger, T., 2018. Exploring the role of management in the coproduction of ecosystem services from Spanish wooded rangelands. Rangel. Ecol. Manage. 71, 549–559. https://doi.org/10.1016/j.rama.2017.09.001.

Van Hecken, G., Bastiaensen, J., 2010. Payments for ecosystem Services in Nicaragua: do market-based approaches work?: payments for ecosystem Services in Nicaragua. Dev. Change 41, 421–444. https://doi.org/10.1111/j.1467-7660.2010.01644.x.

Vialatte, A., Barnaud, C., Blanco, J., Ouin, A., Choisis, J.-P., Andrieu, E., Sheeren, D., Ladet, S., Deconchat, M., Clément, F., Esquerré, D., Sirami, C., 2019. A conceptual framework for the governance of multiple ecosystem services in agricultural landscapes. Landsc. Ecol. 34, 1653–1673. https://doi.org/10.1007/s10980-019-00829-4.

Winkler, K.J., Garcia Rodrigues, J., Albrecht, E., Crockett, E.T.H., 2021. Governance of ecosystem services: a review of empirical literature. Ecosyst. People 17, 306–319. https://doi.org/10.1080/26395916.2021.1938235.

FSC
www.fsc.org
MIX
Papier | Fördert gute Waldnutzung
FSC® C083411

Kolibri 360 GmbH
Ferdinand-Jühlke-Straße 7
99095 Erfurt, Deutschland
produktsicherheit@kolibri360.de